G. Svehla

Wilson and Wilson's Comprehensive Analytical Chemistry, Volume X

Elsevier Scientific Publishing Company, Amsterdam, 1980

ERRATA

p. 29 The equation in Sect. VII should read

p. 31 line 2 of Sect. XVIII. "cyanide" should read "cyanine".

p. 61 line 1. "have been forn" should read "date from".

p. 282 ref. 707. 1966 should read 1961.

Amendment to Chap. 2

It is regretted that the sources of certain material on Robert Boyle [1, 2], Edward Jorden [3], Richard Kirwan [1, 2], the touchstone [4] and atomic spectroscopy [5] were omitted.

It should be noted that the first reference to the term "chemical analysis" in a letter to Frederick Clodius by R. Boyle in 1654 [6] was first reported in the Opening Lecture at Euroanalysis III, Dublin, 1978.

The use of Prof. Thorburn Burns' private library, including the Works of Robert Boyle, is gratefully acknowledged.

REFERENCES

1 D. Thorburn Burns, Anal. Lett., 12A (1979) 89.
2 D. Thorburn Burns, in D. Farrell, D. Thorburn Burns, D. Brown and D. MacDaeid (Eds.), Euroanalysis III, Reviews on Analytical Chemistry, Dublin, 1978, Applied Science, Barking, 1979.
3 D. Thorburn Burns, Anal. Proc., 16 (1979) 219.
4 D. Thorburn Burns, Anal. Proc., 18 (1981) 146.
5 D. Thorburn Burns, Anal. Proc., 12 (1975) 155.
6 T. Birch, The Works of the Honourable Robert Boyle, Vol. 6, J. and F. Rivington et al., London, new edition, 1772, p. 54.

COMPREHENSIVE ANALYTICAL CHEMISTRY

PUBLISHED IN CO-EDITION
WITH SNTL - PUBLISHERS OF TECHNICAL LITERATURE, PRAGUE

DISTRIBUTION OF THIS BOOK IS BEING HANDLED
BY THE FOLLOWING PUBLISHERS

FOR THE U.S.A. AND CANADA
ELSEVIER SCIENCE PUBLISHING COMPANY, INC.
52, VANDERBILT AVENUE
NEW YORK, NEW YORK 10017

FOR THE EAST EUROPEAN COUNTRIES, CHINA, NORTHERN KOREA,
CUBA, VIETNAM AND MONGOLIA
SNTL - PUBLISHERS OF TECHNICAL LITERATURE,
PRAGUE

FOR ALL REMAINING AREAS
ELSEVIER SCIENTIFIC PUBLISHING COMPANY
1, MOLENWERF
P.O. BOX 211
1000 AE AMSTERDAM, THE NETHERLANDS

Library of Congress Cataloging in Publication Data

Kopanica, Miloslav.
   Kinetic methods in chemical analysis.

   (Wilson and Wilson's comprehensive analytical
chemistry ; v. 18)
   Bibliography: p. 227, 438
   Includes index.
   1. Chemistry, Analytic. 2. Chemical reaction, Rate
of. I. Stará, Věra, ing. II. Eckschlager, Karel.
Application of computers in analytical chemistry.
1983. III. Title. IV. Series: Comprehensive analytical chemistry ; v. 18.
QD75.W75 vol. 18 [QD75.2] 543s [543] 82-16259
ISBN 0-444-99685-0

ISBN 0-444-99685-0 (VOL. XVIII)
ISBN 0-444-41735-4 (SERIES)

© 1983 MILOSLAV KOPANICA, VĚRA STARÁ, KAREL ECKSCHLAGER
TRANSLATION © 1983 MADELEINE ŠTULÍKOVÁ

PRINTED IN CZECHOSLOVAKIA

COMPREHENSIVE ANALYTICAL CHEMISTRY

ADVISORY BOARD

Contributors to Volume XVIII

KINETIC METHODS IN CHEMICAL ANALYSIS

M. Kopanica, D.Sc.

J. Heyrovský Institute of Physical Chemistry and Electrochemistry, Czechoslovak Academy of Sciences, Prague

V. Stará, Ph.D.

J. Heyrovský Institute of Physical Chemistry and Electrochemistry, Czechoslovak Academy of Sciences, Prague

APPLICATION OF COMPUTERS IN ANALYTICAL CHEMISTRY

K. Eckschlager, D.Sc.

Institute of Inorganic Chemistry, Czechoslovak Academy of Sciences, Prague

I. Horsák, Ph.D.

Institute of Inorganic Chemistry, Czechoslovak Academy of Sciences, Prague

Z. Kodejš, Ph.D.

Institute of Inorganic Chemistry, Czechoslovak Academy of Sciences, Prague

Z. Ksandr, D.Sc.

Institute of Chemical Technology, Prague

M. Matherny, D.Sc.

Technical University, Košice

I. Obrusník, Ph.D.

Institute of Nuclear Research, Řež

S. Wičar, Ph.D.

Institute of Analytical Chemistry, Czechoslovak Academy of Sciences, Brno

Wilson and Wilson's

COMPREHENSIVE ANALYTICAL CHEMISTRY

EDITED BY

G. SVEHLA, PH.D., D.Sc., F.R.S.C.

Reader in Analytical Chemistry
The Queen's University of Belfast

VOLUME XVIII

Kinetic Methods in Chemical Analysis

Application of Computers in Analytical Chemistry

ELSEVIER SCIENTIFIC PUBLISHING COMPANY
AMSTERDAM   OXFORD   NEW YORK

1983

WILSON AND WILSON'S

COMPREHENSIVE ANALYTICAL CHEMISTRY

VOLUMES IN THE SERIES

Preface

In Comprehensive Analytical Chemistry, the aim is to provide a work which, in many instances, should be a self-sufficient reference work; but where this is not possible, it should at least be a starting point for any analytical investigation.

It is hoped to include the widest selection of topics that is possible within the compass of the work, and to give material in sufficient detail to allow it to be used directly, not only by professional analytical chemists, but also by those workers whose use of analytical methods is incidental to their work rather than continual. Where it is not possible to give details of methods, full reference to the pertinent original literature is made.

Volume XVIII contains two major contributions. The first on Kinetic Methods in Chemical Analysis describes a novel field, which has progressed rapidly in the last decade. Selectivities and sensitivities of some of the catalytic and enzymatic methods often surpass those obtainable with spectroscopic and radiochemical methods. The second major chapter is on the Application of Computers in Analytical Chemistry, surveying this fast-growing field. All the authors come from well-known Czechoslovakian institutions.

Dr. C. L. Graham of the University of Birmingham, England, assisted in the production of this volume; his contribution is acknowledged with many thanks.

July 1982   G. Svehla

Contents

Kinetic Methods in Chemical Analysis

by

MILOSLAV KOPANICA
VĚRA STARÁ

Contents

13

List of Symbols

A, B, C, D	substance
$[A]$	concentration of substance A
$[A]_0$	initial concentration of substance A
$[A]_t$	instantaneous concentration of substance A (at time t)
$[A]_\infty$	concentration of substance A after completion of the reaction
A	absorbance, frequency factor
C	catalyst
E	enzyme
E	redox potential
E^0	standard redox potential
E_A	activation energy
Γ	function
G	function
I	inhibitor
I	ionic strength
J_S	flow of substance S
K	constant
K_M	the Michaelis constant
K_{MX}	concentration stability constant of complex MX
L	ligand
M, M^{v+}	ion of metal M
Ox	oxidation agent
P	reaction product
P	parameter
R	reagent, initial substance, substituent
Red	reducing agent
S	substrate
S	signal
T	temperature
X	polydonor type ligand
Z	precipitation number

DCTA	diaminecyclohexanetetraacetic acid
EDTA	ethylenediaminetetraacetic acid
EGTA	ethyleneglycolbis(2-aminoethylether)tetraacetic acid
NTA	nitrilotriacetic acid
NAD	nicotinamide-adenine dinucleotide
NADH	the reduced form of NAD
NADP	nicotinamide-adenine dinucleotide phosphate
NADPH	the reduced form of NADP
PAR	4-(2-pyridylazo)resorcinol
XO	xylenol orange
a, b	stoichiometric coefficients
e^-	electron
e_{aq}^-	hydrated electron
i	current
i_d	diffusion current
k	rate constant
n	reaction order, coordination number
n_A	number of moles of substance A
p	conversion coefficient
v	reaction rate
v_0, v'	initial rate
v_{max}	maximum reaction rate
$\overleftarrow{v}$	slowed enzymatic reaction
t	time
$t_{1/2}$	half-time
x	amount of substance reacted at time t
z	ionic charge
α, β	order with respect to A, B
γ	activity coefficient
ε	molar (linear) absorption coefficient
ε_r	relative permittivity
$\varkappa$	constant
μ	dipole moment
ν_A	stoichiometric factor for substance A
ϱ	empirical constant
σ	empirical constant
ξ	reaction degree

18

Introduction

Kinetic methods of chemical analysis utilize the dynamic properties of reaction systems. The theory is based on the laws of reaction kinetics. The quantiative relationships that follow from these laws determine the dependence of the reaction rate on the concentrations of the reactants. From the analytical point of view, kinetic methods represent a departure from the traditional view that the concentrations of substances in solutions to be determined must be in chemical equilibrium.

One of the advantages of the kinetic approach to the solution of analytical problems is the possibility of determining traces of many organic substances simply and rapidly, utilizing the catalytic properties of metal ions in solution. Kinetic methods are as sensitive as luminescence and activation analysis methods [2]. However, catalytic reactions are not important only in inorganic analysis. For example, the specific catalytic properties of enzymes have led to the development of a branch of clinical analysis based on the kinetics of enzyme reactions. The kinetic approach has proven to be very powerful in analyses of mixtures of substances which are chemically very similar without separations, because of their different rates of reaction with a single reagent. It is also important that all kinetic methods lend themselves to automation.

Kinetic methods of analysis represent practical application of the laws of reaction kinetics which have been known for a long time. Possibilities of kinetic solution of analytical problems must have been known even at an early stage in the development of physico-chemical methods. Nonetheless, kinetic methods, with very few exceptions, were not studied by analysts, either for routine analysis or in research.

Research in quantitative kinetic analysis began in the mid thirties, with the first research paper of Sandell and Kolthoff [1]. By the mid sixties, the first monograph by Yatsimirskii [2] appeared in the Russian language, followed by an extended English [3] and Russian [2] edition. An extensive monograph of Mark and Rechnitz [4] dealt generally with the role of kinetics in analytical chemistry. The field has been reviewed biannually in the journal Analytical Chemistry since 1964 [5–13], and a number of other reviews appeared as well [14–23]. Even undergraduate and postgraduate textbooks [24–26] took the subject up quite early.

It should be pointed out that the rapid development of kinetic methods has been largely facilitated by the high sophistication of contemporary electronics and measuring and automation techniques. The standard of measuring techniques affects the attitude of the analyst to an analytical method. Not very long ago, kinetic methods performed by standard techniques represented, in an optimum case, reading of values on a meter at certain time intervals. Under these conditions, a ten percent error was considered acceptable. The introduction of instruments continually recording the time dependence of the concentration of a substance improved the quality of kinetic measurements and led to greater use in analytical chemistry.

The formation and development of kinetic methods is an inseparable part of modern analytical chemistry. Great demands are placed on the precision, sensitivity, rapidity and possible automation of analytical methods. This necessitates progress in physico-chemical methods, employing the most varied chemical, physico-chemical and physical properties of substances for their determination. As chemical reactions form the basis of most analytical methods, it is inconceivable that the dynamic character of chemical reactions would remain unused for analytical purposes. As has been shown recently, kinetic methods often permit the solution of analytical problems more simply and effectively than is possible using equilibrium methods.

20

1 Kinetic Aspects of the Analytical Use of Chemical Reactions

Modern analytical chemistry exploits a wide range of methods based on the chemical, physico-chemical and physical properties of substances. Most often, the test material is suitably transferred into solution, a reagent is added and a required component is determined directly or first separated and then determined. The basis of these methods is thus a chemical reaction.

Study of chemical reactions is based on the interpretation of the results of measurements made during the course of the reaction or under static conditions before the reaction or after its completion. Chemical thermodynamics are the theoretical basis of the study of chemical systems under static conditions, provided that the reaction system is in equilibrium; chemical kinetics deal with the laws of the dynamics of chemical reactions.

Not all chemical reactions meet the requirements of analytical chemistry. One of the basic rules quoted in all textbooks on analytical chemistry is that only those reactions are applicable for analytical purposes which proceed quantitatively, unambiguously and, wherever possible, rapidly. These requirements naturally follow from the character of thermodynamic, equilibrium methods. In equilibrium methods the test material is transferred into solution, a reagent is added and the concentration of a required component is determined after completion of all reactions. From the physico-chemical point of view the concept of "equilibrium method" is not quite accurate, as it may lead to the conclusion that the measurement is always carried out at equilibrium and thus that only reversible chemical reactions are used. Actually, equilibrium methods involve all methods of measurement under static conditions, i.e. both at equilibrium and after completion of the reaction. Equilibrium methods include classical analytical methods, such as gravimetry and titrimetry and a number of physico-chemical methods.

Study of reactions of inorganic and organic substances has shown that relatively few reactions fully meet the requirements of analytical chemistry. Many real chemical systems involve side reactions, equilibrium is sometimes slowly attained or the reaction product is not formed quantitatively. The equilibrium methods of chemical analysis then fail and the advantages of the kinetic approach become evident.

The dynamics of reaction systems have been extensively studied and many quantitative relationships generally describe the dependence of the reaction rate on the concentrations of the components of the reaction system. Thus the idea of using such relationships for the determination of a component by measuring the reaction rate naturally emerges. For the determination of the reaction rate the time changes in the concentration of a reactant or a product must be found. In contrast to equilibrium methods, kinetic analytical methods require that the determination be carried out during the reaction, before attainment of equilibrium. From this principle a number of kinetic methods of analysis have been developed and their advantages over equilibrium methods are demonstrated below on various types of analyses.

In determination of trace concentrations of substances that have catalytic properties, the catalyst concentration can be found from the difference in the rates of the so-called indicator reaction in the presence and absence of the catalyst. The methods are highly sensitive and very simple, because all components except the catalyst are present in concentrations that are readily determinable. These methods are suitable for traces of metals and in clinical analysis for determinations involving enzymatic reactions. Enzymes are highly selective biochemical catalysts and thus kinetic methods involving them are extremely selective and incomparably superior to equilibrium methods.

Another analytically important group of methods measure the changes with time in the concentrations of reactants or products at the beginning of the reaction, i.e. the initial reaction rate is measured, when only 3 to 5% of the reactant has been converted. In this initial stage the concentration of the product is very small and thus side reactions involving the product that might affect the reaction mechanism are suppressed. The small concentration of the product and the small change in the reactant concentration at the beginning of the reaction also enable a certain simplification of the relationships describing the dependence of the initial reaction rate on the reactant concentrations. It then follows that the initial rate is initially independent of time and is proportional to the initial concentration of the reactants, thus becoming an analytically important quantity. Kinetic methods based on the measurement of the initial reaction rate then enable a determination, even if the reaction is slow or ambiguous. The methods have found the widest use in clinical analysis (enzyme-catalyzed reactions).

Mixtures of chemically very similar substances must often be analyzed, e.g. substances from a homologous series, isomers or metals from a single group of the periodic system. Equilibrium methods then require prior quantitative separation of the substances. Thermodynamic constants characterizing the equilibria of similar substances with a single reagent are different, but the measurement in equilibrium does not permit sufficient analytical differentiation of the components in the mixture. However, the kinetics of these reactions may differ considerably. The reaction rate depends on the structure of the reactant; even small changes in the structure may strongly affect it and the components can then be analytically differentiated. These experimentally simple methods are termed differential kinetic methods.

These examples show that the use of the kinetic approach in analytical chemistry is fully justified, chiefly because a much wider selection of reactions can be used than in equilibrium methods. Thus analytical chemistry is provided with new possibilities of determining inorganic and organic substances.

2 Kinetics of Chemical Reactions

2.1 Principal Concepts

Kinetic methods of analytical chemistry are based on relationships quantitatively describing the dynamics of reaction systems and permitting the determination of the reaction mechanism. Kinetic solution of analytical problems requires exact knowledge of the dependence of the concentration of a reaction component on time and of the reaction mechanism. Chemical kinetics deal with these problems. Thus the part of chemical kinetics that quantitatively describes and explains the reaction mechanism, permitting interpretation of the results of measurement and calculation of the test substance concentration, forms the theoretical foundations of the kinetic methods of analytical chemistry.

Let substance A react with substance B, with formation of products D and G

$$aA + bB \rightarrow dD + gG \qquad (2\text{-}1)$$

Reaction rate v is defined as the change with time in the amount or concentration of the components in reaction (2-1). If the reaction takes place in solution, which is true in kinetic methods of analysis, the reaction rate is defined as the number of moles consumed or formed in a unit volume per unit time. Reactants are consumed, products formed and the rate of reaction (2-1) is thus given by

$$v = -\frac{1}{a}\frac{d[A]}{dt} = -\frac{1}{b}\frac{d[B]}{dt} = \frac{1}{d}\frac{d[D]}{dt} = \frac{1}{g}\frac{d[G]}{dt} \qquad (2\text{-}2)$$

where t is time. Eq. (2-2) is the mathematical definition of the reaction rate. It follows from experiments that the reaction rate is a certain function of the concentrations of all the components of the reaction system,

$$v = f([A], [B], [D], [G]) \qquad (2\text{-}3)$$

25

Eq. (2-3) is the general form of the kinetic equation, which is the relationship determining the changes in the concentrations of substances participating in the reaction with time. The kinetic equation can be formulated for any reaction system only on the basis of experimental data (changes with time in the system component concentrations).

When a reaction has time-independent stoichiometry, the extent of reaction can be defined as

$$\Delta \xi = \Delta n_i / v_i \tag{2-4}$$

where Δn_i is the change in the amount of any reactant or product and v_i is stoichiometric coefficient of the chemical species.

Naturally, this definition is quite general and independent of external conditions. However, for treatment of experimental data that lead e.g. to the determination of the reaction mechanism or to analytical applicability of kinetic phenomena, equations based on a less general definition of the reaction rate are more suitable.

For many chemical reactions the kinetic equations have a relatively simple form:

$$v = k[A]^\alpha [B]^\beta \tag{2-5}$$

The sum of the exponents, $(n = \alpha + \beta)$, is the overall reaction order, whereas exponents α and β specify the reaction order with respect to substances A and B, respectively. The reaction order must be found experimentally for each system; it is expressed by small integral positive numbers or zero and sometimes by simple fractions. Constant k in Eq. (2-5) is the rate constant with the dimension, $[c]^{-(n-1)} [t]^{-1}$. It is most often expressed in units of $(dm^3 \, mol^{-1})^{n-1} \, s^{-1}$. The rate constant represents the reaction rate at unit concentrations of the reactants.

The concept of reaction order must be differentiated from that of the reaction molecularity, which is determined by the number of particles whose interaction leads to chemical conversion. The reaction molecularity is described only by numbers 1, 2 or 3. Only elemental nuclear disintegrations exhibit monomolecular character, whereas chemical reactions are generally bimolecular. Trimolecular reactions are relatively rare, because simultaneous collision of three particles energetically and sterically capable of interaction has low probability.

2.2 Kinetic Equations

Kinetic equations quantitatively describe the dependence of the reaction rate on the concentrations of the components of the reaction system and thus are of prime importance in kinetic analytical methods. Table 1 summarizes the kinetic equations for simple irreversible reactions in the differential and integral form.

For various types of complex reactions the kinetic equations must be derived; the derivation procedures that are closely related to the determination of the reaction mechanism can be found in monographs on chemical kinetics [27 – 29].

2.3 Factors Affecting the Rate of Chemical Reactions in Solution

The generally accepted theory explaining why chemical reactions occur is the Eyring theory of absolute reaction rates [27, 28, 29]. According to this theory, a chemical reaction occurs when the molecules of the reactants approach one another until a certain critical configuration is formed, in which the molecules form a single entity — the activated complex. The activated complex is an unstable formation which disintegrates either to form the reaction products or the initial reactants; the rate of this disintegration determines the reaction rate.

Employing this concept of chemical reactions, it can be seen from comparison of a reaction in the gaseous and the liquid phase that the rate of the reaction in solution can be affected by other factors, due to a higher degree of condensation of the reactants and solvent-solute interactions. These effects are strongest in reactions of ions, because ions are solvated and interactions resulting from electrostatic interionic forces occur. Thus it can be expected that the rate of a reaction in the gaseous phase will not be identical with that of the same reaction in solution.

Experimental studies of reactions of ions have actually shown that rate constants determined in the common manner vary during the reaction and depend on the initial concentration of the reactants and on the amount of neutral salts in the solution. Further study has revealed that an experimental rate constant can be separated into two factors: one is the actual constant and the other is given by a combination of the activity coefficients of the

TABLE 1

Kinetic equations for simple irreversible reactions

Order $n\ \alpha\ \beta$	Differential form of kinetic equation	Integral form of kinetic equation	Half time	Dimension of constant k
0 0 0	$-\dfrac{d[A]}{dt} = k$	$[A]_0 - [A]_t = kt$	$\dfrac{[A]_0}{2k}$	$mol\,l^{-1}\,s^{-1}$
1 1 0	$-\dfrac{d[A]}{dt} = k[A]$	$\ln\dfrac{[A]_0}{[A]_t} = kt$	$\dfrac{\ln 2}{k}$	s^{-1}
2 2 0	$-\dfrac{d[A]}{dt} = k[A]^2$	$\dfrac{1}{[A]_t} - \dfrac{1}{[A]_0} = kt$	$\dfrac{1}{k[A]_0}$	$l\,mol^{-1}\,s^{-1}$
2 1 1	$-\dfrac{d[A]}{dt} = k[A][B]$	$\dfrac{1}{[B]_0 - [A]_0}\ln\dfrac{[A]_0[B]_t}{[B]_0[A]_t} = kt$	$\dfrac{1}{k[B]_0 - [A]_0}\ln\left(2 - \dfrac{[A]_0}{[B]_0}\right)$	—
3 3 0	$-\dfrac{d[A]}{dt} = k[A]^2$	$\dfrac{1}{2}\left(\dfrac{1}{[A]_t^2} - \dfrac{1}{[A]_0^2}\right) = kt$	$\dfrac{3}{2k[A]_0^2}$	$l^2\,mol^{-2}\,s^{-1}$
3 2 1	$-\dfrac{d[A]}{dt} = k[A]^2[B]$	$\dfrac{1}{[B]_0 - [A]_0}\left(\dfrac{1}{[A]_t} - \dfrac{1}{[A]_0}\right) +$ $+ \dfrac{1}{([B]_0 - [A]_0)^2}\ln\dfrac{[B]_0[A]_t}{[A]_0[B]_t} = kt$	—	—
$n\ n\ 0$ $n \neq 1$	$-\dfrac{d[A]}{dt} = k[A]^n$	$\dfrac{1}{n-1}\left(\dfrac{1}{[A]_t^{n-1}} - \dfrac{1}{[A]_0^{n-1}}\right) = kt$	$\dfrac{2^{n-1}-1}{(n-1)\,k[A]_0^{n-1}}$	$l^{n-1}\,mol^{-(n-1)}\,s^{-1}$

reactants. The equation

$$\frac{dx}{dt} = k[A][B] = k^0 \frac{\gamma_A \gamma_B}{\gamma_{(AB)^{\ddagger}}} [A][B] \tag{2-6}$$

holds for simple second order reactions, where γ_A, γ_B and $\gamma_{(AB)^{\ddagger}}$ are the activity coefficients of substances A and B and of activated complex $(AB)^{\ddagger}$, respectively, and k^0 is the rate constant of the reaction at infinite dilution when $\gamma_A = \gamma_B = \gamma_{(AB)^{\ddagger}} = 1$. According to Eq. (2-6) the rate constant is given by

$$k = k^0 \frac{\gamma_A \gamma_B}{\gamma_{(AB)^{\ddagger}}} \tag{2-7}$$

Because an overwhelming majority of reactions employed in kinetic analytical methods take place in aqueous solutions and often (in methods based on catalytic effects) these reactions involve ions, it is natural that the factors affecting the reaction rate must be taken into account in kinetic analyses. A basic requirement for a successful kinetic measurement is maintenance of a constant temperature.

The rate of a chemical reaction can be increased by an increase in the temperature of the reacting mixture. The dependence of the rate constant on the temperature is given by the Arrhenius equation,

$$K = A \exp\left(-\frac{E_A^{\ddagger}}{RT}\right) \tag{2-8}$$

where A and $E_A^{\ddagger}$ are empirical constants, R is the gas constant and T is the absolute temperature (K). The Arrhenius equation expressed by relationship (2-8) was derived empirically and is rigorously obeyed by simple reactions. A linear dependence of $\ln k$ on $1/T$ follows from Eq. (2-8), the slope being determined by the ratio, $E_A^{\ddagger}/R$. The $\log k$ vs $1/T$ dependence is usually plotted (Fig. 1); the $E_A^{\ddagger}$ value can then be determined from the equation

$$\tan \beta = \frac{E_A^{\ddagger}}{2.303R} \tag{2-9}$$

Constants A and $E_A^{\ddagger}$ in Eq. (2-8) are actually not empirical. Constant $E_A^{\ddagger}$ has the dimension of energy and is termed activation energy and A is the frequency factor; their physical significance follows from the reaction rate theory [27].

The Arrhenius equation, Eq. (2-8), is obeyed exactly for reactions in the gaseous phase over a wide temperature range. Precise measurement of reactions in solution have shown the validity of the equation

$$K = \frac{k}{h} \exp\left(\frac{\Delta S_A^{\ddagger}}{R}\right) \exp\left(-\frac{\Delta H_A^{\ddagger}}{RT}\right) \tag{2-10}$$

where k is the Boltzmann constant $(1.3806 \times 10^{-23} \text{ JK}^{-1})$, h is Planck constant $(6.6262 \times 10^{-34} \text{ Js})$, $\Delta S_A^{\ddagger}$ is the entropy of activation and $\Delta H_A^{\ddagger}$

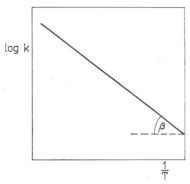

Fig. 1. The plot of the Arrhenius equation.

the enthalpy of activation. The original Arrhenius equation (2-8) thus somewhat simplifies the reaction conditions in solution. This simplification is, however, so small that use of Eq. (2-8) is quite justified when common measuring techniques are employed. If the experimental data exhibit a significant deviation from Eq. (2-8), then the studied reaction has a complex character and the experimental rate constant is a function of a number of rate constants for partial reaction systems.

The effect of solvent and of neutral salts on the reaction rate is equally important. The solvent effect on the rate of a reaction between substances A and B is given by the Kirkwood equation [29]:

$$\ln k = \ln K_0 + \frac{e^2}{2\varepsilon kT}\left[\frac{z_A^2}{r_A} + \frac{z_B^2}{r_B} - \frac{(z_A + z_B)^2}{r_{(AB)^{\ddagger}}}\right] +$$

$$+ \frac{3}{4\varepsilon kT}\left[\frac{\mu_A^2}{r_A^3} + \frac{\mu_B^2}{r_B^3} - \frac{\mu_{(AB)^{\ddagger}}^2}{r_{(AB)^{\ddagger}}^3}\right] \tag{2-11}$$

30

where k_0 is the rate constant of the reaction in a medium of infinite relative permittivity, ε is the solvent relative permittivity, μ_A, μ_B and $\mu_{(AB)^{\ddagger}}$ are the dipole moments of substances A, B and the activated complex, respectively, z_A and z_B are the charges on species A and B, respectively, r_A, r_B and $r_{(AB)^{\ddagger}}$ are the radii of species A, B and the activated complex $(AB)^{\ddagger}$, respectively, and e is the charge of the proton $(1.602 \times 10^{-19}\ C)$. Eq. (2-11) indicates that the logarithm of the rate constant is inversely proportional to the solvent relative permittivity and that, with increasing value of the latter, generally

a) the rate of reaction of two ions with the same charge increases,
b) the rate of reaction of two ions with opposite charges decreases and
c) the rate of reaction of two neutral species increases.

The dependence of the rate constant for reaction of substance A with substance B on the ionic strength I is given by

$$\ln k = \ln k_0 + z_A z_B I + \left(b_A + b_B - b_{(AB)^{\ddagger}}\right)I \tag{2-12}$$

where k_0 is the reaction rate constant in the absence of salts and b_A, b_B and $b_{(AB)^{\ddagger}}$ are empirical coefficients. According to Eq. (2-12) with increasing ionic strength

a) the rate of the reaction of two ions with the same charge increases,
b) the rate of the reaction of two ions with opposite charges decreases,
c) the rate of the reaction of two neutral species that form a polar product changes negligibly and
d) the rate of the reaction of a neutral species with an ion also changes negligibly.

3 Catalytic Reactions

The most important component in a catalytic reaction is the catalyst. Catalytic reactions are often systems of two successive reactions. In the first step the catalyst reacts with one reactant with formation of an intermediate, which is decomposed in the second step liberating the catalyst; the catalyst is thus cyclically reproduced in the reaction system. Hence the catalyst concentration is constant during the reaction and the rate of catalytic reactions is mostly proportional to it.

On the basis of this concept, the catalyst is defined as a substance that increases the rate of a chemical reaction but does not affect the equilibrium [30]. As the equilibrium constant of a reversible reaction equals the ratio of the rate constants for the forward and backward reactions, it is clear that the catalyst affects the rates of both the reactions equally. More rigorously, the catalyst can be defined as a substance that affects the reaction course by allowing it to proceed through steps with a lower activation energy without affecting the resultant chemical equilibrium.

3.1 Kinetic Equations for Catalytic Reactions

The kinetic equation for a catalytic reaction will be derived according to Laidler [29] from the reaction scheme

$$C + S \underset{k_{-1}}{\overset{k_1}{\rightleftharpoons}} X + (Y) \tag{3-1}$$

$$X + R \overset{k_2}{\longrightarrow} P + C + (Z) \tag{3-2}$$

where C is the catalyst, S is a reactant that is often termed the substrate, X is an intermediary complex, R is the other reactant, P is the main reaction product and Y and Z are possible side reaction products. The resultant rate of the overall process depends on rate constants k_1, k_{-1} and k_2. If the rate

33

of the backward reaction (3-1) is higher than that of reaction (3-2), $k_{-1} \gg$ $\gg k_2$, it can be assumed that equilibrium is first established, characterised by the relationship

$$\frac{[X][Y]}{[C][S]} = K_A = \frac{k_1}{k_{-1}} \tag{3-3}$$

where K_A is the equilibrium constant of reaction (3-1). In other words, the rate-determining step is reaction (3-2) and so the overall reaction rate is given by

$$\frac{d[P]}{dt} = -\frac{d[S]}{dt} = k_2[R][X] \tag{3-4}$$

For the equilibrium concentrations of substances S and C we can write

$$[C] = [C]_0 - [X] \tag{3-5a}$$

$$[S] = [S]_0 - [X] \tag{3-5b}$$

The subscript 0 refers to initial concentrations.

In most catalytic reactions the catalyst concentration is very low, so that $[S]_0 \gg [C]_0$; the equilibrium concentration $[S]$ can then be considered approximately equal to the initial concentration $[S]_0$, because $[X]$ cannot be larger than $[C]_0$. Substituting Eqs. (3-5a) and (3-5b) into Eq. (3-3) and solving for $[X]$, the relationship determining the concentration of complex X is obtained,

$$[X] = \frac{K_A[C]_0[S]_0}{K_A[S]_0 + [Y]} \tag{3-6}$$

On substitution of this relationship into Eq. (3-4) the relationship determining the overall rate of product formation is obtained,

$$\frac{d[P]}{dt} = -\frac{d[S]}{dt} = \frac{k_A K_A[C]_0[S]_0[R]}{K_A[S]_0 + [Y]} \tag{3-7}$$

for this so-called "equilibrium case".

If, on the other hand, reaction (3-2) is faster than reaction (3-1), i.e. for $k_2 \gg k_{-1}$ and $k_2 \gg k_1$, equilibrium according to Eq. (3-3) is not attained, because the intermediary complex X immediately reacts further according to reaction (3-2). The concentration of complex X in the reaction system is

34

then very small and can be calculated from the stationary state principle:

$$\frac{d[X]}{dt} = 0 = k_1[C][S] - k_{-1}[X][Y] - k_2[X][R] \qquad (3\text{-}8)$$

Using Eqs. (3-5a) and (3-5b), the kinetic equation determining the rate of formation of the product can be derived as

$$\frac{d[P]}{dt} = -\frac{d[S]}{dt} = \frac{k_1 k_2 [C]_0 [S]_0 [R]}{k_1([C]_0 + [S]_0) + k_{-1}[Y] + k_2[R]} \qquad (3\text{-}9)$$

for the so-called "steady-state" case.

For practical purposes, equations (3-7) and (3-9) can often be simplified to

$$\frac{dx}{dt} = K\alpha_c[C]_0([R]_0 - x) \qquad (3\text{-}10)$$

Here x is the decrease in the concentration of R, the substance which is monitored during the process, α_c is a function containing all the other concentration terms (except $[C]_0$, $[R]_0$ and x), and K is the term which involves all the rate and equilibrium constants. dx/dt is the rate of the decrease in the concentration of R, in other words, the rate of the reaction. During one measurement the value of α_c can be regarded as constant, provided that the experimental conditions are properly adjusted.

Eq. (3-10) holds rigorously when the reaction rate in the absence of the catalyst is zero, i.e. when the reaction does not proceed at all in the absence of the catalyst. However, a great majority of catalytic reactions do proceed (even if slowly) in the absence of a catalyst, as can be expressed by the equation,

$$S + R \xrightarrow{\ k_3\ } P + Z \qquad (3\text{-}11)$$

The rate of this reaction, termed the indicator reaction according to Yatsimirskii [3], is given by

$$\left(\frac{dx}{dt}\right)_{\text{uncat}} = k_3[S]_0([R]_0 - x) \qquad (3\text{-}12)$$

This relationship holds rigorously only when $[S] > [R]$.

As reaction (3-11) proceeds simultaneously with the catalytic reaction, the overall rate of the catalytic reaction is obtained from the sum of Eqs.

(3-10) and (3-12),

$$\left(\frac{dx}{dt}\right)_{\text{overall}} = \left(\frac{dx}{dt}\right)_{\text{cat}} + \left(\frac{dx}{dt}\right)_{\text{uncat}} =$$

$$= K\alpha_c[C]_0([R]_0 - x) + k_3[S]_0([R]_0 - x) \qquad (3\text{-}13)$$

As will be shown in Chapter 8, Eq. (3-13) is the principal relationship for determination of the test substance — the catalyst.

3.2 The Catalysis Mechanism

Study of the kinetics of catalytic reactions has shown that three types of reaction mechanism are possible [30]. The first type includes the above systems of successive reactions. These are homogeneous reactions involving molecules, ions or radicals and the reaction rate is determined by the rate of the decomposition of the intermediate. This group includes reactions catalyzed by acids and bases, analytically important redox reactions catalyzed by ions and analytically equally important enzyme-catalyzed reactions.

The second type of catalytic process involves heterogeneous catalytic reactions, in which the reaction rate is influenced at the interface. These include e.g. electrode reactions with catalytic effects on the electrode surface that have also been used as analytical determinations (see Chapter 7).

The third type of catalytic reaction is the catalytic chain reaction, in which initiation of the chain process is catalytically accelerated. These reactions can also be used analytically.

3.2.1 MECHANISMS OF ANALYTICALLY IMPORTANT CATALYTIC REACTIONS

It has been pointed out above that analytically important catalytic reactions are redox processes catalyzed by certain ions (the first type of catalytic reaction mechanism). Redox reagents can basically be classified into two groups, namely, oxidants acting through their d-orbitals (such as Cu^{2+}, Ag^{2+}, Mn^{3+}, Fe^{3+}, VO_2^+, CrO_2^{2-}, MoO_2^{2-}, MnO_4^-, OsO_4) and similar reductants (e.g. Cr^{2+}, Fe^{2+}, Ti^{3+}, V^{3+}, VO^{2+}), and oxidants acting through their s- and p-orbitals (e.g. H_2O_2, NO_3^-, ClO_3^-, BrO_3^-, $S_2O_3^{2-}$, ClO_4^-, I_2) with similar reductants (Cl^-, Br^-, I^-, HS^-, Sn^{2+}, SO_3^{2-}, N_3^-, phenols,

$C_2O_4^{2-}$). Reactions of oxidants acting through their s- and p-orbitals are slower than those of the first group and thus are advantageous as indicator reactions for determining catalytically active substances. Reactions of this type have been employed for determination of some 50 elements [2]. The methods are based on the measurement of the rates of various reactions of inorganic and organic substances. Although these reactions are varied, certain common features following from their mechanism have been found and used for more detailed classification.

Analytically useful catalytic reactions can be divided into two groups, depending on whether the ion acting as the catalyst changes its oxidation state or not during the reaction. Reactions in which the oxidation state of the ion-catalyst changes are marked by a high sensitivity. The mechanism of reactions of this type involves the steps

$$Red + C^{(n+1)+} \rightarrow P + C^{n+} \tag{3-14}$$

$$C^{n+} + Ox \rightarrow C^{(n+1)+} + Z \tag{3-15}$$

where $C^{(n+1)+}$ and C^{n+} are the catalyst in the two oxidation states, Red and Ox are the reactants in the redox reaction and P and Z are the reaction products. In these reactions, ions of the transition elements and halides usually function as catalysts.

The reaction mechanism described by Eqs. (3-14) and (3-15) is valid only under two conditions. Red must not react directly with Ox. The direct reaction is thermodynamically more favourable than the catalytic reaction but certain kinetic effects (e.g. charge transfer is possible only with transfer of some group or the redox reaction is not complementary) may hinder or completely prevent the direct reaction. Secondly, inequality $E_{Ox}^0 > E_C^0 > E_{Red}^0$ must be valid, i.e. the standard redox potential of the $C^{(n+1)+}/C^{n+}$ system must be more positive than that of the P/Red system and more negative than that for the Ox/Z system. Reactions of this type which are most frequent are catalytic oxidations of acrylamines, phenols and some organic dyes [31 – 33]. Their advantage is easy monitoring by spectrophotometric methods.

The catalyst interaction with the reactant is usually substitution of water molecules in the cation coordination sphere by molecules of amines or phenols; a charge-transfer complex is often formed [50]. The existence of such a complex was found e.g. in the oxidation of p-phenetidine with the

37

pentavalent vanadium ions as the catalyst [33]. The charge-transfer complex of V(V) with phenetidine is stable at temperatures below 5 °C; at higher temperatures an oxidation process occurs with formation of an arylamine radical and ions of tetravalent vanadium.

The complex formation is affected by the hydrogen ion concentration and thus the catalyst activity also depends on it.

The choice of suitable oxidants for analytical methods based on these catalytic effects is chiefly determined by the above condition, i.e. the validity of inequality $E^0_{Ox} > E^0_{Red}$. This condition is usually readily satisfied. It is much more difficult to find a catalyst satisfying the condition that $E^0_{Ox} > E^0_C > E^0_{Red}$, because the number of suitable $C^{(n+1)+}/C^{n+}$ systems is rather limited. The use of a suitable complexing agent that decreases the redox potential may help here. The greatest problem is the necessity that there be no direct reaction of Red with Ox. For this reason, many effective oxidants, such as potassium permanganate, ions of tetravalent cerium or those of tervalent cobalt and manganese, cannot be used.

Some reactions from this group are very simple, e.g.

$$V^{3+} + Fe^{3+} \rightarrow V^{4+} + Fe^{2+} \tag{3-16}$$

where transfer of two electrons is made possible by the presence of another ion capable of existence in at least two stable oxidation states. Reaction (3-16) is catalyzed by the system Cu^+/Cu^{2+} or Ag^+/Ag^{2+}. The mechanism of reaction (3-16) involves the reaction steps,

$$V^{3+} + Cu^{2+} \rightarrow V^{4+} + Cu^+ \tag{3-16a}$$

$$Cu^+ + Fe^{3+} \rightarrow Fe^{2+} + Cu^{2+} \tag{3-16b}$$

A number of redox reactions are known in analytical chemistry in which the catalytic activity of metal ions is utilized. Electron transfer occurs mostly by successive or chain reactions in which the catalyst participates. A typical example of such an analytically important reaction is the oxidation of oxalic acid by potassium permanganate catalyzed by Mn^{2+} ions. Simultaneous transfer of two electrons which might occur, e.g. in the reaction,

$$Tl^+ + 2 Ce^{4+} \rightarrow Tl^{3+} + 2 Ce^{3+} \tag{3-17}$$

is impossible [36], because such a reaction would require simultaneous interaction of three particles, which is statistically improbable. Transfer of two electrons can occur simultaneously with transfer of an atom or of a group of atoms, e.g. in the reaction

$$O_2SO^{2-} + OCl^- \rightarrow O_2SO_2^{2-} + Cl^- \tag{3-17a}$$

In electron-transfer reactions, unstable intermediates may also be formed during the individual reaction steps. An example is the analytically important oxidation of tervalent arsenic compounds by ceric ions, catalyzed by iodide ions. The catalysis mechanism [37] is expressed by the reactions,

$$I^- + Ce^{4+} \rightarrow Ce^{3+} + I \tag{3-18a}$$

$$I + As^{3+} \rightarrow (I-As)^{3+} \tag{3-18b}$$

$$(I-As)^{3+} + Ce^{4+} \rightarrow Ce^{3+} + I^- + As^{5+} \tag{3-18c}$$

where $(I-As)^{3+}$ is a reaction intermediate.

In catalytic reactions of the above type, an indirect catalysis mechanism is relatively rare; an example is the oxidation of ammonia by hydrogen peroxide, catalyzed by cupric salts. In this reaction Cu^{2+} ions react with hydrogen peroxide with formation of hydroxyl radicals that then oxidize ammonia; a similar mechanism was found for the oxidation of iodide by hydrogen peroxide, catalyzed by iron compounds [38].

For oxidations of inorganic substances, oxidants with higher redox potentials are suitable and thus the use of potassium permanganate and chromate, ceric salts, etc. is common. For the same reason these reactions are usually catalyzed by the ions of metals that behave as strong oxidants in their highest oxidation states, e.g. Co^{3+}, Ni^{3+}, Ag^{2+}, Mn^{3+}, Cr^{6+} or Os^{8+}. As the redox potentials of reagents and catalysts are rather high, complexing agents must often be used to decrease the redox potential of the catalyst or a reagent to satisfy the condition, $E_{Ox}^0 > E_C^0 > E_{Red}^0$. From the analytical point of view it is advantagenous that catalytic reactions leading to oxidation of inorganic substances are relatively simple and, in contrast to oxidants of organic substances, are not accompanied by side reactions.

The other large group of catalytic reactions comprises the reactions in which the oxidation state of the ion − catalyst − remains unchanged. These include catalytic oxidations by hydrogen peroxide in acidic medium. Catalytic activity is exhibited by cations with high positive charges that readily form peroxocomplexes (e.g. Fe(III), Ti(IV), Hf(IV), Th(IV), V(V), Nb(V), Ta(V), Cr(VI), Mo(VI) and W(VI)). These unstable complexes decompose to yield OH^-, OH^+ or O_2H^+ radicals, which then act as oxidants in the process itself. Hydroxyl radicals are analogously formed in catalytic oxidations of organic compounds by hydrogen peroxide [34, 35] and effect the oxidation.

This group of catalytic reactions also includes catalytic decompositions, hydrolysis and catalytic substitution reactions of organic and inorganic substances [15]. These reactions are important analytically, because they are often catalyzed by elements from the main groups of the periodic system whose ions have no vacant d-orbitals and thus cannot catalyze redox processes. The mechanism of catalysis then involves the effect of the catalyst on the polarization of the bonds in the reactants or on the orientation of the reactants, thus making the reaction possible. Then an important step in the catalysis is often chelation; the ions of the catalyst form stable chelates with the substrates. An example of the first catalysis type is hydrolysis of monoalkyl phosphates, $ROPO_3^{2-}$. These substances alone do not hydrolyze in alkaline media. In the presence of Cu^{2+} or Mg^{2+} ions neutral chelates are formed, the negative charge on the substrate is neutralized and hydrolytic decomposition can begin. The second catalysis type can be demonstrated on decarboxylation of oxaloacetic acid, with the mechanism described by the reaction scheme

$$\longrightarrow CH_3COCOOH + Cu^{2+} \tag{3-19}$$

Here chelates with both the substrate and the reaction product are formed. However, because the (product-Cu(II)) chelate has low stability, the copper is displaced from it by the substrate and the catalytic cycle is repeated.

From the analytical point of view, catalytic substitution reactions of complexes are also important (for details see Chapter 6).

40

3.2.2 ACTIVATION OF CATALYTIC REACTIONS

The sensitivity of an analytical method based on the measurement of the rate of a catalytic reaction is higher, the greater the difference between the rates of the catalytic and the indicator reactions. Various methods have been tested for increasing the rate of the catalytic reaction, e.g. an increase in the temperature of the reaction system. The only really effective method is the use of activators of catalytic reactions. The activator is defined as a substance that alone does not affect the reaction rate, but increases it considerably when combined with a suitable catalyst [39].

The activator effect depends on the given catalytic mechanism and may appear in any reaction step. If the catalyst directly interacts with the substrate with formation of a substrate-catalyst intermediate, $S-C$, the presence of an activator may affect the formation of $S-C$ or successive reactions leading to the formation of the reaction product. In both cases it is necessary that an activator-catalyst complex, $A-C$, be formed. This complex may form hydrogen bonds with the substrate and thus orient the substrate and the catalyst into positions suitable for the formation of intermediate $C-S$, depending on the activator structure. In this way oxidation reactions catalyzed by pentavalent vanadium compounds are activated (glycerol is the activator in this reaction [40]), or oxidations of aromatic compounds catalyzed by Cu^{2+} ions (OH^- ions act as the activator [41]).

The catalysis mechanism of homogeneous redox reactions is described by general Eqs. (3-14) and (3-15) and was demonstrated e.g. for the oxidation of sulphanilic acid by potassium peroxodisulphate catalyzed by Ag^+ ions [42]. The rate-determining step is reaction (3-15). If the reaction takes place in the presence of a complexing agent for silver ions, with the stability of the Ag^{2+} complex higher than that of the Ag^+ complex, the acceleration of the $Ag^+ \rightarrow Ag^{2+}$ process (reaction (3-15)) can be expected, leading to acceleration of the overall reaction. The validity of this assumption has been verified experimentally and 2,2'-bipyridine has been chosen as a suitable complexing agent [43, 44]. The activator here accelerates the reactions leading to catalyst regeneration. An increase in the rate of a catalytic reaction may, under optimum conditions, substantially increase the sensitivity of a kinetic analytical method; for example, in the determination of silver based on the catalytic oxidation of sulphanilic acid, addition of 2,2'-bipyridine increases the sensitivity 5000 times [45]. Another analytically

41

advantageous application of activators is based on the fact that the presence of a suitable complexing agent leads not only to faster catalytic regeneration, but also to suppression of the effect of other catalysts that might be present, resulting in an improvement of the selectivity of the method. However, experimental data are lacking for a detailed evaluation of the activator effect in this respect.

3.3 Catalytic Reactions with an Induction Period

Some catalytic reactions are characterized by an induction period, which is the time interval that elapses from reactant mixing to the appearance of the product and during which the reaction apparently does not take place. This phenomenon was first observed by Landolt [46] on the reaction of iodate with sulphite in an acidic medium. The reaction product with excess iodate ions is iodine, which can be monitored visually. However, iodine is formed only after a certain incubation period. This phenomenon was termed the *Landolt effect* after its discoverer and can be observed with many slow reactions, e.g. with redox processes involving halogens in various oxidation states. In acid-base and complexation reactions, the Landolt effect has been observed less frequently.

Reactions exhibiting the Landolt effect can be described by the reaction scheme

$$A + B \xrightarrow{k_1} D \tag{3-20a}$$

$$D + E \xrightarrow{k_2} P \text{ or } A \tag{3-20b}$$

where $k_2 > k_1$. It follows from the difference in the rates of the two reactions that the presence of the product D in the reaction system can be detected only when substance E (the Landolt reagent) has been entirely consumed in reaction (3-20b). If reaction (3-20a) can be accelerated by a catalyst, then the time elapsed from the beginning of the reaction to the appearance of product D is a measure of the catalyst concentration. Direct proportionality between the catalyst concentration and the reciprocal induction period $(1/t_i)$ was found empirically [47] and has been utilized analytically.

The validity of the relationship

$$[C] = \text{const.} \; (1/t_i) \tag{3-21}$$

was verified theoretically by Svehla [14]. However, his exact solution in-dicated deviations from linearity which are nevertheless so small that the error does not exceed 6%.

An advantage of kinetic methods based on the Landolt effect is simplicity of the procedure and instrumentation. Their sensitivity is high (from 0.1 to 1.0 $\mu g \cdot ml^{-1}$).

3.4 Oscillating Chemical Reactions

These reactions [51] are very interesting theoretically; the intermediate concentrations and thus also the reaction rate vary periodically. The study of oscillating reactions is very important in clarification of the mechanisms of some biological processes, e.g. muscle contractions, regulation of cell division, general biological movement on cell and subcell surfaces, nerve processes and "biological clocks". Analytically utilizable reactions of this type involve e.g. catalytic oxidation of citric, malonic or succinic acid by BrO_3^- ions [52–54]. These reactions can be catalyzed by the ions of metals exchanging a single electron during the reaction and with a standard redox potential between 0.9 and 1.6 V, e.g. the Ce^{4+}/Ce^{3+} and complex Fe^{2+}/complex Fe^{3+} systems.

In the oxidation of malonic acid catalyzed by cerium ions, the reaction can be monitored relatively simply, because the catalyst reacts according to the equations,

$$Ce^{3+} \xrightarrow{\quad HBrO_3 \quad} Ce^{4+} \tag{3-22}$$

$$Ce^{4+} \xrightarrow{\quad malonic\ acid \quad} Ce^{3+} \tag{3-23}$$

and changes in the Ce^{4+} ion concentration can readily be monitored spectrophotometrically or by measuring the non-equilibrium redox poten-tial of the Ce^{4+}/Ce^{3+} system. The measurement shows that the Ce^{4+} concentration varies periodically. Reaction (3-22) is autocatalytic and an intermediate formed during the reduction of BrO_3^- ions acts as the auto-catalyst. In parallel with reaction (3-23) bromoderivatives of malonic acid are formed and decomposed, leading to liberation of Br^- ions that hinder reaction (3-22) [55].

The periodicity of the process can be clarified as follows: The reaction system contains a certain concentration of Ce^{4+} ions. During reaction (3-23) Br^- ions are formed that slow down reaction (3-22). The concentration of Br^- ions in the system depends on the rate of reaction (3-23) and on the rate of the reaction in which Br^- ions are consumed through reaction with excess BrO_3^- ions. If the concentration of Br^- ions is sufficiently large, reaction (3-22) is stopped. When the concentration of Ce^{4+} ions, which decreases due to reaction (3-23), attains the lower limiting value, the Br^- ion concentration sharply decreases. Hence reaction (3-22) is strongly accelerated and the concentration of Ce^{4+} ions increases. When the Ce^{4+} ion concentration reaches the upper limiting value, the Br^- ion concentration rapidly increases, leading to a slow-down of reaction (3-22). The whole cycle is then repeated.

From the analytical point of view it is important that the number of cycles per time unit is proportional to the initial reactant concentrations [55]. However, this dependence does not have unlimited validity; with decreasing reagent concentrations the cycle frequency decreases. Therefore, the cycle frequency is measured over a short time interval after the beginning of the reaction. The cycle frequency strongly depends on the temperature and, according to Zhabotinskii [55], the temperature must be maintained constant within $\pm 0.04\ °C$ to ensure precise results. An advantage is the simplicity and precision of the determination of substances using oscillating reactions. So far, only the reaction of malonic acid has been proposed for analytical applications. The determination of ruthenium is based on the fact that the presence of Ru(III) and (IV) compounds increases the period frequency in the oxidation of malonic acid catalyzed by cerium ions. Down to 10^{-2} µg Ru per ml can be determined with a relative standard deviation of 0.03 [56]. The determination is not disturbed by small concentrations of Pt, Rh and Ir, which is advantageous.

3.5 Reactions Catalyzed by Organic Catalysts

Reactions of organic substances catalyzed by organic compounds have not yet been widely used in analytical chemistry. Organic catalysts affect the rate of redox reactions in a similar manner to inorganic catalysts. Redox reactions of organic substances can be reversible, irreversible or partially reversible with successive irreversible inactivation of the product.

A partially reversible reaction is described by the scheme,

$$\text{Ox} \ + \ e^- \ \rightleftharpoons \ \text{Red}_1 \ \rightarrow \ \text{Red}_2 \tag{3-24}$$

$$\text{Red} \ - \ e^- \ \rightleftharpoons \ \text{Ox}_1 \ \rightarrow \ \text{Ox}_2 \tag{3-25}$$

An example of catalysis with a reversible redox system is the reaction of divalent titanium with iodine, catalyzed by 2-hydroxy-2-aminophenazine [48]; the catalyst is effective down to a concentration of 10^{-7} mol. l^{-1}.

Catalysis involving a partially reversible redox system consists of a fast reversible reaction with electron transfer, followed by irreversible reactions of the intermediates formed either together, with the solvent or with dissolved oxygen. Some thio-substances react in this way, e.g.

$$2\,\text{RSH} \ \rightleftharpoons \ \text{RS}-\text{RS} \ + \ 2\,\text{H}^+ \ + \ 2\,e^- \tag{3-26}$$

The redox potential of substance RSH depends solely on the concentration of the reduced form, which makes it possible to follow concentration changes of RSH. It is assumed that radical RS˙ is formed during the electron transfer,

$$\text{RS}^- \ \rightleftharpoons \ \text{RS}^˙ \ + \ e^- \tag{3-27}$$

The reversibility of reaction (3-27) is a condition for the catalytic activity of substances containing $-\text{SH}$ or $=\text{S}$ groups. These substances have a strong effect on the rate of the indicator reaction of iodine with sodium azide and thus can be determined with relatively high sensitivity [49].

3.6 Selectivity of Catalytic Reactions

As catalysts present even at very low concentrations affect the rates of catalytic reactions, trace amounts of catalytically active substances can readily be determined by measuring the rate of catalytic reactions. Table 2 summarizes the elements that can be determined on the basis of their catalytic activity and the corresponding minimum determinable concentrations. It can be seen that over 40 elements can be determined with a high sensitivity. These advantages of catalytic methods are often emphasized, but less attention is paid to the fact that these catalysts are not as specific as enzymes. For example, the oxidation of iodide with hydrogen peroxide is catalyzed

TABLE 2

Survey of the elements which can be determined using their catalytic activity.

The numbers beside the symbols of the elements denote the detection limit expressed as the negative logarithm of the corresponding concentration in g ml^{-1} (according to [2] and [57]).

Ia	IIa	IIIb	IVb	Vb	VIb	VIIb	VIIIb	Ib	IIb	IIIa	IVa	Va	VIa	VIIa
H														
Li	Be9									B	C	N	O	F10
Na	Mg6									A16	Si6	P8	S10	C17
K	Ca6	Sc	Ti7	V10	Cr9	Mn10	Fe9 Co12 Ni9	Cu11	Zn9	Ga	Ge7	As5	Se9	Br8
Rb	Sr	Y	Zr7	Nb7	Mo10	Tc	Ru11 Rh10 Pd9	Ag10	Cd6	In8	Sn	Sb8	Te10	I10
Cs	Ba	La	Hf7	Ta7	W10	Re9	Os10 Ir11 Pt7	Au8	Hg9	Tl	Pb8	Bi7	Po	At
Fr	Ra	Ac	Th7	Pa	U9									

by Ti, Zr, Hf, Th, Nb, Cr, Mo and W, and the frequently used reaction of arsenic (III) with cerium (IV) is catalyzed by Hg, Ag, I^-, Ni and Ti.

Consider the indicator reaction

$$A + B \rightarrow P \tag{3-28}$$

which takes place in the presence of catalyst C at a rate defined by the equation

$$\frac{d[P]}{dt} = k'[A]^{p'}[B]^{q'} + k[A]^p[B]^q[C] \tag{3-29}$$

where k' and k are the rate constants for the uncatalyzed and catalyzed reactions, respectively and symbols p', q', p and q are, in a simple case, the stoichiometric coefficients. If several different catalysts $C_1, C_2, ..., C_n$ participate in reaction (3-28), then the reaction rate can be described by the equations [58],

$$v_1 - v_{10} = k_{11}\pi_{11}[C_1] + k_{12}\pi_{12}[C_2] + ... + k_{1n}\pi_{1n}[C_n] \tag{3-29a}$$
$$\vdots$$
$$v_n - v_{n0} = k_{n1}\pi_{n1}[C_1] + k_{n2}\pi_{n2}[C_2] + ... + k_{nn}\pi_{nn}[C_n] \tag{3-29b}$$

where $v_1, ..., v_n$ are the rates of the catalyzed reaction, $v_{i0} = k'[A]^{p'i}[B]^{q'i}$ is the rate of the uncatalyzed reaction and $\pi_{ik} = [A]^{pik}[B]^{qik}$, where $i, k = 1, ...$ ..., n. For the activity of a certain catalyst to be selective, the activities of the individual catalysts must differ considerably. Under various reaction conditions, $k_{ii} > k_{ik}$ or the individual catalysts must react according to different reaction mechanisms, $p_{ii}, q_{ii} > p_{ik}, q_{ik}$. Present knowledge on the catalytic mechanism is insufficient and thus only the sum of the two effects can be considered. In evaluating the selectivity of a catalytic reaction, we can then only rely on an empirical rule, according to which the selectivity of a catalytic reaction is higher when the reaction catalysts are not chemically similar. For successful application of catalytic reactions in analysis, they must often be suitably modified.

3.6.1 MODIFICATION OF THE SELECTIVITY OF CATALYTIC REACTIONS

The selectivity of a catalytic reaction can be improved if the character of the catalyst-reactant interaction can be suitably affected. The simplest way is variation of the pH of the reaction medium, the concentration of the reacting substances and the temperature.

As the catalytic activity appears only in a narrow hydrogen ion concentration range for most catalysts, the selectivity of a catalytic determination can be improved by a suitable change in this concentration. The reaction between iodide and hydrogen peroxide is catalyzed by zirconium and hafnium. Because only certain hydroxo-complexes of the metals are catalytically active and are formed at different pH values, the maximum catalytic effect of zirconium appears at pH 1.1 and that of hafnium at pH 2. This effect was used by Yatsimirskii in a catalytic determination of hafnium and zirconium in the presence of one another [59].

A change in the concentration of H^+ ions can also influence the formation of the catalyst-substrate or catalyst-activator complex. For example, copper catalyzes the oxidation of phenols and arylamines in an alkaline medium. In acidic media the stability of the Cu-phenol complexes decreases considerably, the catalytic effect of copper is suppressed and arylamines are oxidized catalytically. This effect can be used for the catalytic determination of iron in the presence of excess copper by measuring the rate of oxidation of phenols catalyzed by iron in acidic media, which is unaffected by the presence of excess copper [60].

The selectivity of catalytic reactions can also sometimes be influenced by a change in the reactant concentrations. As has been shown by Worthington and Pardue [61], the reaction between Ce(IV) and As(III) is chiefly catalyzed by osmium when the concentration of As(III) is high and that of Ce(IV) low. When the concentration ratio is reversed, the reaction is almost selectively catalyzed by ruthenium.

Attempts have been made to affect the sensitivity of catalytic reactions by a change in the temperature. Wolff and Schwing [62] have found that oxidations with bromate are catalyzed by hexavalent molybdenum, tungsten and chromium to various degrees at various temperatures. However, the differences in the activities of these catalysts at various temperatures were insufficient for precise simultaneous determination of the three metals.

As the selectivity of catalytic reactions strongly depends on the character of the catalyst-substrate reaction, possibilities of regulating the selectivity by modification of the substrate, which is possible with organic substrates, have also been studied. Tailor-made substrates for analytical purposes have not yet been developed but certain prospects can be found in the works of Dolmanova et al. [63]. The authors studied catalytic oxidations of substituted phenols by hydrogen peroxide and showed that the catalytic activity of copper is especially high for para derivatives, whereas ortho-substituted phenols are suitable substrates for oxidations catalyzed by nickel and cobalt.

When the catalyst forms inert complexes with the substrate (the platinum metals), the character of the catalyst coordination sphere is important for modification of the complexation reaction between the substrate and the catalyst. Müller et al. [64] have found that the reaction of Mn(III)-pyrophosphate with p-benzoquinone is catalyzed by ruthenium and palladium. The chlorocomplexes of these metals have the same catalytic effects, whereas in sulphate solutions the activity of ruthenium is increased and that of palladium decreased. Analogously, the reaction between luminol and periodate is catalyzed by rhodium only in sulphate solutions, while iridium catalyzes this reaction even in the presence of chloride [65].

3.6.2 IMPROVEMENT OF THE SELECTIVITY BY MASKING AND SEPARATIONS

Because the selectivity of catalytic reactions cannot always be improved in the above ways, masking agents and the use of trace analysis separation techniques are often useful.

Masking agents that are often used in catalytic determinations involve EDTA, CN^-, F^-, citrate, sulphosalicylic acid and others. Sometimes the indicator reaction itself is the source of a masking agent. For example, the reaction between iodide and hydrogen peroxide is catalyzed by niobium and tantalum. If, however, hydrogen peroxide is added to the sample before the determination, niobium is masked and tantalum can be determined in its presence. The iodide-azide indicator reaction is also selective, because sodium azide masks many interfering metals.

The use of activators that function as masking agents is frequently advantageous. For example, the reaction of p-phenetidine with chlorate is catalyzed by vanadium (V), iron (III) and copper (II). Citric acid acts as an activator in this reaction and masks ferric and cupric ions. Similar properties are exhibited by other activators, such as 2,2'-bipyridyl, ethylenediamine and sulphosalicylic acid [17, 45].

If a suitable masking agent is not available, the selectivity of a catalytic reaction must be improved by separation of interfering catalysts. Trace analysis separation techniques are suitable; liquid-liquid extraction has the greatest importance, followed by ion-exchange and chromatographic methods. The matrix effect must be examined thoroughly. For example, thin-layer and paper chromatography are rarely useful, because the matrix often affects the catalytic reactions. In extraction separations the simplest technique is to be preferred, i.e. back-extraction or destruction of the organic solvent should be avoided. It is advantageous to carry out the catalytic determination directly in the extract or in the extract dissolved in a mixed solvent, e.g. water − acetone − chloroform [66], water − dioxan − nitrobezene [67], water − acetone − n-butanol [68] or water − ethanol − methylisobutylketone [69].

The following conditions must be satisfied to enable an extraction − catalytic determination [58]:

a) the indicator reaction must proceed in an organic or mixed solvent and the catalyst activity must be retained in this medium,

b) an extraction system must be available for specific separation of the test metal and the extraction agent, if extracted itself, must not interfere in the catalytic reaction.

The first condition is often difficult to meet, because the catalytic activity mostly depends on the formation of a complex between the catalyst and a component of the indicator reaction. Catalysis can occur when the coordination sphere of the metal in this complex is not fully saturated or, which is very rare, when the coordination-saturated complex reacts according to an inner sphere mechanism. However, in extraction separation the metal to be separated must be present as a coordination-saturated complex, i.e. it cannot be extracted in its catalytically active form. Therefore, the conditions of the extraction-catalytic determination must be selected so that the catalytically active complex is formed only after the extraction, which is attained by a ligand-exchange reaction or by dissociation. Extractions of very stable or inert complexes are unsuitable for catalytic determinations, because their ligand-exchange reactions are very slow or non-existent.

50

The above mixed solvents are suitable media for some catalytic reactions and the determinations are equally precise as in aqueous solutions. A certain disadvantage of mixed solvents is the limited solubility of common extraction agents in polar media and the above mixed solvents sometimes exhibit a tendency to phase separation. Therefore the optimum method involves catalytic determination in an organic solvent in which the solubility of the extract is high.

So far, only a few indicator reactions proceeding in organic solvents are known. According to Otto et al. [58], suitable indicator reactions are the oxidation of bromopyrogallol red by $S_2O_8^{2-}$ in the presence of 1,10-phenanthroline in nitrobezene (determination of silver [67]), the oxidation of sulphanilic acid by hydrogen peroxide in pyridine (determination of copper [70]), the oxidation of 1-naphthylamine by bromate in chloroform (determination of molybdenum [66]) or the oxidation of p-phenetidine by hydrogen peroxide in the presence of 1,10-phenanthroline in methylisobutylketone (determination or iron [69]).

Extraction-catalytic methods are very attractive analytically, because they enable analyses of very complex materials. Further development can be expected in this field.

4 Enzyme Reactions

Enzymes are macromolecular protein catalysts that make possible many complex reactions in living organisms at normal temperatures. It is important that these catalysts are also active in vitro and thus can be used in analytical chemistry. Enzymes are highly active catalytically (much more so than inorganic catalysts) and often very selective, which is analytically important. Common catalysts only affect the rate of the indicator reaction. The specificity of enzyme catalytic effects results in a particular reaction course. Thus only a single reaction of the substrate takes place, even if several reactions are thermodynamically possible. The rate of this reaction is increased to such a degree by the enzyme that the other thermodynamically feasible reactions do not interfere. For example, urease catalyzes only the hydrolysis of urea, glucosidase catalyzes the oxidation of only two sugars from sixty oxidizable carbohydrates. Therefore it is logical and sometimes necessary to determine enzymes or other substances participating in enzyme reactions kinetically. This fact has long been known. Wider use of kinetic methods in clinical and biochemical laboratories was, however, originally prevented by distrust stemming from tedious and imprecise procedures and from difficulties in obtaining pure enzymes, which were also rather expensive.

These difficulties have been overcome by sophisticated measuring techniques, by the availability of sufficiently pure enzymes at an acceptable price and of immobilized enzymes that enable prolonged use and storage without loss of activity.

4.1 The Kinetic Equation for Enzyme-Catalyzed Reactions

A simple enzyme-catalyzed reaction is characterized by the reaction scheme [71, 72],

$$E + S \underset{k_{-1}}{\overset{k_1}{\rightleftharpoons}} X \overset{k_2}{\longrightarrow} E + P \qquad (4\text{-}1)$$

where E is the enzyme, S is the substrate, X is an intermediate and P is the product. Because of the high catalytic activity of enzymes, the rate of enzyme reactions is measurable using common experimental techniques at enzyme concentrations of about 10^{-8} to 10^{-10} mol.l^{-1} and at substrate concentrations above 10^{-6} mol.l^{-1}. Under these conditions $[S]_0 \gg [E]_0$, hence the intermediate concentration is very low and consequently the stationary state principle can be used for its calculation. Assuming the validity of this principle, it holds for reaction scheme (4-1) that

$$\frac{d[X]}{dt} = 0 = k_1[E][S] - k_{-1}[X] - k_2[X] \tag{4-2}$$

It follows from Eq. (4-2) that

$$[X] = \frac{k_1[E][S]}{k_{-1} + k_2} \tag{4-3}$$

Because it holds for the rate of product formation that $d[P]/dt = k_2[X]$, it follows from Eq. (4-3) that

$$\frac{d[P]}{dt} = \frac{k_1 k_2 [E][S]}{k_{-1} + k_2} \tag{4-4}$$

To derive the kinetic equation for reaction (4-1), it is advantageous to express the reaction rate as a function of the initial concentrations of the substrate and the enzyme. The relationships between the instantaneous and initial concentrations of the enzyme and substrate are given by

$$[E] = [E]_0 - [X] \tag{4-5a}$$

$$[S] = [S]_0 - [X] \tag{4-5b}$$

As $[X]$ is always smaller than $[E]_0$ and $[E]_0 \ll [S]_0$, the relationship

$$[S]_0 - [X] = [S]_0 \tag{4-5c}$$

is valid with sufficient precision. Substituting Eqs. (4-5a), (4-5b) and (4-5c) into Eq. (4-2),

$$\frac{d[X]}{dt} = k_1[S]_0 ([E]_0 - [X]) - k_{-1}[X] - k_2[X] = 0 \tag{4-2a}$$

is obtained, whose solution for $[X]$ is

$$[X] = \frac{k_1 [S]_0 [E]_0}{k_{-1} + k_2 + k_1 [S]_0} \tag{4-3a}$$

On substitution of Eq. (4-3a) into the relationship $d[P]/dt = k_2[X]$, an equation determining the rate of reaction (4-1) is obtained,

$$\frac{d[P]}{dt} = \frac{k_1 k_2 [E]_0 [S]_0}{k_{-1} + k_2 + k_1 [S]_0} \tag{4-4a}$$

The stationary state principle is used in the study of virtually all enzyme reactions. For example, for the reaction system,

$$E + S \underset{k_{-1}}{\overset{k_1}{\rightleftharpoons}} X \underset{k_{-2}}{\overset{k_2}{\rightleftharpoons}} E + P \tag{4-6}$$

the solution of the kinetics using the stationary state principle leads to the kinetic equation [24],

$$-\frac{d[S]}{dt} = \frac{d[P]}{dt} = \frac{(k_1 k_2 [S] - k_{-1} k_{-2} [P]) [E]_0}{k_1 S + k_{-2} [P] + k_2 + k_{-1}} \tag{4-7a}$$

This equation, analogous to Eq. (4-4a), is usually formulated as

$$-\frac{d[S]}{dt} = \frac{(v_{max}/K_M) [S] - (v_p/K_p) [P]}{1 + ([S]/K_M) + ([P]/K_p)} \tag{4-7b}$$

where

$$v_{max} = k_2 [E]_0 \tag{4-8}$$

$$v_p = k_{-1} [E]_0 \tag{4-9}$$

$$K_p = \frac{k_{-1} + k_2}{k_1} \tag{4-10}$$

$$K_M = \frac{k_{-1} + k_2}{k_{-2}} \tag{4-11}$$

v_{max} is the maximum reaction rate and K_M is the Michaelis constant. Quantities K_M and K_p are not equilibrium constants, but constants having significance only under stationary state conditions.

Eq. (4-1) is only a theoretical reaction mechanism. It has been found that in enzyme reactions at least two intermediates are formed, so that the reaction mechanism is expressed by the scheme

$$E + S \underset{k_{-1}}{\overset{k_1}{\rightleftharpoons}} X_1 \underset{k_{-2}}{\overset{k_2}{\rightleftharpoons}} X_2 \underset{k_{-3}}{\overset{k_3}{\rightleftharpoons}} P + E \qquad (4\text{-}12)$$

where X_1 and X_2 are intermediates $E-S$ and $E-P$, respectively. The kinetic equation again has the form of Eq. (4-7b), but the definitions of the Michaelis constant, K_M, and the maximum rate, v_{max}, are now

$$K_M = \frac{k_2 k_3 + k_{-1} k_3 + k_{-1} k_{-2}}{k_1 (k_{-2} + k_2 + k_3)} \qquad (4\text{-}13)$$

$$v_{max} = \frac{k_2 k_3 [E]_0}{k_3 + k_{-2} + k_2} \qquad (4\text{-}14)$$

For measurement under the initial conditions it holds that $[P] = 0$ and the initial reaction rate v_0 of reaction (4-6) is given by

$$v_0 = -\left(\frac{d[S]}{dt}\right)_{init} = \frac{k_2 [E]_0 [S]_0}{K_M + [S]_0} \qquad (4\text{-}15)$$

Eq. (4-15) is graphically represented by the dependence of the rate of an enzyme-catalyzed reaction on the initial concentration of the substrate, given in Fig. 2. It follows from this dependence that Michaelis constant K_M is the substrate concentration at which the initial reaction rate equals one half of the maximum rate v_{max}. The Michaelis constant is thus a measure of

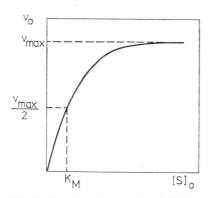

Fig. 2. The dependence of the initial rate of an enzyme-catalyzed reaction on the initial concentration of the substrate at a constant enzyme concentration.

the decomposition of the enzyme-substrate complex and its value indicates how complete the reaction is. The smaller the K_M value, the more quantitative is the reaction.

The relationship between the initial (v_0) and maximum (v_{max}) rates of reaction (4-6) is given by

$$v_0 = \frac{v_{max}[S]_0}{K_M + [S]_0} \tag{4-16}$$

from which it follows that, at $v_0 = v_{max}/2$, $K_M = [S]_0$. The reciprocal of Eq. (4-16),

$$\frac{1}{v_0} = \frac{K_M}{v_{max}[S]_0} + \frac{1}{v_{max}} \tag{4-17}$$

shows that the dependence of $1/v_0$ on $1/[S]_0$ is linear with a slope of K_M/v_{max} and an intercept on the $1/v_0$ axis of $1/v_{max}$. From this dependence the Michaelis constant is determined graphically [73] as in Fig. 3.

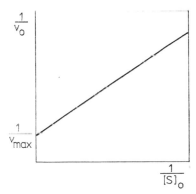

Fig. 3. Dependence of the reciprocal initial rate of an enzyme-catalyzed reaction on the reciprocal initial concentration of the substrate.

4.2 Factors Affecting the Rate of Enzyme Reactions

4.2.1 ACTIVATORS

It has been shown experimentally that the rate of reactions catalyzed by some enzymes increases in the presence of certain substances called activators. Cations of some metals, anions of some organic acids and proteins usually act as activators. The presence of an activator not only accelerates enzyme-catalyzed reactions, but sometimes even makes them possible. The mechanism of enzyme activation can be represented by the scheme

$$E(\text{inactive}) + \text{activator} \rightleftharpoons E(\text{active}) \tag{4-18}$$

$$E(\text{active}) + S \rightleftharpoons P \tag{4-19}$$

At low activator concentrations the initial reaction rate is proportional to the activator concentration and at higher concentrations it becomes con-

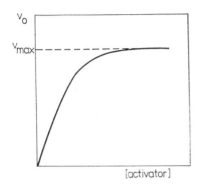

Fig. 4. Dependence of the initial rate of an enzyme catalyzed reaction on the activator concentration. (The enzyme and substrate concentrations are constant.)

centration-independent; the dependence is depicted in Fig. 4. The linear portion of this dependence corresponds to very low activator concentrations with most reactions and thus it can be used for the determination of traces of some metal ions that function as enzyme activators [74].

4.2.2 INHIBITORS

The rate of enzyme-catalyzed reactions is decreased considerably by small concentrations of some substances termed inhibitors. This phenomenon is caused either by the formation of an enzyme-inhibitor complex or of an enzyme-substrate-inhibitor complex. The deceleration of an enzyme reaction can be reversible, i.e. the enzyme attains the original activity on removal of the inhibitor. The mechanism can be represented by the scheme,

$$E + S \underset{k_{-1}}{\overset{k_1}{\rightleftharpoons}} ES \tag{4-20}$$

$$ES \underset{k_{-2}}{\overset{k_2}{\rightleftharpoons}} E + P \tag{4-21}$$

$$E + I \underset{k_{-3}}{\overset{k_3}{\rightleftharpoons}} EI \tag{4-22}$$

where I is the inhibitor. Using the stationary state principle, the relationship

$$v_0 = \frac{k_2[E]}{1 + \left(\dfrac{k_{-1} + k_2}{k_1[S]}\right)\left(1 + \dfrac{k_3[I]}{k_{-3}}\right)} \tag{4-23}$$

can be derived for the initial reaction rate. On substitution from Eq. (4-11) and substitution of the relationship

$$K_1 = k_{-3}/k_3 \tag{4-24}$$

the equation assumes the form

$$v_0 = \frac{v_{max}}{1 + (K_M/[S])(1 + [I]/K_I)} \tag{4-25}$$

According to Eq. (4-25), the presence of the inhibitor causes an apparent increase in constant K_M that can be expressed by factor $(1 + [I]/K_I)$.

The mechanism of irreversible inhibition is described by the reaction scheme

$$E + S \rightleftharpoons ES \tag{4-26}$$

$$E + I \rightleftharpoons EI \tag{4-27}$$

$$EI + S \rightleftharpoons EIS \tag{4-28}$$

$$I + ES \rightleftharpoons EIS \tag{4-29}$$

$$ES \rightleftharpoons E + P \tag{4-30}$$

As the enzyme-inhibitor-substrate complex is not decomposed further, an increase in the inhibitor concentration leads to a decrease in the initial rate. The dependence of initial rate v_0 on the inhibitor concentration is given by

59

the equation

$$v_0 = \frac{\dfrac{v_{max}}{1 + [I]/K_M}}{1 + K_M/[S]} \qquad (4\text{-}31)$$

At low inhibitor concentrations the dependence of the decrease of the initial rate on the inhibitor concentration is linear (Fig. 5) and can be used as a calibration curve for the kinetic determination of inhibitors.

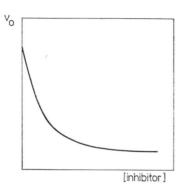

Fig. 5. Dependence of the initial rate of an enzyme-catalyzed reaction on the inhibitor concentration. (The enzyme and substrate concentrations are constant.)

4.2.3 HYDROGEN ION CONCENTRATION

The catalytic activity of enzymes generally depends on the hydrogen ion concentration; the maximum activity is usually attained in solutions with pH = 7 ± 1. When the pH of the solution containing an enzyme differs considerably from the above value, the enzyme is denatured, its activity decreases and the initial reaction rate also decreases. Therefore, the pH must be controlled in kinetic methods involving enzyme reactions. The optimal pH, at which the given enzyme has maximum activity, is found experimentally and must be maintained. It should be borne in mind that virtually all kinetic methods for the determination of enzymes and substrates employ the measurement of the rate of an indicator reaction, in which the product of a reaction of the test component reacts with a certain reagent. For example the determination of sucrose is based on the following reactions:

60

1) sucrose is converted into glucose using the invertase enzyme; 2) glucose is converted into hydrogen peroxide by oxidase; 3) the concentration of the H_2O_2 formed is determined spectrophotometrically or electrochemically versus time. Because each of these reactions has a different pH optimum, it is necessary to invert sucrose at pH 4.6, to oxidize the glucose and to measure the H_2O_2 concentration at pH 8.9. Thus it is clear that a reaction must be selected for indication of the course of the studied reaction which takes place at a pH as close as possible to the optimal pH for the given enzyme reaction (here the sucrose inversion must be carried out at as high a pH as possible and the glucose oxidation at as low a pH as possible).

4.2.4 TEMPERATURE

Even if the mechanism of enzyme reactions is very complex, the validity of the Arrhenius relationship has often been verified experimentally for narrow temperature intervals. The dependence of $\log v_{max}$ or $\log K_M$ on $1/T$ is linear when one of the steps in the enzyme reaction is rate-determining and the measured dependence determines the activation energy of this step. A non-linear temperature dependence can result from several factors, such as thermal denaturing of the enzyme, which sometimes may occur even at lower temperatures. All enzymes are denatured at high temperatures and the enzyme reactions are substantially decelerated.

The temperature must be kept constant in kinetic analyses based on enzyme reactions. Reproducible results are obtained, similar to all other methods of kinetic analysis, when the temperature is maintained within $\pm 0.1\ °C$.

The effect of the ionic strength obeys the general rules described in Section 2.3.

4.3 Principles of the Analytical Use of Enzyme Reactions

All substances participating in enzyme reactions can be determined by kinetic methods used for catalytic reactions. These methods may be applied for enzymes and all substances which affect the catalytic activity of an enzyme like activators and inhibitors. Further the kinetic determination of substrates can be also made, based on the measurement of the initial rate.

4.3.1 DETERMINATION OF SUBSTRATES

Determination of substrates is based on the validity of Eq. (4-15); the dependence of the initial rate of an enzyme reaction on the initial concentration of the substrate is depicted in Fig. 2. If the substrate concentration is small, then $[S] \ll K_M$ and Eq. (4-15) can be written in the form

$$-\left(\frac{d[S]}{dt}\right)_{init} = \frac{k_2[E]_0[S]_0}{K_M} \qquad (4\text{-}32)$$

Hence the initial rate is proportional to the initial concentration of the substance at a constant enzyme concentration.

An example of the use of Eq. (4-32) for analytical purposes is the determination of glucose. Glucosidase enzyme catalyzes the reaction

$$\beta\text{-D-glucose} + H_2O + O_2 \rightarrow \text{D-glutamic acid} + H_2O_2 \qquad (4\text{-}33)$$

The reaction is indicated colorimetrically or fluorimetrically (employing an indicator reaction of the hydrogen peroxide formed with a dye), by measuring the decrease in the oxygen content with an oxygen electrode, by measuring changes in the potential of the enzyme reaction system during the reaction, etc.

4.3.2 DETERMINATION OF ENZYMES

As in the previous case, the kinetic determination of enzymes is based on the validity of Eq. (4-15). The dependence of the initial reaction rate on the substrate initial concentration for various concentrations of the enzyme is depicted in Fig. 6, from which it follows that, at a certain substrate concentration (greater than that given by point A in Fig. 6), the initial reaction rate no longer depends on the concentration of the substrate and is proportional to the concentration of the enzyme. This linear dependence makes it possible to determine the enzyme by differential kinetic methods (Section 8.2.1.1) — by the constant time or constant concentration methods or by the tangent method.

A suitable example is a simple kinetic determination of lipase, acylase or chymotrypsine. Some of the fluorescein esters are used as substrates. These esters do not fluoresce, but they hydrolyze to form fluorescein in the presence of the test enzyme [75]. The initial reaction rate is found from the time dependence of the fluorescence intensity.

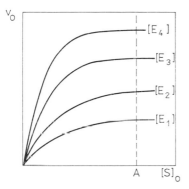

Fig. 6. Dependence of the initial rate of an enzyme reaction on the initial concentration of the substrate at various concentrations of the enzyme.
$[E_1] < [E_2] < [E_3] < [E_4]$

4.3.3 DETERMINATION OF THE ACTIVATORS AND COENZYMES

The principle of the determination of activators follows from the role of these substances in enzyme reactions. Activators convert inactive or poorly active enzymes into the active form. The enzyme activity increases with increasing concentration of the activator up to complete activation. With increasing enzyme activity, the initial rate of the enzyme reaction increases (Fig. 4) and is proportional to the activator concentration for small activator concentrations. Therefore, the increase in the concentration of a product of the enzyme reaction is measured as a function of time at the beginning of the reaction.

An example is the determination of traces of magnesium. Isocitricodehydrogenase enzyme (ICDH) catalyzes the oxidation of isocitric acid to oxalsuccinic acid and the reaction takes place in the presence of a coenzyme, nicotinamide adenine dinucleotide phosphate (NADP), only when ICDH is activated by Mg^{2+} ions. The initial rate is monitored spectrophotometrically, because the reduced form of the coenzyme, NADPH, has an absorption maximum at 340 nm [74, 76].

As the activity of many enzymes depends on the presence of another enzyme, a coenzyme, the initial reaction rate is then proportional to the coenzyme concentration. Therefore, coenzymes can be kinetically determined analogously to activators.

4.3.4 DETERMINATION OF INHIBITORS

If an enzyme reaction is decelerated by an inhibitor, the initial reaction rate decreases with increasing inhibitor concentration (Fig. 5). This dependence is linear for low inhibitor concentrations and can be utilized for analytical purposes. The deceleration of the reaction (in %) is plotted vs the inhibitor concentration, as shown in Fig. 7. The deceleration of an enzyme reaction (%) is found from the formula

$$\bar{v} = 100 \, (v - v_{\mathrm{I}})/v$$

where v and v_{I} are the reaction rates in the absence and presence of the inhibitor, respectively.

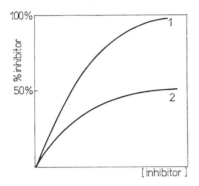

Fig. 7. Dependence of the deceleration of an enzyme reaction on the inhibitor concentration.
1 — irreversible inhibition; *2* — reversible inhibition.

With irreversible inhibition, the dependence depicted in Fig. 7 is linear up to a deceleration of 60—70%, so that good calibration graphs for determination of inhibitors are obtained. Ions of various transition metals, anions of some organic acids and many other organic substances act as inhibitors.

An example of determination of an inhibitor is the determination of mercury based on hydrolysis of amygdaline catalyzed by glucosidase, producing CN^- ions whose concentration can be determined electrochemically by measuring potential difference between a silver and a platinum electrode immersed in the test solution [77]. The ΔE value obtained is proportional to the CN^- concentration and hence also to the enzyme concentration. When traces of mercury are present, the reaction is slowed-down and from the plot of ΔE vs time the initial rate of the hydrolysis of amygdaline is found for mercury concentrations from 5×10^{-7} to 1×10^{-5} mol.1^{-1}.

4.3.5 GENERAL REMARKS ON THE METHODS OF ENZYME ANALYSIS

It must be borne in mind that work with enzymes in the laboratory requires some special precautions compared to other common reagents. Enzymes are readily deactivated at increased temperatures and thus all analytical operations must be carried out below 30 °C. Enzymes should be stored in a refrigerator at a temperature of 2 to 5 °C and some even well below 0 °C. Because enzymes are stabilized by salts at high concentrations, it is recommended that they be stored as suspensions in ammonium sulphate in a refrigerator; the activity of some enzymes then remains unchanged for several months.

Enzymes are also deactivated at pH values above 9 and below 5; hence reagents for pH adjustment must be added close to the vessel walls dropwise with stirring. Other factors, such as contact with organic solvents at normal temperatures and the effect of light and atmospheric oxygen, also lead to enzyme denaturation. Therefore, the manufacturer's recommendations should be observed.

In contrast to common analytical reagents, which are chemically pure substances (99 – 100%), the purity of good enzyme preparations is 1 – 5%. Therefore, a suitable unit must be defined for quantitative expression of the enzyme purity and activity. According to the guidelines of the Commission for Clinical Chemistry of IUPAC drawn up in 1959, the "international unit" (I.U.) was accepted, which represents an amount of the enzyme that converts 1 μmole of the substrate per minute under the optimal conditions (substrate concentration, pH, ionic strength, buffer, etc.).

4.4 Immobilized Enzymes

As has already been pointed out, the original objections to the use of enzymes in analytical practice stemmed chiefly from economical aspects. Enzymes are not cheap and cannot be regenerated after use without a loss in activity. A great development was the preparation of immobilized enzymes [78, 79], based on the study of living systems in which enzymes are present in soluble forms and also bound to various biostructures. Therefore, a method of bonding the enzyme to a suitable insoluble material was sought.

Basically, there are four methods for enzyme immobilization:

a) physical or chemical adsorption on various insoluble materials;

b) protein cross-linking of the enzyme to itself or to another type of protein;

c) entrapment within a gel matrix;

d) covalent bonding to an insoluble carrier (glass, cellulose, ion-exchange resin, etc.).

In these ways immobilized enzymes are obtained that function as heterogeneous specific catalysts, retaining all the typical properties of soluble enzymes. From a practical point of view, it is important that immobilized enzymes can be removed from the reaction mixture at any time (e.g. by centrifugation or filtration). Immobilized enzymes can be used for several analyses, they can be stored at normal temperatures and their activity remains constant for a long time. Many immobilized enzymes, mostly covalently bonded to insoluble carriers, e.g. cellulose, polyamide, polystyrene, etc., are marketed [79].

The catalytic activity of immobilized and free enzymes is the same, but their kinetic properties are somewhat different, because immobilized enzymes function as heterogeneous catalysts. Diffusion of the substrate to the enzyme bonding site substantially affects the kinetic parameters of the enzyme system. The surface of the carrier with the bound enzyme is covered with a stationary layer of the solvent, in which a concentration gradient of the substrate is developed, depending on the intensity of stirring.

Bound enzymes are saturated at higher substrate concentrations than free enzymes [80]. Therefore, the Michaelis constants are quoted for immobilized enzymes. This increase is usually related to changes in the substrate and in the carrier, to diffusion effects and sometimes to changes in the enzyme configuration. In some cases, however, the K_M values for immobilized and soluble enzymes are the same. For example, the K_M values for urease (substrate urea) are 10.0 mM for soluble enzyme and 7.60 mM for the immobilized enzyme; on the other hand, the K_M value, 0.448 mM, is valid for both soluble and immobilized invertase (sucrose substrate) [89]. The curves of the dependence of the enzyme activity on the pH can be affected by carrier ionizable groups and thus free enzymes have somewhat different optimal pH values than immobilized enzymes. The rate of thermal denaturation of immobilized enzymes is often less than that for the free enzyme. However, the differences in the kinetic behaviour of immobilized enzymes do not affect their possible analytical use.

Immobilized enzymes have been used analytically since about 1965. Guilbault et al. [81] prepared choline esterase immobilized by building it into a starch gel; the starch solution containing the enzyme, glycerol and a buffer was carefully introduced into urethane foam. The gel was dried in vacuo, forming an insoluble layer from which circular plates about 2 cm in diameter could be cut. This immobilized enzyme has proven advantageous in flow-through measurements. Guilbault and co-workers have described various analytical procedures involving immobilized choline esterase. For example, inhibitors, choline esterases, can be determined fluorimetrically [82]. A naphthyl acetate solution is passed through a plate containing immobilized choline esterase placed in a tube. The enzyme catalyzes hydrolysis of naphthyl acetate to highly fluorescing 2-naphthol. If inhibitors are present, the reaction is decelerated and the fluorescence intensity decreases.

The enzyme immobilized by the above method has also been used for electrochemical measurements [81]. A disc containing the enzyme was placed between two platinum electrodes of the same diameter, provided with holes for passage of the test solution. If the solution contains acetyl-thiocholine iodide, enzyme hydrolysis occurs producing thiocholine. On passage of a constant current between the electrodes ($\sim 2\ \mu A$), the thiocholine is electrochemically oxidized on the anode and the reaction can be followed by monitoring the potential difference between the two electrodes in dependence on time. The presence of inhibitors leads to a decrease in the hydrolysis rate and a change in the slope of the potential-time curve. These analyses yield very good results, but the apparatus is rather complicated, as one or more pumps are required for maintaining the flow of the solution through the immobilized enzyme. From this point of view, enzymes immobilized in polyacrylamide gels [83, 84] have been found more advantageous. These gels are prepared from N,N-methylenebisacrylamide, which, when mixed with a suitable buffer and the enzyme, is polymerized by adding riboflavine. The gel formed is dispersed mechanically to form small particles and fractions with uniform grain sizes are separated by sieving. These particles are packed into columns (similar to ion-exchange columns), through which the substrate solution in a suitable buffer is passed and the reaction product is determined in the eluate. It has been experimentally verified that enzymes immobilized in polyacrylamide gels are sufficiently stable to enable storage for up to 3 months without a decrease in their activity. It has also been shown that the concentration of the products of the enzyme reaction after passage of the substrate through the immobilized enzyme is proportional to the substrate concentration [84].

Another development in the analytical use of immobilized enzymes is the determination of glucose in blood, described by Updike and Hicks [85]. The test solution is pumped through a micro-tube containing immobilized glucose oxidase and variations in the oxygen concentration are monitored in the eluate with an oxygen electrode.

Polyacrylamide or polyvinylalcohol gels containing various immobilized enzymes can also be prepared in the form of membranes, the catalytic activity and selectivity of the enzymes being preserved, similar to immobilized enzymes in the form of discs, plates, beads, etc. Membranes with immobilized enzymes can simply be prepared e.g. by evaporating polyvinylalcohol gel with an enzyme on a glass plate, briefly irradiating with UV light and peeling the membrane from the glass.

Combination of ion-selective electrodes with membranes containing immobilized enzymes produced the "enzyme electrodes". The first such electrode was prepared by Updike and Hicks [86], who covered the active element of an oxygen electrode with a membrane containing immobilized glucose oxidase. If this electrode is immersed in a solution containing glucose and oxygen, the two substances diffuse through the membrane with the enzyme. Part of the oxygen is, however, consumed in the enzyme-catalyzed oxidation of the glucose, in proportion to the glucose concentration in the solution. Glucose can thus be determined in this way. Combination of membranes with immobilized enzymes with suitable ion-selective electrodes leads to relatively simple sensors for biochemical and biomedical [87, 88] measurements. For more details on enzyme electrodes, see Chapter 9.

4.4.1 QUANTITATIVE TREATMENT OF REACTIONS INVOLVING ENZYMES IMMOBILIZED IN ARTIFICIAL MEMBRANES

It can evidently be assumed that the kinetic behaviour of the substrate and product in an enzyme-catalyzed reaction (4-1) differs in dependence on whether the reaction takes place in homogeneous solution or the substrate and the product pass through a membrane with an immobilized enzyme. This assumption has been experimentally verified by comparing the kinetic parameters under these different conditions [90, 91]. It is difficult to express general relationships describing time changes in the concentration in reaction systems involving immobilized enzymes, because the reaction is complicated by diffusion and transport of substances across the phase boundary. The concept of apparent rate constant, which simply determines the overall reaction rate related only to changes in the substrate concentration in solution, obviously does not suffice for formulation of a general quantitative relationship.

The kinetics of a reaction catalyzed by an enzyme immobilized in an

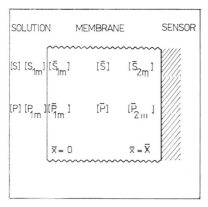

Fig. 8. Scheme of transport of the substrate and the product through the membrane towards the sensor.

artificial membrane was quantitatively treated by Blaedel et al. [98]. The authors dealt with certain simplified systems where the results obtained could be experimentally verified. Let an ion-selective electrode covered with a membrane containing an immobilized enzyme be immersed in a solution (see Fig. 8). The concentrations of the substrate and the product in solution are denoted as [S] and [P], respectively; the respective concentrations inside the membrane are denoted as $[\bar{S}]$ and $[\bar{P}]$ and indices 1 m and 2 m denote the concentrations of the substances at the solution/membrane and membrane/sensor interfaces. $\bar{X}$ is the membrane thickness and $\bar{x}$ the distance from the solution/membrane interface.

Because the sensor responds to concentrations $\bar{S}_{2m}$ or $\bar{P}_{2m}$, an equation must be found that defines these concentrations. The following equations have been derived for fluxes of the substrate and the product inside the membrane, $\bar{J}_{S_{1m}}$ and $\bar{J}_{P_{1m}}$,

$$\bar{J}_{S_{1m}} = -\bar{a}\bar{D}_S \frac{[\bar{S}_{2m}] - [\bar{S}_{1m}]\cosh \bar{a}\bar{X}}{\sinh \bar{a}\bar{X}} \tag{4-34}$$

$$\bar{J}_{P_{1m}} = \bar{a}\bar{D}_S \frac{[\bar{S}_{2m}] - [\bar{S}_{1m}]\cosh \bar{a}\bar{X}}{\sinh \bar{a}\bar{X}} -$$

$$-\frac{\bar{D}_S}{X}([\bar{S}_{2m}] - [\bar{S}_{1m}]) - \frac{\bar{D}_P}{X}([\bar{P}_{2m}] - [\bar{P}_{1m}]) \tag{4-35}$$

where

$$\bar{a} = \sqrt{(v_{max}/\bar{K}_M \bar{D}_S)}$$

In these equations $\bar{D}_S$ and $\bar{D}_P$ are the diffusion coefficients of the substrate and the product inside the membrane and $\bar{K}_M$ is the Michaelis constant. Eqs. (4-34) and (4.35) are valid provided that the concentration of the substance is small, $[\bar{S}] \ll [\bar{K}_M]$. On attaining the stationary state,

$$J_{S_{1m}} = \bar{J}_{S_{1m}} \tag{4-36}$$

$$J_{P_{1m}} = \bar{J}_{P_{1m}} \tag{4-37}$$

where J_S and J_P are fluxes of the substrate and the product in solution.

It follows from the condition of equilibrium in a stationary state that the concentrations of all the substances in solution and inside the membrane are related by distribution ratios σ_S and σ_P:

$$\frac{\bar{S}_{1m}}{S_{1m}} = \frac{\bar{S}_{2m}}{S_{2m}} = \sigma_S \tag{4-38}$$

$$\frac{\bar{P}_{1m}}{P_{1m}} = \frac{\bar{P}_{2m}}{P_{2m}} = \sigma_P \tag{4-39}$$

As the measurement is carried out in a stirred solution, the effect of laminar flow on the transport of the substances across the interface must also be considered. The material transport can then be expressed according to Treybal [93] using empirical transfer coefficient p. For the fluxes of the substrate and the product across the interface, it then holds that

$$J_{S_{1m}} = -p_S([S_{1m}] - [S]) \tag{4-40}$$

$$J_{P_{1m}} = -p_P([P_{1m}] - [P]) \tag{4-41}$$

Combination of Eqs. (4-34)–(4-41) yields equations determining the values of $[\bar{S}_{2m}]$ and $[\bar{P}_{2m}]$ which are monitored by the ion-selective electrode:

$$[\bar{S}_{2m}] = \cfrac{1}{\cfrac{\cosh \bar{a}\bar{X}}{\sigma_S} + \cfrac{\bar{a}\bar{D}_S}{p_S} \sinh \bar{a}\bar{X}} [S] \tag{4-42}$$

$$[\bar{P}_{2m}] = \sigma_P[P] + \cfrac{\cfrac{\bar{a}\bar{D}_S \sigma_P}{p_P} \sinh \bar{a}\bar{X} + \cfrac{\bar{D}_S}{\bar{D}_P} \cosh \bar{a}\bar{X} - \cfrac{\bar{D}_S}{\bar{D}_P}}{\cfrac{\cosh \bar{a}\bar{X}}{\sigma_S} + \cfrac{\bar{a}\bar{D}_S}{p_S} \sinh a\bar{X}} [S] \tag{4-43}$$

The validity of these relationships can be verified experimentally. Potential E of an ion-selective electrode is given by

$$E = E' + c \log([\text{ion}] + B) \tag{4-44}$$

where E' is the potential involving the values of the interface potentials and the effect of the activity coefficients and can be determined by measurement on a solution with a known composition. Constant c equals 59 mV for an ideal electrode and B is the term expressing the interferences, depending on the quality and the concentration of the other ions in the solution (e.g. the buffer components).

If the measurement is performed before the formation of product P in solution, i.e. for $[P] = 0$ in Eq. (4-34), it is possible to substitute Eq. (4-34) into Eq. (4-44), obtaining the relationship

$$E = E' + c \log \bar{G} + c \log([S] + B/\bar{G}) \tag{4-45}$$

71

where

$$\bar{G} = \frac{\dfrac{\bar{a}\bar{D}_s\sigma_s}{p_P}\sinh \bar{a}\bar{X} + \dfrac{\bar{D}_s}{\bar{D}_P}\cos \bar{a}\bar{X} - \dfrac{\bar{D}_s}{\bar{D}_P}}{\dfrac{1}{\sigma_s}\cosh \bar{a}\bar{X} + \dfrac{\bar{a}\bar{D}_s}{p_s}\sinh \bar{a}\bar{X}} \qquad (4\text{-}46)$$

The dependence of E on the concentration of urea, calculated from Eq.(4-45) for an NH_4^+-sensitive electrode provided with a membrane with immobilized urease, is given in Fig. 9. The dependence is linear over a wide range of substrate concentrations $(10^{-4}$ to 10^{-2} M). It follows from Eq. (4-45) that the electrode response depends on the amount of enzyme in the membrane; 45 mg urease per ml of gel is a limiting value and at higher concentrations of the enzyme in the membrane the $E - [CO(NH_2)_2]$

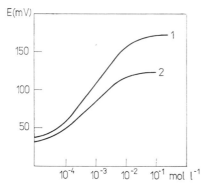

Fig. 9. Dependence of the response of an NH_4^+-sensitive electrode covered with a membrane containing immobilized urease on the urea concentration.
1 — 45 mg urease in 1 ml membrane gel; *2* — 24 mg urease in 1 ml membrane gel.

curve is identical with the curve (1) in Fig. 9. The theoretical dependences given in Fig. 9 were experimentally verified by measuring the response of an NH_4^+-sensitive electrode with a membrane containing immobilized urease [92] and of a CN^--sensitive electrode covered with a membrane containing immobilized β-glucosidase [94].

Thus it has been theoretically verified that the amount of reaction product is proportional to the substrate concentration in solution (Eq. (4-43)) provided that the substrate concentration is small $[(S < K_M)]$ and that the substance can be determined rapidly using a suitable sensor for the product. If a sensor is not available, small discs (with a diameter of about 1 cm) can be cut from the membrane with the immobilized enzyme, soaked in a suitable buffer and activator, placed in a vessel containing the substrate and the other reaction components and the solution stirred. At certain time intervals small samples (about 1 ml) are taken and the concentration of the product is determined in them by a suitable method.

5 Uncatalyzed Reactions

Until recently it was generally felt that uncatalyzed reactions were of little importance for kinetic analysis. If analytical methods for the determination of a substance based on catalyzed and uncatalyzed reactions are compared, it is found that this is correct. The advantages of catalytic reactions for determination of small concentrations of catalysts are obvious. Furthermore, probably no analyst would use a kinetic method based on an uncatalyzed reaction for determination of a substance when a suitable equilibrium method is available, because the kinetic method has poorer precision. On the other hand, in many determinations of organic substances, kinetic methods are more advantageous or even the only methods available. For example, ketones are determined by titration with hydroxylammonium chloride. Acetylacetone cannot be determined in this way, because its reaction with the titrant takes more than 50 hours [95]. Study of the kinetics of this reaction has shown that the only possible analytical use of this reaction is the measurement of its initial rate [96]. Another example is the determination of phenols or cresols based on their bromination. The equilibrium method is tedious and requires special equipment [95], but the kinetic method based on the measurement of the time required for the formation of monobrominated derivatives of the test substances is much simpler and no special apparatus is needed [97].

As can be seen from the above examples, kinetic methods based on uncatalyzed reactions have rather limited use for determination of a single substance, but are useful when no suitable equilibrium methods are available.

Kinetic methods are very important for analyses of mixtures of chemically similar substances — differential kinetic methods developed recently have found wide use, especially in organic analysis, but also in analyses of mixtures of metals on the basis of differences in the rates of substitution reactions of some complexes (see Chapter 6). In analyses of mixtures of similar organic substances, different rates of reaction with a single reagent due to their different structures are utilized.

5.1 The Effect of the Structure of the Reactants on the Reaction Rate

The fact that the structure of organic compounds affects the reaction rate was found empirically. Detailed study of many reactions of organic substances has shown that the reaction rate or the rate constant is a measurable parameter which is related to the structure of chemically similar substances reacting with a single reagent. For example, Hammett [99] compared the

TABLE 3

Relative values of the rate constants for esterification of carboxylic acids, RCOOH

Acid	k_{rel}
CH_3COOH	1.00
C_2H_5COOH	0.83
$C_6H_5CH_2COOH$	0.56
iso-C_3H_7COOH	0.54
$BrCH_2COOH$	0.54
$(CH_3)_2CBrCOOH$	0.037
tert.-C_4H_9COOH	0.025
$Cl_2CHCOOH$	0.0175
$(C_6H_5)_2CHCOOH$	0.0153
$Br_2CHCOOH$	0.0139
Cl_3CCOOH	0.0101
CH_3CBr_2COOH	0.0066
Br_3CCOOH	0.0037

rate of the esterification of various carboxylic acids in ethanol and found that the rate constants differ by as much as three orders of magnitude. Their relative values are given in Table 3, the rate constant for acetic acid being denoted as 1.00. Naturally, the great differences in the rate of esterification of various organic acids lend themselves to kinetic analysis of mixtures of these acids. A kinetic method [100] is based on esterification with diphenyl-diazomethane. There are also great differences in the rates of reactions other than esterification, involving variously substituted organic compounds and a single reagent (e.g. hydrolysis, saponification, condensation, etc.), which have a certain dependence on the structure of the organic compounds. Therefore, efforts have been made to find a general explanation of the laws governing the reactions of organic compounds and to apply kinetic differences to chemical analyses. Relationships between the reactivity and the structure of compounds are among the most important research topics in organic chemistry. So far, the treatment is not general enough to permit derivation of quantiative relationships describing the dependence of the reactivity on the compound structure. Only empirical equations are available, obtained on the basis of a treatment of vast experimental material [98, 101].

These equations express the dependence of experimental parameters, such as the rate or an equilibrium constant, on some independently variable factors that cannot be expressed numerically, e.g. substituents, solvents, etc. Therefore, they are generalizations of experimental data, indicating that certain substituents similarly affect the reactivity of various compounds in various reactions.

For substitution in a side chain of an aromatic compound,

$$\text{(5.1)}$$

where R is a substituent in any position with respect to the side chain, it holds, according to Hammett [102], that

$$\log (k/k_0) = \varrho\sigma \tag{5-2}$$

where k and k_0 are the rate constants for the substituted and unsubstituted substances (when hydrogen replaces R in Eq. (5-1)), respectively, and σ and ϱ are empirical constants. Quantity σ is the substituent constant and depends on the character and position of substituent R, whereas ϱ depends on the reaction type, conditions and the character of the side chain X. Constant σ is positive for groups capable of accepting electrons and negative for those capable of donating electrons and is a quantitative measure of the polar effect of substituent R. Constant ϱ is a measure of the reaction sensitivity toward polar effects.

Eq. (5-2) is called the Hammett equation and is only valid for substituents in meta and para positions. Selected values of σ are given in Table 4. If constants σ and ϱ are known, the relative rates can be calculated for series of reactions. According to Jaffé [103], the knowledge of 400 values of constant ϱ and 100 values of constant σ permits prediction of about 40,000 rate constants, which demonstrates the importance of the Hammett equation. However, it should be borne in mind that these calculations are limited by the conditions for the validity of the Hammett equation. The calculation of the relative rates for a pair of substances, 1 and 2, employs the equation

$$\log \left(k_2/k_1\right) = \left(\sigma_2 - \sigma_1\right) \varrho \tag{5-3}$$

The Hammett equation is valid only for aromatic compounds. In seeking

TABLE 4

The values of empirical constants according to ref. [103]

R	σ_m	σ_p
CH_3	−0.069	−0.170
CH_3CH_2	−0.070	−0.151
$CH(CH_3)_2$	−0.068	−0.151
$C(CH_3)_3$	−0.10	−0.197
C_6H_5	+0.06	−0.01
$COCH_3$	+0.376	+0.502
COC_6H_5	−	+0.459
CN	+0.56	+0.66
COO	−0.1	0.0
$COOH$	+0.35	+0.406
$COOCH_3$	+0.321	+0.385
NH_2	−0.16	−0.66
$N(CH_3)_3$	+0.88	+0.82
NO_2	+0.710	+0.778
O	−0.708	−1.0
OH	+0.121	−0.37
F	+0.337	+0.062
Cl	+0.373	+0.227
Br	+0.391	+0.232
I	+0.352	+0.276
SH	+0.25	+0.15
SCH_3	+0.15	0.0
$SOCH_3$	+0.52	+0.49

Subscripts m and p denote substitution in meta and para positions, respectively.

an analogous relationship for aliphatic compounds, marked steric effects must be considered. Taft [104] studied these effects and formulated an empirical equation, formally analogous to the Hammett equation. In contrast to the Hammett equation that holds only for some compounds, the Taft equation is valid for substances with any structure, but only for certain reaction types. The Taft equation is applied in the form

$$\log (k/k_0) = \varrho^*\sigma^* \tag{5-4}$$

where σ^* is a constant for a polar substituent. Its value was determined by Taft from the measurement of the rate of hydrolysis of aliphatic esters in acidic and alkaline media, using the relationship

$$\sigma^* = \frac{1}{2.48} \left[\log (k/k_0)_B - \log (k/k_0)_A \right] \tag{5-5}$$

where $(k/k_0)_B$ and $(k/k_0)_A$ are the ratios of the hydrolysis rate constants in basic and acidic media, respectively. The validity of Eq. (5-5) is based on the assumption that the steric effects for hydrolysis of an ester are the same in acidic and basic media and that the difference in the reaction rates can be used as a measure of the polar effects. As the hydrolysis proceeds differently in basic and acidic media, Taft [105] assumed that acid hydrolysis is controlled exclusively by steric effects. The equations,

$$\log (k/k_0)_B = 2.48\sigma^* + E_{S_B} \tag{5-6}$$

and

$$\log (k/k_0)_A = E_{S_A} \tag{5-7}$$

then hold.

Here E_S is the steric coefficient expressing the intensity of the steric effects and subscripts A and B denote the reaction in acidic and basic media, respectively. The values of some polar and steric constants calculated by Taft [104] are given in Table 5.

The validity of the Hammett and Taft equations has been verified many times and they can be used for prediction of the reaction rates of variously substituted compounds. It should be added that the equations do not include other factors that also may affect the reaction rate, e.g. the effect of characteristic groups neighbouring a substituent studied or the effect of conformation (for example, the rates of the same reactions of threo- and erythro-compounds are usually considered different [106]). As these effects have not yet been quantitatively expressed, predictions of reaction rates from the Hammett and Taft equations must be accepted with reserve and the reaction mechanism and the polar and steric effects of substituents must be known.

Differential kinetic methods developed for analyses of mixtures of chemi-

TABLE 5

Polar and steric constants for aliphatic compounds according to refs. [104, 105]

R	σ^*	E_s
H	0.49	1.24
C_6H_5	0.60	—
CH_3	0.0	0.0
C_2H_5	−0.10	−0.07
C_3H_7	−0.12	−0.36
iso-C_3H_7	−0.19	−0.47
tert.-C_4H_9	−0.30	−1.54
CH_3COO	2.0	—
CH_3CO	1.65	—
$ClCH_2$	1.05	−0.24
$BrCH_2$	1.0	−0.27
ICH_2	0.85	−0.37
cyclo-C_6H_{11}	−0.15	−0.79
cyclo-C_5H_9	−0.20	−0.51

cally similar substances are based on the differences in the reaction rates with a certain reagent (for details see Section 8.3.2). These methods are suitable for analyses of substances belonging to a homologous series or for mixtures of isomers or polymers. In an ideal case, the Hammett equation should provide information on whether substances to be analyzed will react with a reagent at different rates. For a pair of such substances it can be written that

$$\log (k_1/k_2) = \varrho(\sigma_2 - \sigma_1) + (E_{s_2} - E_{s_1}) = \varrho \, \Delta\sigma + E_s \tag{5-8}$$

Obviously, ratio k_1/k_2 must not equal unity if a differential kinetic method is to be used. The $\Delta\sigma$ value is generally constant for two given substances and cannot be altered by variation of the reaction conditions or a change of the reagent. Hence, ratio k_1/k_2 can be affected only by variation of constants ϱ and E_S. The value of constant ϱ depends on the reaction conditions and can be affected e.g. by varying the ionic strength, changing the solvent, etc., whereas constant E_S can be affected only by a change of the reagent. Theoretically, it should be possible to find reaction conditions and a reagent for which the k_1/k_2 ratio attains an optimal value for a given differential kinetic method. However, the values of the empirical constants are often lacking and the empirical equations do not include all the possible effects and therefore the decisive criterion for the choice of the reagent and of the reaction conditions in differential kinetic methods still depends on experimental results. Theoretical solution depends on further development of theoretical organic chemistry.

5.2 Oxidation-Reduction Reactions

Oxidation-reduction (redox) processes, important in analytical chemistry, can be represented by the scheme

$$Ox_1 + Red_2 \xrightleftharpoons{k_{12}} Red_1 + Ox_2 \tag{5-9}$$

where k_{12} is the experimentally determinable rate constant. In contrast, rate constants k_1 and k_2 for the reaction steps (isotope exchange reactions),

$$Ox_1 + v_1e^- \xrightleftharpoons{k_1} Red_1 \tag{5-10}$$

$$Red_2 \xrightleftharpoons{k_2} Ox_2 + v_2e^- \tag{5-11}$$

cannot usually be determined experimentally. A general treatment of the mechanism of redox processes is not yet available. For example, Michaelis [107] assumed that reactions of the type (5-9) proceed through a series of successive steps always involving exchange of a single electron; however, the validity of this assumption has not been fully experimentally verified. The explanation given by Taube [108] fits the experimental results much better. The author assumes the formation of an activated complex during the redox reaction, which enables electron transfer. The correctness of this assumption was recently verified by interpretation of the experimental data obtained from many redox reactions. At present the opinion prevails that all redox reactions can be divided into two groups according to the type of activated complex formed [134]. Electron transfer either takes place from one primary bonding system to the other, or an electron is transferred within a single system. In the former case an outer-orbital activated complex is formed, e.g. in the reduction of $Co(NH_3)_6^{3+}$ by hydrated Cr^{2+} ions [135]; in the latter case an inner-orbital activated complex is involved, e.g. in the reduction of $Co(NH_3)_5H_2O^{3+}$ by hydrated Cr^{2+} ions [136]*).

For the theory of redox reactions, the relationships derived by Marcus [109] are probably of the greatest importance, as they express the dependence of the rate constant of the redox reaction (5-9) on the rate constants of isotope exchange reactions (5-10) and (5-11) and the dependence of the rate constant of the homogeneous redox reaction (5-9) on the rate of the same reaction taking place as a heterogeneous process on a metal electrode. These dependences are expressed by the equations

$$k_{12} = (k_{11}k_{22}K_{12}f_{12})^{1/2} \tag{5-12}$$

$$(k_{chem}/Z_{chem})^{1/2} \cong k_{el}/Z_{el} \tag{5-13}$$

where K_{12} is the equilibrium constant of reaction (5-9), Z is the collision number (the number of collisions of the reacting particles per unit volume per unit time) and subscripts *chem* and *el* denote the homogeneous chemical and electrode reaction, respectively. Term f_{12} in Eq. (5-12) is defined by

$$\ln f_{12} = \frac{(\ln K_{12})^2}{4 \ln (k_{11}k_{22}/Z^2)} \tag{5-14}$$

Theoretical rate constants for redox reactions involving the formation of an activated complex can be calculated from Eq. (5-12). Equilibrium constant K_{12} is determined from the standard redox potentials and rate constants k_{11} and k_{22} can be obtained experimentally from radiometric measurements [110]. The values thus calculated agree very well with the experimental rate constants [111–113].

Eq. (5-13) was verified by Candlin and co-workers [114, 115] who compared the rates of the oxidation of Eu(II), V(II) and the complex

---

*) Square brackets are omitted when writing the formulae of simple coordination compounds to avoid confusion with expressions for equilibrium concentrations.

of Cr(III) with bipyridine by various complexes of Co(III) (e.g. $Co(NH_3)_6^{3+}$, $Co(NH_3)_5F^{2+}$, etc.) with the rate of electrochemical reduction of these Co(III) complexes determined polarographically [116]. It follows from Eq. (5-13) that the rate constant ratio, k_{chem}/k_{el}, must also be constant for a series of similar compounds oxidized by the same reagent in solution (k_{chem}) and reduced electrochemically at a constant electrode potential (k_{el}). In the works cited this constancy was verified experimentally. However, it must be pointed out that the Marcus equations are not generally valid for all types of redox reactions, but hold only for redox reactions of hydrated or weakly complexed metal ions in which electron transfer is effected through the formation of an activated complex.

5.3 Fast Reactions

Fast reactions cannot be excluded from the discussion of the use of un-catalyzed reactions in kinetic analysis. Their rate cannot be measured by common analytical measuring techniques (e.g. spectrophotometry or electroanalytical methods) and they were earlier considered as taking place instantaneously. The need to study fast reactions stems from efforts to elucidate the mechanism of chemical reactions in solution that consist of a series of fast reaction steps. In the interaction of two uncharged particles in solution the maximum possible reaction rate is determined by the second-order rate constant with a value of the order of $10^9 \, l \, . \, mol^{-1} \, s^{-1}$, whereas in the reaction of two ions with unit positive and negative charges the reaction rate is higher and the rate constant can be as large as $10^{11} \, l \, . \, mol^{-1} \, s^{-1}$. The fastest reaction in solution is neutralization,

$$H^+ + OH^- \rightarrow H_2O \tag{5-15}$$

whose rate from left to right has a rate constant of $1.4 \, . \, 10^{11} \, l \, . \, mol^{-1} \, s^{-1}$ at 25 °C [117].

Because mixing of small volumes of two solutions normally takes at least one second, it is obvious that special measuring techniques must be employed for monitoring of concentration changes of substances participating in fast reactions. For reactions with half-times down to 10^{-3} s, flow-through methods have been developed [118]. The initial substances are placed in two cylindrical vessels with pistons. In the beginning of the reaction the two solutions are driven at a constant velocity into a mixing chamber of very small volume, in which the solutions are homogeneously mixed within less than one millisecond. The mixture flows from the mixing chamber through a tube in which the concentration of the required component is measured at a certain distance from the chamber. The principle of the method is depicted in Fig. 10. At a constant flow-rate, the composition of the mixture at a certain place in the tube is constant. Time interval t that elapses between

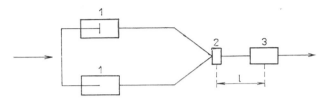

Fig. 10. Principle of the measurement of reaction rate by the flow-through method. 1 — vessels with pistons, containing the reactants; 2 — mixing chamber; 3 — point at which the concentration of a component of the reaction system is measured.

mixing the reactants in the mixing chamber and measuring the concentration at point 3 (Fig. 10) depends on the flow-rate Q_V and distance l between the mixing chamber and measuring point 3. If the cross-sectional area of the tube is S, then

$$Q_V = l . S/t \qquad (5\text{-}16)$$

or

$$t = l \frac{S}{Q_V} \qquad (5\text{-}17)$$

Because S and Q_V are constant, t is proportional only to distance l. Therefore, the reaction course can be followed by measuring the concentration of a component under constant conditions for various values of l.

The flow-through method has been variously modified; the above technique is called the continuous flow method and permits the measurement of the rate of reactions with half-times down to 10^{-3} s [119]. A very frequent modification that also has found use in analytical chemistry [120, 121] is the stopped-flow method [122]. Here distance l between the mixing chamber and the measuring point is constant and as small as possible. The measurement is carried out in dependence on time, because the flow is stopped when the solution reaches the measuring point after passage through the mixing chamber. For this purpose, another vessel with a piston is included. The stopped-flow method can be used for reactions with half-times down to 10 ms. The concentration is mostly measured spectrophotometrically, but electrochemical measurements are also possible.

For reactions reaching equilibrium within a few milliseconds the flow-through methods cannot be used and special methods must be employed. These are based on the following reasoning: If the reaction system is in equilibrium and this equilibrium is perturbed in a defined way, it can be assumed that the same reaction that led to the original equilibrium will return the system back to equilibrium conditions. By monitoring the kinetics of the reaction leading to equilibrium re-establishment, the problem of solution mixing, which is practically insoluble with very fast reactions, can be avoided. This idea forms the basis of various relaxation methods [123, 124], that enable measurement of the rate for reactions with half-times down to 10^{-9} s. It is evident that, in contrast to flow-through methods, methods based on perturbation of the equilibrium are applicable only to reversible reactions.

In the study of fast reactions the existence of hydrated electrons was experimentally verified [125]. Hydrated electrons e_{aq}^- are formed on passage of ionizing radiation (e.g. γ-rays or accelerated electrons) through water. They are unstable particles with life-times in water determined by the reactions

$$e_{aq}^- + e_{aq}^- + 2\,H_2O \xrightarrow{\;k_1\;} H_2 + 2\,OH^- \tag{5-18}$$

$$e_{aq}^- + H_3O^+ \xrightarrow{\;k_2\;} H + H_2O \tag{5-19}$$

$$e_{aq}^- + H_2O \xrightarrow{\;k_3\;} H + OH^- \tag{5-20}$$

where $k_1 = 1 \times 10^{10}$ l.mol^{-1}.s^{-1}, $k_2 = 2.3 \times 10^{10}$ l.mol^{-1}.s^{-1} and $k_3 = 16$ l.mol^{-1}.s^{-1} [126, 127], but they have a well-defined absorption spectrum with a maximum at 720 nm, so that their reactions can be experimentally followed (the molar (linear) absorption coefficient of a hydrated electron is 1.5×10^4 l.mol^{-1}.s^{-1} at 720 nm). The hydrated electron is the single strongest chemical reductant capable of existence in water.

The pulse radiolysis method was used to measure the rate constants for reactions of hydrated electrons with various organic substances [128 – 130] and it has been found that oxidants react with hydrated electrons at a rate approaching the diffusion limit ($k \sim 10^{10}$); other substances react rather slowly with hydrated electrons. Various complexes with the same central atom react with hydrated electrons at various rates [130]. Among organic substances, those containing nitrogen mostly react fastest with hydrated electrons [128, 131, 132].

On this basis, Hart and Fielden [133] have shown that any organic or inorganic compound that reacts with hydrated electrons at a rate higher than that of decomposition of the hydrated electrons can be determined by measuring the reaction rate. Hydrated electrons are generated by X-ray pulses in solutions with pH 11 and saturated with hydrogen; the decrease in the hydrated electron concentration during the reaction is monitored spectrophotometrically. The rate constants need not be known; a calibration curve must be constructed for each test substance (provided that it is available pure). The method is very sensitive and permits the determination of substances at concentrations down to 10^{-7} mol . 1^{-1} with a relative error of $\pm 10\%$.

6 Reactions of Complexes

Complexes and complexing agents are important in both classical and instrumental chemical analysis. The rate of their reactions varies considerably; formation of some complexes is characterized by rate constants of the order of 10^9 s^{-1} and substitution reactions of some complexes are so slow that their half-times amount to tens of hours. A great variety of catalytic effects is typical of reactions of complexes and can be utilized analytically.

6.1 Kinetics of Complex Formation in Aqueous Solutions

Complex formation is generally a very fast process. In the reaction of a solvated metal ion with a complexing agent, simple addition never occurs, but the ligand replaces a solvent molecule in the cation coordination sphere,

$$MS_n + L \rightarrow MS_{(n-1)}L + S \tag{6-1}$$

and the reaction can proceed until the coordination sphere of ligand L is completely occupied. In Eq. (6-1) M denotes the cation, S the solvent molecule, L the ligand and n is the coordination number.

In complex formation in aqueous solutions, the reaction rate-determining step is the substitution of water molecules in the cation inner coordination sphere by complexing agent molecules and usually takes less than 1 millisecond. Detailed study of such fast processes was, naturally, made possible only by the development of some new instrumental techniques, such as nuclear magnetic and electron spin resonance or relaxation methods. These methods have provided data on the basis of which knowledge can be systematized and metal ions classified into three groups [137].

The first group contains cations with water molecules bound by labile

87

bonds in their inner coordination sphere; they are therefore substituted by ligands in virtually any interaction of the two particles. The substitution rate is limited only by diffusion and is not very dependent on the metal ion quality. This group includes the alkali metals and some alkaline earth elements.

The second group contains cations with small ionic radii and high charge, i.e. Be^{2+} and most tervalent cations, such as Al^{3+}, Fe^{3+}, La^{3+} or In^{3+}. Water molecules are strongly bound in the inner coordination sphere of these ions and thus the substitution process is relatively slow. Because of the small radius and high charge of the ion, internal hydrolysis may occur, i.e. the water molecule decomposes to the OH^- group bound by the cation and to H^+ bound more strongly by the ligand molecule. Therefore, the rate of complex formation for cations of this group depends on the reagent basicity. For example, a ligand that is a good proton acceptor hastens the hydrolysis.

The third group comprises the ions of the transition elements and the main group of metals and is characterized by substitution rates specific for each metal ion and virtually independent of the ligand bonding strength. The substitution rate thus depends on the charge, size and electron configuration of the cation and is independent of the character of the ligand. It decreases with increasing charge and decreasing cation size. In formation of polydentate complexes (e.g. with EDTA), multiple substitution takes place and two limiting cases can occur. If the entering ligand is bonded more strongly than water molecules, its presence makes the remaining water molecules labile in the inner coordination sphere of the metal ion. In the opposite case, the solvent molecules are stabilized. Therefore, the "local" charge density at the bonding site of the ligand molecule is more important for the overall substitution rate than the total charge of the complexing agent [138, 139].

6.2 Catalytic Effects in Reactions of Complexes

Numerous reactions involving complexing agents or complexes exhibit various catalytic effects. In complex formation reactions catalytic effects caused by the ligand, the metal ion and the complex have been encountered and studied. Catalytic effects occuring in substitution reactions of chiefly polydentate complexes and in coordination chain reactions are analytically important.

A catalytic effect of the ligand appears in reaction systems with slow reaction between metal ion M and ligand X,

$$M + X \xrightarrow{k_1} MX \tag{6-2}$$

and fast reaction between metal ion M and ligand Y,

$$M + Y \xrightarrow{k_2} MY \tag{6-3}$$

complex MX being thermodynamically more stable than complex MY. If less stable complex MY also reacts with ligand X, mixed complex MXY is temporarily formed,

$$MY + X \xrightarrow{k_3} MXY \xrightarrow{k_4} MX + Y \tag{6-4}$$

In reactions (6-2) to (6-4), $k_1 - k_4$ are the appropriate rate constants for which it holds that $k_2, k_3, k_4 > k_1$ (the charges are omitted for the sake of simplicity). These reactions schematically represent the formation of complex MX catalyzed by ligand Y. The actual reaction mechanism will, of course, be more complex, e.g.

$$M(H_2O)_6 + Y \rightarrow M(H_2O)_6 Y$$
$$\downarrow$$
$$M(H_2O)_5 Y \xrightarrow{X} M(H_2O)_5 YX \longrightarrow M(H_2O)_5 X + Y$$
$$\downarrow$$
$$M(H_2O)_4 Y \xrightarrow{X} M(H_2O)_4 YX \tag{6-5}$$

6.3 Substitution Reactions of Complexes

Study of substitution reactions of complexes has clarified many questions of complex formation mechanisms and stability of complexes and some of these reactions have also been used for indirect determination of some metals. Substitution reactions of complexes are also often accompanied by catalytic effects. It has been known for a long time that some metals accelerate the ligand replacement of water molecules, e.g. with some complexes of cobalt with amines [140, 141]. From the analytical point of view, substitution reactions of complexes, in which ligand exchange occurs and which are catalyzed by metal ions, are more important. Typical reactions of this type are the substitution reactions of hexacyanoferrate with water [142, 143], bipyridine [144], or phenanthroline [145], represented in a simplified form as

$$[Fe(CN)_6]^{4-} + X \rightarrow [Fe(CN)_5X]^{3-} + CN^- \tag{6-6}$$

where X is the molecule of water or other complexing agent. Reaction (6-6) is catalyzed by all metal ions that form stable complexes with cyanide (e.g. mercury, silver, gold) and can be used for determination of these metals [3].

6.3.1 SUBSTITUTION REACTIONS OF POLYDENTATE COMPLEXES

Substitution reactions of polydentate complexes have a more complex mechanism. Reactions of complexes of EDTA (ethylenediaminetetraacetic acid, H_4Y) and its analogues are well known [146]. Three types of substitution reaction of polydentate complexes are generally possible.

First, a ligand can be exchanged between two metals,

$$ML + M' \rightleftharpoons M'L + M \tag{6-7a}$$

Second, a metal can be exchanged between two ligands,

$$ML + X \rightleftharpoons MX + L \tag{6-7b}$$

Third, complete exchange between two complexes, one of which is polydentate, can occur,

$$ML + M'X \rightleftharpoons M'L + MX \tag{6-7c}$$

(Charges are omitted for the sake of simplicity.)

Reaction systems (6-7a) to (6-7c) hold only for systems involving ligands capable of forming only complexes with a metal-to-ligand ratio of 1 : 1. The kinetics of these reactions have been studied in detail [146–149] and it has been found that several coordination bonds in the complex are broken (six in EDTA complexes) with gradual formation of unstable intermediates. Therefore, these reactions are usually slow. If re-formation of the initial complex can be suppressed, e.g. by increasing the hydrogen ion concentration in the reaction system with resultant protonation of the ligands liberated, the overall rate of the substitution reaction increases.

With substitution reactions between zinc ions and the nickel EDTA complex (NiY^{2-}),

$$NiY^{2-} + Zn^{2+} \xrightleftharpoons[k_{Ni}^{ZnY}]{k_{Zn}^{NiY}} ZnY^{2-} + Ni^{2+} \tag{6-8}$$

the overall reaction rate is given by [14]

$$k = k_{Zn}^{NiY}[Zn^{2+}] + k_d^{NiY} \tag{6-9}$$

where k is the overall rate constant for reaction (6-8) and k_d^{NiY} is the rate constant for the dissociation of the Ni-EDTA complex. The $k_{Zn}^{NiY}[Zn^{2+}]$ term in Eq. (6-9) is constant if excess zinc ions are present in the reaction system. The reaction of cupric ions with the Ni-EDTA complex,

$$NiY^{2-} + Cu^{2+} \xrightarrow{k_{Cu}^{NiY}} CuY^{2-} + Ni^{2+} \tag{6-10}$$

is much faster than reaction (6-8) [149]. The reaction of Zn^{2+} ions with the Cu-EDTA complex,

$$CuY^{2-} + Zn^{2+} \rightarrow ZnY^{2-} + Cu^{2+} \tag{6-11}$$

is also very fast [150]. These facts suggest that the rate of the zinc ion reaction with the Ni-EDTA complex (reaction (6-8)) might be increased by adding traces of Cu^{2+} ions. According to reactions (6-10) and (6-11), the CuY^{2-} complex is rapidly formed, immediately reacts with zinc ions to form the ZnY^{2-} complex and the Cu^{2+} ions are liberated. Cupric ions are thus not consumed and can be considered as a catalyst.

The rate of the catalyzed reaction is given by [149]

$$k^* = k_{Zn}^{NiY}[Cu^{2+}] + k_{Cu}^{NiY}[Zn^{2+}] \tag{6-12}$$

provided that the rate of dissociation of the Ni-EDTA complex is neglected, which is acceptable, and that the pH of the reacting system is close to 5. As zinc ions are present in a large excess, the second term on the right-hand side of Eq. (6-12) is constant and the reaction rate is proportional to the Cu^{2+} ion concentration. The plot of rate constant k^* vs the Cu^{2+} ion concentration is a straight line whose slope determines rate constant k_{Cu}^{NiY} and whose intercept with the vertical axis determines rate constant k_{Zn}^{NiY}. Hence copper can be determined on this basis using a suitable catalytic method.

6.3.2 COORDINATION CHAIN REACTIONS

Detailed study of substitution reaction (6-7c) involving a polydentate complex led to discovery of catalyzed chain reactions. Margerum et al. [151, 152] have found that complex ligand exchange occurs in the reaction of Cu-EDTA with Ni-trien (trien = triethylenetetramine),

$$NiT^{2+} + CuY^{2-} \rightarrow CuT^{2+} + NiY^{2-} \tag{6-13}$$

where T is trien. Reaction (6-13) is very slow, the presence of traces of any ligand (EDTA or trien, but not of the complex) increases the reaction rate and the presence of free metal ions decreases it. As a number of important kinetic parameters for the complexes involved in reaction (6-13) are known, it was assumed that the exchange proceeds through a series of steps with chain character:

$$NiT^{2+} \underset{k_{-1}}{\overset{k_1}{\rightleftharpoons}} Ni^{2+} + T \tag{6-13a}$$

$$T + CuY^{2-} \underset{k_{-2}}{\overset{k_2}{\rightleftharpoons}} CuT^{2+} + Y^{4-} \tag{6-13b}$$

$$Y^{4-} + NiT^{2+} \overset{k_3}{\longrightarrow} NiY^{2-} + T \tag{6-13c}$$

$$Y^{4-} + Ni^{2+} \overset{k_4}{\longrightarrow} NiY^{2-} \tag{6-13d}$$

This mechanism shows that the dissociation of the Ni-trien complex produces free trien (initiation of the reaction), which reacts with the Cu-EDTA complex and forms one of the products, CuT^{2+}, with simultaneous liberation of the EDTA ligand. The latter reacts with the Ni-trien complex forming the other product, NiY^{2-}, and liberating another trien that reacts further according to reaction (6-13b). Reactions (6-13b) and (6-13c) thus represent the propagation cycle and step (6-13d) is the termination reaction of the process. If this reaction mechanism is valid, then the rate of substitution reaction (6-13) is determined by the rate of the propagation step (6-13b) or (6-13c), i.e.

$$\frac{d[CuT^{2+}]}{dt} = k_2[T][CuY^{2-}] - k_{-2}[Y][CuT^{2+}] \tag{6-14}$$

It was found experimentally [16] that, in the presence of excess Ni-trien complex, the rate of reaction (6-13) is given by

$$\frac{d[CuT^{2+}]}{dt} = k_0[CuY^{2-}][NiT^{2+}]^{1/2} \tag{6-15}$$

and in the presence of excess Cu-EDTA complex by

$$\frac{d[CuT^{2+}]}{dt} = k_0'[NiT^{2+}]^{3/2} \tag{6-16}$$

where k_0 and k_0' are the experimental rate constants obtained from spectrophotometric monitoring of the concentration of the Cu-trien complex in dependence on time.

The assumed reaction mechanism was verified employing the stationary state principle. For the rate of reaction (6-13) the relationship

$$\frac{d[CuT^{2+}]}{dt} = k_2 \left(\frac{k_1}{k_{-1}}\right)^{1/2} [CuY^{2-}][NiT^{2+}] \tag{6-17}$$

was derived for excess Ni-trien complex and

$$\frac{d[CuT^{2+}]}{dt} = \left(\frac{k_1}{k_4}\right)^{1/2} k_3[NiT^{2+}]^{3/2} \tag{6-18}$$

for excess Cu-EDTA complex, which is in very good agreement with experimental dependences (6-15) and (6-16). This chain mechanism has also been demonstrated for other complexes, e.g. for substitution reactions between the Cu-EDTA and Zn-pentane (N,N,N',N'-tetrakis(2-aminoethyl)ethylenediamine) complexes [153], between the Cu-EDTA and Ni-tetrene (tetraethylpentamine) complexes [148] or between the Cu-TTHA (triethylenetetraminehexaacetic acid) and Ni-trien complexes [154]. The chain mechanism, however, was not found e.g. for the reactions of the Cu-EDTA complex with Ni complexes of ammonia, glycine or ethylenediamine; free ligand then does not affect the rate of the substitution process [155].

The possibilities of analytical use of coordination chain reaction (6-13) follow from the fact that trace amounts of added ligand considerably increase the reaction rate. Addition of free ligand suppresses the kinetic effects of the initiation and termination reactions ((6-13a) and (6-13d)) and the rate of the product formation is determined by the rate of the propagation cycle. Shortly after ligand addition the system attains stationary state when the rates of reactions (6-13b) and (6-13c) are equal, so that it holds that

$$k_T^{CuY}[T][CuY^{2-}] = k_Y^{NiT}[Y^{4-}][NiT^{2+}] \qquad (6\text{-}19)$$

where k_T^{CuY} is the rate constant for the triene reaction with the Cu-EDTA complex and k_Y^{NiT} is that for the EDTA reaction with the Ni-triene complex. It is evident from Eq. (6-19) that the $[Y^{4-}]/[T]$ ratio is constant,

$$\frac{[Y^{4-}]}{[T]} = \frac{k_T^{CuY}[CuY^{2-}]}{k_Y^{NiT}[NiT^{2+}]} \qquad (6\text{-}20)$$

The value of this ratio depends on the solution pH, because the rate constants are also pH-dependent. To simplify the expression for the rate of the substitution reaction, it is advantageous that the EDTA concentration is made equal to that of triene, because then the reaction rate can be expressed by a first order equation, as follows from Eq. (6-19). In the following equations, symbol Y^{4-} is replaced by Y' for the sake of simplicity, because variously protonated EDTA ions are present in dependence on the solution pH (HY^{3-}, H_2Y^{2-}, etc.). Experimentally determined optimum conditions for monitoring reaction (6-13) involve a solution pH of about 8 and approximately a 20-fold excess of CuY^{2-} over Ni^{2+}. Then $k_T^{CuY} > k_Y^{NiT}$, hence $[Y] \gg [T]$ and therefore $[Y']$ remains constant during the reaction. The reaction is monitored spectrophotometrically at 550 nm (the absorption maximum of the Ni-trien complex) [152]. The rate of reaction (6-13) is given by

$$\frac{d[CuT^{2+}]}{dt} = -\frac{d[NiT^{2+}]}{dt} = k_0[NiT^{2+}] \qquad (6\text{-}21)$$

with experimental rate constant k_0 defined by

$$k_0 = k_Y^{NiT}[Y'] \qquad (6\text{-}22)$$

If no ions forming complexes with EDTA are present in the test solution, then $[Y']$ in equation (6-22) corresponds to the amount of EDTA $[Y']_p$ actually added to the reaction system. Then the plot of Eq. (6-22), i.e. the dependence of k_0 on $[Y']_p$, is a straight line passing through the origin (Fig. 11, straight line a). If the studied reaction system contains traces of

94

metal ions, then the plot is straight line *b* and the $[Y']_p$ value at point *B* gives the amount of EDTA required to form complexes with the metal impurities. Thus straight line *b* corresponds to the blank experiment. If the reaction system also contains a small amount of a test ion (e.g. Zn^{2+}), then the k_0 vs $[Y']_p$ dependence is represented by straight line *c* and the $[Y']_p$ value at point *C* gives the total amount of EDTA required to complex the impurities and the Zn^{2+} present.

In the determination the blank is performed first, then the solution is divided into several aliquots and to each of them a different amount of

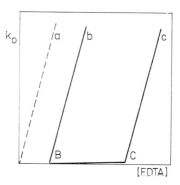

Fig. 11. Dependence of rate constant k_0 on the amount of EDTA added.
a — theoretical dependence (no impurities present); *b* — blank (the system is contaminated by traces of metal ions); *c* — sample.

EDTA is added. These solutions are mixed with the reaction mixture (a mixture of the two complexes and a buffer) and the k_0 values are measured. Extrapolating the k_0 vs $[Y']_p$ dependence to zero rate of reaction (6-13) yields point *C* (Fig. 11). Basically, this is a titration of a metal ion that acts as an inhibitor of the EDTA catalyzed reaction. In this way trace concentrations (10^{-6} to $10^{-9}M$) of metals forming stable complexes with EDTA can be determined.

For effective use of substitution reaction (6-13), the rate of propagation reactions (6-13b) and (6-13c) must, according to Margerum and Steinhaus [152], considerably exceed the rate of dissociation of the initial complexes. The lowest concentration determinable according to reaction (6-13) is given by the concentration of the free ligand that is introduced into the reaction system through dissociation of the Cu-EDTA or Ni-trien complex; it is five times greater than the concentration given by the dissociation. For the method to be sufficiently sensitive and precise the ions of the metals to be determined must form sufficiently stable complexes with EDTA. If a metal is to be determined in a concentration range from 10^{-6} to 10^{-8}M, the conditional stability constant of its complex with EDTA must exceed 10^6. The values of conditional stability constants strongly depend on the pH, which can also be used for masking of some metals. Common masking agents known from complexometry can also be used.

To accelerate this method of determination, an automatic analyzer has been constructed [156], which permits the determination of metals in a concentration range from 10^{-6} to 10^{-8}M. One determination takes about 10 minutes and the error is $\pm 2 \times 10^{-8}$M.

Coordination chain reactions can also be used for indirect determinations when the test substance causes a change in the concentration of EDTA in the reaction system. In this way e.g. traces of oxygen can be determined [157], using its reaction with $Co(CN)_5^{3-}$ [158] in the presence of EDTA.

6.3.3 SUBSTITUTION REACTIONS OF BINUCLEAR COMPLEXES

Among reagents forming mono- and binuclear complexes, reactions of complexes of triethylenetetraminehexaacetic acid (TTHA) have been mainly analytically studied [159–161]. Various types of substitution reactions of TTHA complexes were recently studied, namely, a) reactions of complexes with metal ions, b) reactions of complexes with other ligands and c) reactions between two complexes. Reactions of mononuclear complexes are discussed in Section 6.3.1. Reactions of TTHA complexes with other ligands have the same mechanism as analogous reactions of EDTA complexes, as follows from the study of the kinetics of the reaction of the Ni-TTHA complex with CN⁻ ions [162] and of TTHA with Ni-tetrene [163]. On the other hand, substitution reactions of TTHA with other metal ions substantially differ from analogous substitution reactions of mononuclear complexes. A typical example is the reaction of the binuclear Cu/TTHA complex with Ni(II) ions [164, 165]. The reaction proceeds in two steps. In the first, the mixed binuclear Cu-Ni-TTHA complex is formed,

$$Cu_2X^{2-} + Ni^{2+} \rightarrow CuNiX^{2-} + Cu^{2+} \tag{6-23}$$

and the formation of the Ni_2X^{2-} product,

$$CuNiX^{2-} + Ni^{2+} \rightarrow Ni_2X^{2-} + Cu^{2+} \tag{6-24}$$

is the rate-determining step (X is the TTHA anion).

In this reaction scheme, reaction intermediate $CuNiX^{2-}$ is in mobile equilibrium with free Ni^{2+} ions, the Cu-TTHA mononuclear complex, free Cu^{2+} ions and the Ni-TTHA mononuclear complex, so that step (6-24) can be expressed by a simplified reaction scheme,

$$
\begin{array}{l}
CuX^{4-} + Ni^{2+} \\
\quad \updownarrow \\
CuNiX^{2-} \qquad\qquad + Ni^{2+} \longrightarrow Ni_2X^{2-} + Cu^{2+} \\
\quad \updownarrow \\
NiX^{4-} + Cu^{2+}
\end{array}
\tag{6-25}
$$

Because Ni^{2+} ions are present in a great excess over the mixed binuclear complex in this reaction scheme, the formation of CuX^{4-} particles is strongly suppressed (the dashed arrow in Eq. (6-25)) and the resultant reaction product, Ni_2X^{2-}, is chiefly formed by addition of nickel to the NiX^{4-} particles.

Similar to the reactions of EDTA complexes, it was found that the reaction of the Cu-TTHA mononuclear complex with Ni-tetrene has chain character. The rate of this reaction can also be increased by adding traces of one of the two ligands [166]. On the other hand, the chain mechanism has not been reliably demonstrated for the reaction of the Cu-TTHA binuclear complex with Ni-tetrene. In the beginning of this reaction, the concentration of the Cu-TTHA binuclear complex rapidly decreases owing to the rapid formation of the Cu-Ni-TTHA mixed binuclear complex, whose formation prevents the chain mechanism. Hence the accelerating effect of the ligand is strongly suppressed and the reaction is unsuitable for analytical purposes.

6.4 Analytical Use of the Modifying Effect of Complexing Agents on the Rate of Catalytic Reactions

Catalytic reactions involving very small concentrations of metal ions as catalysts are important for the determination of trace concentrations of metals (see Section 10.1). It is evident that complexing agents binding metal ions (catalysts) affect the rate of catalytic reactions. The possibility of analytically utilizing these modifying effects of complexing agents was pointed out by Yatsimirskii in his monograph [3]. Mottola [167] recently returned to this problem and studied the effect of complexing agents on the rate of oxidation of various organic compounds catalyzed by some metal ions. The modifying effects are of three kinds, determining their possible analytical use.

The first kind is an inhibition effect of the ligand; the reaction rate decreases because the catalyst is bound in a stable complex with the ligand. The most important analytical applications of this effect are catalytic end-point titrations [168, 169], where the titrant is a catalyst of the indicator reaction and the titrand is its inhibitor. The inhibitor is consumed during the titration and the first traces of the free reagent-catalyst appear at the equivalence point, leading to an immediate increase in the rate of the indicator reaction. A classical example of a catalytic end-point titration is the determination of traces of silver. Silver ions inhibit the oxidation of tervalent arsenic by ceric ions and iodide ions catalyze this reaction. Hence the indicator reaction of As^{3+} with Ce^{4+} exhibits virtually zero rate in the presence of Ag^+. If the Ag^+ ions are titrated by a potassium iodide standard solution, the rate of the indicator reaction sharply increases at the equivalence point. An advantage is the fact that the reaction of the titrant with the indicator is not stoichiometric, but a catalytic effect is used which appears immediately on attaining the equivalence point. Hence the indication of catalytic end-point titrations is much more sensitive than that using indicators.

The second kind of ligand modifying effect on the rate of catalytic reactions is the catalytic effect of the complex formed by the reaction of the ligand with the catalyst — a metal ion. These reactions can be used for determination of traces of metal ions (catalysts) and of traces of complexing agents. An example of analyses of this type is the determination of low concentrations of nitrilotriacetic acid (NTA) [170], by measuring the rate of oxidation of malachite green by potassium periodate catalyzed by Mn^{2+} ions. On addition of NTA the relatively stable Mn-NTA complex is formed, causing a substantial increase in the indicator reaction rate. The constant concentration method (see Section 8.2.1) is advantageous for analytical use of these effects.

The third kind of ligand effect on the rate of catalytic reactions is transient increase in the reaction rate, caused by the catalyst-ligand complex that enables the reaction to take place with a lower activation energy and thus with a rate higher than in the presence of the catalyst alone. It is typical of these reactions that the catalyst is not cyclically renewed, but the complex formed is either decomposed or converted into an inactive form (e.g. by a change in the valence of the central ion) and consequently the increase in the reaction rate is only transient. Monitoring of this effect thus requires good apparatus, especially a rapid mixing device and rapid signal recorder which monitors time changes in the concentration of a reaction component and thus also the reaction rate. This effect has not yet been extensively used in analyses. One of the few examples is the determination of small amounts of oxalic acid in urine [171], based on an increase in the rate of the reaction of the complex of ferrous ions with 1,10-phenanthroline (ferroin) with hexavalent chromium ions, caused by the formation of $Cr(IV)$, $Cr(V)$ and $Cr(VI)$ complexes with oxalic acid. These complexes are reduced to the complex of tervalent chromium with oxalic acid, which is inactive.

6.5 Differential Kinetic Methods Based on Reactions of Complexes

2/55/0

Kinetic studies of substitution reactions [146–152] of the complexes of EDTA and related compounds have shown that, during substitution reactions of the complexes of trans-1,2-diaminocyclohexane-N,N.N',N'-tetraacetic acid (DCTA), direct reaction of the DCTA complex with another metal ion does not occur [172–174], while direct exchange of two metals (by reaction type (6-7a)) is possible during substitution reactions with other EDTA type complexes. It has been found that the mechanism of the DCTA complex substitution reaction consists of the following steps

$$\text{M-DCTA} + \text{H}^+ \rightarrow \text{H-DCTA} + \text{M} \tag{6-26}$$

$$\text{H-DCTA} + \text{M}' \rightarrow \text{M}'\text{-DCTA} + \text{H}^+ \tag{6-27}$$

where M and M' are two different metal ions, whose charges are not given here. Reaction (6-26) is the rate-determining step in the reaction system. Its rate, which depends solely on the H^+ ion concentration, is specific for each metal. Differences in the rates of the dissociation of various DCTA complexes are very large; for example, at a particular H^+ ion concentration the dissociation half-time for the Ba-DCTA complex is 10 ms and that for the Ni-DCTA complex 100 years. This kinetic behaviour of the DCTA complexes, which is rather unique, has been explained on the basis of the structure of the DCTA complex molecule [175], where the cyclohexane ring stabilizes the spatial arrangement of the bonding groups, as shown by the following formula

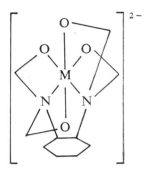

For analytical application of the reaction system described by Eqs. (6-26) and (6-27), Margerum and co-workers [176] used cupric or lead (II) ions as ion M', which hastens the dissociation of the M-DCTA complex by depleting the free ligand, H-DCTA (reaction (6-26)), and permits spectrophotometric monitoring of the reaction (an increase in the concentration of the Cu or Pb complex with DCTA). Under these conditions, the rate of dissociation of the M-DCTA complex equals the rate of formation of the Pb- or Cu-DCTA complex and is given by the relationship,

$$\frac{d[CuCy]}{dt} = k_d[MCy] \tag{6-28}$$

where Cy is the DCTA anion and k_d is the rate constant of the M-DCTA complex dissociation.

As the M-DCTA complex may be present in the analyzed solution as M-DCTA, MH-DCTA or even M(X)-DCTA (where X is a halogen or another monodentate ligand), the dissociation rate is further determined by the rate constants k_H^{MCy}, k_H^{MHCy} and k_H^{MXCy} and the overall rate constant, k_d (Eq. (6-28)) can be defined by the relationship

$$k_d = \frac{k_H^{MCy} + k_H^{MHCy}K_{MHCy}[H^+] + k_H^{MXCy}K_{MXCy}[X^-][H^+]}{1 + K_{MHCy}[H^+] + K_{MXCy}[X^-]} \tag{6-29}$$

where $K_{MHCy} = [MHCy]/[MCy][H^+]$, $K_{MXCy} = [MXCy]/[MCy][H^+]$.

Equation (6-29) is less complicated in practice, as the formation of mixed complexes MXCy is negligible, except when M is Hg^{2+} or when the k_H^{MHCy} and k_H^{MCy} values are comparable. Therefore, knowledge of the k_H^{MCy} values

TABLE 6

Rate constants for dissociation of M-DCTA complexes (25° C, 0.1M $NaClO_4$)

Metal	k_H^{MCy}, $mol^{-1}s^{-1}$	pH
Ba*)	1.1×10^8	7.0—7.6
Sr*)	6.1×10^6	5.5—7.6
Ca*)	4.1×10^5	5.5—7.6
Mg*)	6.3×10^4	5.5—7.6
Mn(II)	3.2×10^2	4.7—6.1
Zn	1.7×10^2	4.5—6.8
La	1.3×10^2	4.0—5.4
Ce(III)	53	4.0—5.4
Cd**)	42	3.8—4.7
Pr	35	4.0—5.4
Pb	23	4.3—5.8
Nd	16	4.0—5.4
Sm	5.1	4.0—5.4
Cu(II)	3.9	2.9—4.8
Co(II)	3.2	3.1—5.5
Eu	2.2	4.0—5.4
Gd	1.3	4.0—5.4
Tb	0.58	4.0—5.4
Y	0.36	4.0—5.4
Dy	0.26	4.0—5.4
Th	0.23	4.0—5.4
Ho	0.13	4.0—5.4
Er	0.053	4.0—5.4
Tm	0.037	4.0—5.4
Yb	0.023	4.0—5.4
Sc	0.019	3.0—3.7
Lu	0.017	4.0—5.4
Al	0.011	2.0—2.8
Ni	3.4×10^{-4}	4.0—5.2

*) 0.5M NaAc, **) 0.1M KNO_3 (according to Margerum [176]).

suffices for the determination of individual metals by the differential kinetic method. The rate constants for various DCTA complexes are given in Table 6, from which it can be seen that there are large differences among these constants. As the k_H^{MCy} values depend on the pH, dissociation half-time values suitable for the measurement can be obtained by pH adjustment. If, for example, the pH of the reaction system is close to 7, Ca, Ba and Sr can be determined by the stopped-flow technique using the rapid dissociation of these complexes ($k_H^{MCy} = 10^5 - 10^8$), as other DCTA complexes dissociate sufficiently slowly at this pH to avoid interference in the determination of the alkaline earths [175]. If the pH of the system is less than 4, the transition metals can be determined; the rapid dissociation of the alkaline earth complexes does not interfere. It follows from the data given in Table 6 that the dissociation rate constants are so close for some complexes that the complexes cannot be differentiated on the basis of their kinetic properties. However, a detailed analysis of the problem has shown that ratios of dissociation constants greater than 2 suffice for successful analyses. With many complexes the values of constants k_d can be experimentally affected e.g. by variation of the temperature by $\pm 10\,^{\circ}C$, by a change in the ionic strength or by a change in the oxidation state of the metal ion.

The determination procedure consists of adding about a 10% excess of DCTA over the total content of all metal ions (approximate titration) to the sample, adding this mixture to a solution of Cu^{2+} or Pb^{2+} ions in a buffer of a suitable pH and recording the absorbance-time dependence.

If Cu^{2+} ions are used as indicator ions (ions M' in reaction (6-27)) for the measurement of the dissociation rate in a mixture of several DCTA complexes formed by ions $M_1, M_2, M_3, ..., M_i$, with dissociation rate constants $k_1, k_2, k_3, ..., k_i$, and if the increase in the Cu-DCTA complex concentration with time is measured, then the dependence of the absorbance on time can be expressed by the relationship

$$\frac{A}{b} = \varepsilon_{CuCy}[DCTA]_{excess} + \varepsilon_{Cu}([Cu]_{total} - [DCTA]_{excess}) +$$

$$+ (\varepsilon_{CuCy} - \varepsilon_{Cu})(1 - e^{-k_1 t})[M_1]_{init} + (1 - e^{-k_2 t})[M_2]_{init} +$$

$$+ ... + (1 - e^{-k_i t})[M_i]_{init} \tag{6-30}$$

where A is the measured absorbance, b is the thickness of the cuvette used

102

and ε are the appropriate molar absorbances. Eq. (6-30) can be solved graphically for the given initial concentration values, $[M_n]_{init}$ [177, 178] if the mixture consists of two metals; for a multicomponent mixture a computer must be employed. The absorbance vs time curves must be recorded for solutions of individual complexes under identical experimental conditions in order to calibrate the dependence (to determine the rate constant values).

The absorbance-time curves can be recorded with common spectrophotometers under suitable acidity conditions for all the metals listed in Table 6, except for strontium and barium; the stopped-flow technique must be employed in analyses for the alkaline earths, because of their rapid dissociation. Mixtures containing from two to seven metals in concentrations of $10^{-5} - 10^{-6}$M were analyzed in this way with errors not exceeding 15%.

The polarographic method is advantageous for monitoring the time-course of substitution reactions, continuously recording the limiting current of a metal ion that is complexed or liberated from a complex, from which the concentration of required components can be directly calculated using the methods of differential kinetic analysis (Section 8.3.2). In analysis of mixtures of calcium and magnesium the substitution reactions of their DCTA complexes with lead (II) ions is used. The decrease in the Pb^{2+} ion concentration recorded polarographically is used for calculation of the calcium and magnesium concentration by the logarithmic extrapolation method [184].

If cadmium ions are used for displacement of calcium or magnesium from their DCTA complexes, adding it in a concentration equal to the sum of the concentrations of the alkaline earth metals, the substitution reaction

$$CaCy^{2-} + MgCy^{2-} + 2\,Cd^{2+} \rightarrow 2\,CdCy^{2-} + Ca^{2+} + Mg^{2+}$$

$$(6-31)$$

takes place as an irreversible second-order reaction which is rather slow, so that special fast reaction techniques are not needed [179]. Because the rates of the substitution reactions of the Ca-DCTA and Mg-DCTA complexes with cadmium ions are very different, reaction (6-31) can be used for differential kinetic determination of calcium and magnesium using linear extrapolation (see Section 8.3.2.2). The sum of the concentrations of the determinands must be known and is found from the amount of cadmium ions liberated from the Cd-EDTA complex by the calcium and magnesium

103

ions according to the reaction

$$Ca^{2+} + Mg^{2+} + 2CdY^{2-} \rightarrow 2Cd^{2+} + CaY^{2-} + MgY^{2-} \quad (6\text{-}32)$$

which is quantitative in 3 to 5M ammonia medium.

The differential kinetic method of analyses of mixtures of two metals based on reactions (6-26) and (6-27) has been modified [180] to enable measurement of the reaction rates without using special measuring methods. In the reaction mixture containing DCTA complexes and cupric ions the concentration of hydrogen ions was continuously increased (by hydrolysis of tert. butyl chloride), simultaneously recording the absorbance of the Cu-DCTA complex being formed and the pH as functions of time. Sigmoid curves are obtained for reaction of a single DCTA complex with Cu^{2+} ions, analogous to titration curves. If several DCTA complexes react with Cu^{2+} ions, the resultant curves shift and the component concentrations can be calculated from these shifts. This method has several advantages: no special fast-reaction measuring techniques are required, the analysis time is not longer when two metals which form DCTA complexes that react with Cu^{2+} ions at very different rates are analyzed, because the reaction rate increases with decreasing pH, and the reaction components can be mixed in the usual way, because the initial reaction mixture has a high pH at which the substitution reactions proceed slowly.

Substitution reactions of EDTA-type complexes have also found use in analyses of binary mixtures of heavy metals using the logarithmic extrapolation method (see Section 8.3.2.1). The reactions between PAR (4-(2-pyridylazo)resorcinol) complexes of heavy metals and EGTA (ethyleneglycol-bis(2-aminoethylether)N,N,N',N'-tetraacetic acid) [181] and between the eriochrome black R (Calcon) complexes and EDTA [182] were used, the substitution reaction being of the following type,

$$MR + Y \rightarrow MY + R \qquad (6\text{-}33)$$

Another interesting example of the use of substitution reactions of complexes in diferential kinetic methods is the determination of mixtures of rare earth metals described by Yatsimirskii et al [183]. In this method, different rates of substitution reactions between the complexes of the xylenol orange metallochromic indicator with the rare earth metal ions and EDTA are employed. This is a (6-7b) type reaction,

$$Ln-XO + Y \rightarrow LnY + XO \qquad (6\text{-}34)$$

where Ln is a lanthanide, Y is the EDTA anion and XO is the xylenol orange molecule (the charges are omitted). Reaction (6-34) is irreversible and fast for all lanthanides — the half times are of the order of hundredths of a second — and all the reactions are first order with respect to the Ln − XO complex. Therefore, the stopped-flow method, coupled with spectrophotometric measurement of the concentration of the Ln − XO complexes, must be employed for monitoring time variations in the concentration of the Ln − XO complex.

The following reactions take place after the addition of EDTA to a reaction mixture containing the xylenol orange complexes of dysprosium, holmium and ytterbium:

$$Dy - XO + Y \xrightarrow{k_1} DyY + XO \tag{6-35a}$$

$$Ho - XO + Y \xrightarrow{k_2} HoY + XO \tag{6-35b}$$

$$Yb - XO + Y \xrightarrow{k_3} YbY + XO \tag{6-35c}$$

The measured absorbance at time t (A_t) is given by the equation

$$A_t = A_1 e^{-k_1 t} + A_2 e^{-k_2 t} + A_3 e^{-k_3 t} \tag{6-36}$$

where A_1, A_2 and A_3 are the absorbance values for the xylenol orange complexes of Dy, Ho and Yb, respectively, corresponding to time t_0 and k_1, k_2 and k_3 are the rate constants of reactions (6-35a) to (6-35c), respectively. On the basis of Eq. (6-36) the following relationships can be formulated for the data obtained by the measurements at three times t_1, t_2 and t_3:

$$A_{t_1} = A_1 e^{-k_1 t_1} + A_2 e^{-k_2 t_1} + A_3 e^{-k_3 t_1} \tag{6-37a}$$

$$A_{t_2} = A_1 e^{-k_1 t_2} + A_2 e^{-k_2 t_2} + A_3 e^{-k_3 t_2} \tag{6-37b}$$

$$A_{t_3} = A_1 e^{-k_1 t_3} + A_2 e^{-k_2 t_3} + A_3 e^{-k_3 t_3} \tag{6-37c}$$

As the values of constants k_1, k_2 and k_3 can be determined experimentally (by measurement in a solution containing only a single Ln − XO complex and EDTA), the A_1, A_2 and A_3 values, proportional to the initial concentrations of the Dy, Ho and Yb complexes with xylenol orange, can be calculated by solving Eqs. (6-37a) to (6-37c). This is thus an application of the method of proportional equations, which is discussed in detail in Section 8.3.2.1.

The above differential kinetic method for the analysis of lanthanoid mixtures yields very good results and is much more precise than equilibrium methods. This method is an example of analysis in which the kinetic approach to the solution of the analytical problem (analysis of a mixture of chemically similar substances) is more advantageous and provides an experimentally relatively simple and sufficiently precise analytical procedure, in all respects surpassing procedures based on measurements performed at equilibrium.

7 Reactions Affecting Electrode Processes

For analytical use of kinetic phenomena accompanying chemical reactions, homogeneous reactions in solution are of the greatest importance; these are chiefly reactions of inorganic systems, catalyzed or uncatalyzed. Reactions connected with processes taking place at electrodes are also used in kinetic analysis. Electrochemical reactions are sometimes affected by the rate of a homogeneous chemical reaction that may precede or follow the actual electrode process. Because of these reactions, e.g., the transport of depolarizer toward the electrode surface in voltammetric measurements is not controlled solely by diffusion, but also by the rate of a preceding or subsequent reaction involving the depolarizer. The current obtained is readily distinguished experimentally from the diffusion current and is termed the kinetic current. It is analytically important that the kinetic currents are proportional to the concentration of the test substance in the solution (similar to the diffusion current).

7.1 Polarographic Kinetic Currents

The theory of polarographic kinetic currents has been thoroughly developed and relationships have been derived describing the dependence of these currents on the rate constant of the accompanying chemical reaction [185, 186]. Kinetic currents appear when either a reaction preceding the electrode process or a reaction parallel to the electrode process occurs in solution.

Preceding chemical reactions may affect the electrode process, e.g. in polarographic reduction of complexes. The ratio of complexed ions M^{2+} and complex MX^- in the solution is constant and is determined by the stability constant of the complex; thus the reaction system is in equilibrium. The equilibrium is disturbed in the close vicinity of the electrode as a result of the reduction of free M^{2+} ions and is re-established by dissociation of the complex,

$$MX^- \underset{k_f}{\overset{k_d}{\rightleftharpoons}} M^{2+} + X^{3-} \tag{7-1}$$

where k_d and k_f are the dissociation and formation rate constants, respectively, for the complex. The measured current corresponding to the reduction of free M^{2+} ions is thus increased by a value determined by the dissociation rate of complex MX^-. Such kinetic currents were observed in the polarographic reduction of cadmium chelates with NTA (nitrilotriacetic acid), EDTA and TTHA (triethylenetetraminehexaacetic acid) [187−189].

The dependence of the kinetic current on the rate of the preceding reaction for complex MX^- is determined by the relationship

$$\frac{i}{i_d - i} = 0.886 \frac{(k_d t)^{1/2}}{(K_{MX^-}[X^{3-}])^{1/2}} \tag{7-2}$$

where i is the kinetic current, i_d is the hypothetical diffusion current which would appear if the dissociation rate were higher than the diffusion rate and K_{MX^-} is the stability constant of complex MX^-. Relationships analogous to Eq. (7-2) have also been derived for other types of reactions accompanying electrode processes [190−192], so that rate constants for many chemical reactions can be determined polarographically. Advantages of this approach are its simplicity and the ability to determine the rate constants of very fast reactions.

Rate constants as high as 10^9 s^{-1} have been determined polarographically, but values above 10^7 s^{-1} are not reliable, because the measurement of kinetic currents for very fast chemical reactions is strongly influenced by the electric double-layer structure [193].

The analytical importance of reactions accompanying polarographic electrode processes can be demonstrated on the amperometric titration of Cr(III) with EDTA [194]. The Cr(III) complex with EDTA is formed very slowly; it takes up to several hours at 20 °C and at concentrations of the two substances of about 10^{-3} mol . l^{-1}. On the other hand, the reaction of Cr^{2+} ions with EDTA is instantaneous from the analytical point of view. At pH 4.7, the half-wave potential of the reduction of Cr^{3+} at a dropping mercury electrode is -0.8 V $(Cr^{3+} \rightarrow Cr^{2+})$ and that of the reduction of

108

the Cr^{3+}-EDTA complex is -1.2 V (Cr^{3+}-EDTA $\rightarrow$ Cr^{2+}-EDTA). Hence a wave corresponding to the reduction $Cr^{3+} \rightarrow Cr^{2+}$ is obtained in a suitable electrolyte containing only a Cr^{3+} salt. On addition of EDTA, the following processes take place:

The concentration of the Cr^{3+}-EDTA complex is immeasurably small (slow reaction). The Cr^{3+} ions are reduced electrochemically,

$$Cr^{3+} + e^- \rightarrow Cr^{2+} \tag{7-3}$$

The Cr^{2+} ions formed immediately react with EDTA,

$$Cr^{2+} + EDTA \rightleftharpoons Cr^{2+}\text{-EDTA} \tag{7-4}$$

If the electrode potential is more positive than   1.2 V, the electrochemical oxidation

$$Cr^{2+}\text{-EDTA} \rightleftharpoons Cr^{3+}\text{-EDTA} + e^- \tag{7-5}$$

takes place [195]. As soon as the electrode potential reaches a value of -1.2 V, the Cr^{3+}-EDTA complex begins to be reduced. Hence addition of EDTA to a solution of Cr^{3+} ions leads to appearance of the wave of the Cr^{3+}-EDTA complex at -1.2 V and to a decrease in the wave of reduction of Cr^{3+} ions at -0.8 V. The solution does not contain a measurable amount of the Cr^{3+}-EDTA complex, as can be readily demonstrated spectrophotometrically. Hence amperometric titration of Cr(III) with EDTA can be carried out, with indication on the basis of the decrease in the limiting current of the Cr^{3+} ion reduction or of the increase in that for the Cr^{3+}-EDTA complex, although virtually no Cr^{3+}-EDTA complex is formed in the solution. These changes in the limiting currents result from chemical reaction (7-4) in which the product of the electrochemical reduction of Cr^{3+} ions (7-3) participates and which leads to the subsequent electrochemical reaction (7-5).

7.2 Catalytic Electrode Processes

Detailed investigation of reactions accompanying electrode processes has shown that the product of the electrode reaction may also react with another substance present in the test solution with regeneration of the initial

109

depolarizer [196, 197]. This process can be schematically represented as

$$A \xrightarrow{\text{el}} B \tag{7-6}$$

$$B \xrightarrow{\text{chem}} A \tag{7-7}$$

If chemical reaction (7-7) is sufficiently fast, then the regenerated depolarizer A again participates in electrode reaction (7-6). An increase in the diffusion current of substance A results and thus this current is termed catalytic, i_{cat}. The total measured current, i_{tot}, is the sum of the catalytic and diffusion currents,

$$i_{tot} = i_{cat} + i_d \tag{7-8}$$

Therefore, reactions parallel to the electrode process are commonly termed catalytic reactions. Owing to the cyclic character of these reactions, the limiting current increases and thus analytical procedures based on catalytic reactions are more sensitive than common polarographic methods. It is also important that analytically advantageous consequences of catalytic electrode reactions appear not only in polarography, but also in voltam- metric, chronopotentiometric, chronoamperometric and chronocoulometric methods [198].

Catalytic currents have most often been observed with reduction as the primary electrode reaction and thus substance Z, with which the product reacts, must exhibit oxidative properties. This mechanism can be described by the scheme,

$$Ox + \nu e^- \rightarrow Red \tag{7-9}$$

$$Red + Z \rightleftharpoons Ox \tag{7-10}$$

The catalytic currents described by equations (7-9) and (7-10) have been observed in polarographic reduction of molybdate or vanadate in the presence of H_2O_2 [199], in reduction of the Cr^{3+}-EDTA complex in the presence of NO_3^- ions [200] and in the reduction of niobium and zirconium in the presence of H_2O_2 [201].

Less frequently, depolarizer Ox is not completely regenerated in reaction cycle (7-9)—(7-10). This situation occurs in the polarographic reduction of

110

uranyl ions in an acidic medium [202],

$$2\,UO_2^{2+}\ +\ 2\,e^-\ \rightarrow\ 2\,UO_2^+ \tag{7-11}$$

$$2\,UO_2^+\ +\ H^+\ \rightarrow\ UO_2^{2+}\ +\ UOOH^+ \tag{7-12}$$

It can be seen that only one half of the reduced hexavalent uranium is regenerated. The catalytic activity of reaction cycle $(7\text{-}11)-(7\text{-}12)$ is decreased by dismutation of the product of the electrode reaction of UO_2^{2+}; however, this catalytic electrode reaction has been used for the determination of uranium [203].

Catalytic electrode reactions can also include the reaction mechanism

$$ML_x(H_2O)_m^{v+}\ +\ ve^-\ \xrightarrow{\ \text{el}\ }\ M^0\ +\ x\,L\ +\ m\,H_2O \tag{7-13}$$

$$x\,L\ +\ M(H_2O)_6^{v+}\ \xrightarrow{\ \text{chem}\ }\ ML_x(H_2O)_m^{v+}\ +\ (6-m)\,H_2O \tag{7-14}$$

Complex $ML_x(H_2O)_m^{v+}$ is reduced at an electrode and liberated ligand L (reaction (7-14)) reacts with the hydrated metal ions with formation of the initial complex, $ML_x(H_2O)_m^{v+}$. Ligand L can thus be considered as a catalyst in the reduction of the $M(H_2O)_m^{v+}$ ions. It is assumed that the critical factor of the whole process is the formation of the complex by the reaction of the metal ion with the ligand adsorbed on the electrode surface [204]. Catalytic effects have been observed with complexes of nickel with amines, e.g. o-phenylenediamine [204−206], p-phenylenediamine [207], pyridine [208, 209] and ethylenediamine [209], for complexes of indium with halides [210] and of cystine with tin and cobalt [211]. Catalytic waves of this type have been used for the determination of low concentrations of p-phenylenediamine $(10^{-7}M)$ in the presence of its isomers that are catalytically inactive [205, 206].

7.3 Hydrogen Catalytic Currents

Catalytic waves of hydrogen observed in polarography result from a decrease in the large hydrogen evolution overvoltage on a mercury electrode caused by certain substances, such as organic substances with characteristic groups acting as proton donors and platinum metal chlorides. The catalytic character of this process follows from the reaction scheme [212],

$$HA_{ads} + e^- \rightarrow A^- + H_{ads} \qquad (7\text{-}15)$$

$$A^- + H^+ \rightarrow HA \qquad (7\text{-}16)$$

Organic substances affecting the hydrogen overvoltage, HA, are weak acids which are adsorbed on the electrode. Most important analytically are the catalytic effects of proteins containing disulphidic or sulphydryl groups, which produce a marked decrease in the hydrogen overvoltage in solutions containing cobalt salts in an ammoniacal buffer [213, 214]. Even very small concentrations of these proteins can be detected from a typical double-wave beyond that of the reduction of divalent cobalt. This effect is widely used in clinical laboratories as the "Brdička reaction" for serological diagnosis of cancer and infectious diseases.

8 Kinetic Methods

8.1 Classification of Kinetic Methods

Analytical methods based on the measurement of time changes in the concentration of certain components of suitable reaction systems are very varied, differing both methodically and in application. The field of kinetic analysis should include the methods of determination of trace concentrations of metals, of organic substances and a large group of clinical and biochemical methods. Some kinetic methods are very simple and require no specific equipment. However, the need to automate analytical procedures and the possibility of automating any kinetic method have led to development of automated kinetic methods that naturally need sophisticated instrumentation.

The classification of kinetic methods adopted in this book is based on two principles. From the analytical point of view, kinetic methods are divided into two large groups — methods for determination of a single substance and methods for analysis of mixtures of substances. Further classification within these groups follows from the kinetic nature of the reactions used, namely, in the first group the reactions are classified into catalyzed and uncatalyzed, in others the classification is based on the reaction order. The classification is schematically shown in Table 7.

8.2 Methods for Determination of a Single Component

Among the kinetic methods suitable for the determination of a single component of the test system, those based on catalytic reactions are undoubtedly of the greatest importance, because they offer the highest sensitivity. Methods employing uncatalyzed reactions are especially important when no suitable equilibrium method is available.

113

TABLE 7

Classification of kinetic methods of chemical analysis

Kinetic methods of chemical analysis

Methods for determination of a single substance

Methods based on uncatalyzed reactions

Methods based on catalyzed reactions

- Constant time methods
- Constant concentration methods

1. Differential methods
2. Integration method
3. Method of tangents
4. Methods based on measurement of induction period
5. Catalytic end-point titrations

Methods for analysis of mixtures of substances

Methods based on great differences in the rate of reaction of the test substances with the reagent

Differential kinetic methods

1. Methods based on first and pseudofirst order reactions
2. Methods based on second order reactions
3. Methods based on pseudozero order reactions

8.2.1 METHODS USING CATALYTIC REACTIONS

8.2.1.1 Differential Methods

Differential methods can be used when a sufficiently sensitive and precise analytical method is available for monitoring small concentration changes of a reactant at the beginning of the reaction. The value of x in Eq. (3-13) is very small compared with $[R]_0$, so that the $([R]_0 - x)$ term can be considered constant and equal to $[R]_0$. Eq. (3-13) then assumes a simplified form,

$$\left(\frac{dx}{dt}\right)_{tot} = f[C]_0 + f' \tag{8-1}$$

which is rigorously valid at the beginning of the reaction when $[S] \gg [C]_0$ and $[R] \gg [C]_0$. The plot of Eq. (8-1) is a straight line expressing the dependence of v_{init} on $[C]_0$ (Fig. 12). The intercept on the vertical axis is

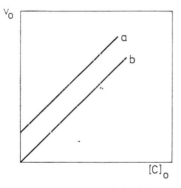

Fig. 12. Dependence of the initial rate of a catalytic reaction on the catalyst concentration.
a — dependence given by Eq. (8—1); b — the same dependence for $f' = 0$.

equal to f' and determines the reaction rate in the absence of the catalyst. When $f' = 0$, the straight line passes through the origin and indicates a reaction that takes place only in the presence of the catalyst.

The relationship describing the dependence of the change in the concentration of the initial substance, x, on the catalyst concentration, C, at the beginning of the reaction,

$$\Delta x = (f[C]_0 + f') \Delta t \tag{8-2}$$

follows from Eq. (8-1). According to Eq. (8-2), this dependence is linear assuming that Δt is constant. If the decrease in the concentration of the initial substance in the indicator reaction, Δx, is measured in the presence of various concentrations of the catalyst, C, at the same time interval after

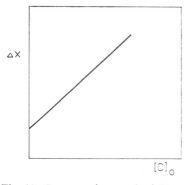

Fig. 13. Constant time method. Dependence of Δx on catalyst concentration C.

the beginning of the reaction, the dependence of Δx on $[C]_0$ is linear. The plot (Fig. 13) can be used as a calibration curve for the determination of the catalyst concentration; functions f and f' need not be determined, because the reaction conditions are the same for the construction of the calibration curve and for the analysis itself. Instead of concentration change Δx, the dependence of any physico-chemical parameter proportional to the concentration can be used. The constant time method is very simple and rapid and can be applied to all catalytic reactions for which the initial rate can be measured. It is unsuitable for reactions with an induction period.

Differences in the impurity content in the standard and test solutions can affect the reaction rate differently and render the analysis imprecise. It is therefore advantageous to perform the measurements on an aliquot of the test solution and then on another aliquot with a known amount of test substance C added [215].

If the test substance concentration in the solution is $[C]_0$ and the decrease in its concentration determined by the constant time method is Δx, then, on addition of a known amount of the substance, $[C]_p$, the resultant test solution concentration is $([C]_0 + [C]_p)$ and the corresponding decrease in the concentration is Δx_p. The relationship between $[C]_0$ and $[C]_p$ is then

$$[C]_0 = \frac{\Delta x}{\Delta x_p - \Delta x} [C]_p \qquad (8\text{-}3)$$

from which $[C]_0$ can be calculated.

This procedure is called the addition method [215]; in this way the precision of the determination is improved and the constant time method can be used even when the reaction rate in the absence of the catalyst is very low compared with the rate of the catalyzed reaction.

Constant Concentration Method

This method is also based on Eq. (8-1) which can be formulated as follows,

$$\frac{1}{\Delta t} = \frac{f[C]_0 + f'}{\Delta x} \qquad (8\text{-}4)$$

It follows from Eq. (8-4) that, at constant Δx, the dependence of $1/\Delta t$ on the catalyst concentration, C, is linear. Therefore, in this method the time is measured that is required for the concentration of the initial substance to attain a preset value. The calibration curve is constructed analogously to that in the constant time method.

The constant concentration method is especially advantageous when the concentration change is measured continuously. As with the constant time method, the constant concentration method cannot be used for reactions with an induction period.

The precision of the determination is improved if the addition method is used. The relationship between $[C]_0$ and $[C]_p$ is then given by,

$$[C]_0 = \frac{\Delta t_p}{\Delta t - \Delta t_p} [C]_p \qquad (8\text{-}5)$$

where Δt_p is the time value obtained by the constant concentration method after addition of a known amount of the test substance $[C]_p$.

The Yatsimirskii [216] Differential Method

The rate of a catalytic reaction is a linear function of the catalyst concentration assuming that the measurement is carried out under the initial conditions. If concentration change Δx is measured spectrophotometrically, the concentration term, Δx, in Eq. (8-1) can be replaced by term ΔA, i.e. the change in the absorbance proportional to Δx. Then, according to Yatsimirskii, the relationship between the catalyst concentration and the measured absorbance can be formulated as follows,

$$\frac{\Delta A}{\Delta t} = K'[C] \tag{8-6}$$

where constant K' includes the rate constant, the molar linear absorption coefficient of the test substance and the cuvette thickness. If a suitable time interval (Δt = const.) is chosen for the measurement, then Eq. (8-6) can be written as

$$\Delta A_t = K'[C] \tag{8-6a}$$

where ΔA_t is the absorbance measured after time Δt. The ΔA_t value is simply measured by mixing all the components of the reaction system in a cuvette, after time Δt (1 to 2 min.) mixing the same components in another cuvette and following the difference in the absorbance, which is constant and equal to ΔA_t for a certain time. A calibration curve is used, which is the dependence of ΔA_t on the catalyst concentration. The method is very simple and rapid but, according to the author, is suitable only for reactions in which the initial substance or the product produce coloured solutions.

However, the differential method has much wider use [249]. In two separate vessels the same indicator reactions take place under identical conditions of temperature and reactant concentrations; one vessel contains a known amount of the catalyst, C_A, and the other the amount to be determined, C_S. A certain quantity proportional to a reactant concentration (absorbance, electrode potential) is measured in the two vessels during the reaction and the difference, ΔP, is recorded as a function of time. At the beginning of the reaction, the measured substance concentration is equal

118

in the two vessels, i.e. ΔP is equal to zero. During the reaction the difference, ΔP, increases to a maximum value, because of the different reaction rates in the two vessels. On completion of the catalytic reaction in both vessels, the concentrations measured are again the same and ΔP is again zero. The ΔP dependence on time has the shape of an irregular maximum, whose height or area is proportional to the catalyst concentration, C_S. In a series of measurements with constant concentration C_A the recorded maximum is positive for $[C_S] > [C_A]$, negative for $[C_S] < [C_A]$ and zero for $[C_S] = = [C_A]$.

8.2.1.2 Integration Methods

If x cannot be neglected in Eq. (3-13) because $[R] \ll [S]$, the methods based on Eq. (8-1) cannot be used. Then Eq. (3-13) is integrated from t_1 to t_2 and from $[R]_1$ to $[R]_2$, yielding

$$\ln \frac{[R]_1}{[R]_2} = (f[C]_0 + f')(t_2 - t_1) \tag{8-7}$$

where $[R]_1$ and $[R]_2$ are concentrations of substance R at times t_1 and t_2, respectively. Time t_1 thus need not correspond to the beginning of the reaction and the method can also be used for reactions with an induction period.

It has been shown [217] that the determination of $[C]_0$ according to Eq. (8-7) is most precise when the measurement is performed at time interval $(t_2 - t_1) = t_e$ at which the $[R]_1/[R]_2$ ratio equals number e (the base of natural logarithms). Eq. (8-7) then assumes the form

$$\frac{1}{t_e} = f[C]_0 + f' \tag{8-8}$$

When using Eq. (8-8), time t_e is determined for various concentrations of the test substance C; the $1/t_e$ vs $[C]_0$ dependence can be used as a calibration curve.

Another method employs modified Eq. (8-7),

$$K^* = \frac{\ln([R]_1/[R]_2)}{t_2 - t_1} = f[C]_0 + f' \tag{8-7a}$$

where K^* is the overall rate constant.

If constant reaction conditions are maintained (ionic strength, temperature, concentration of substance S), the K^* value is only a function of the substance concentration $[C]$ and the dependence of K^* on $[C]_0$ is linear and can be used as a calibration curve.

8.2.1.3 Method of Tangents

As shown in Chapter 2, the integrated forms of the kinetic equations for various types of chemical reaction are equations of straight lines (see Table 1). The slopes of these straight lines are proportional to the reaction rate constants and thus also to the reaction rates. The integrated form of the kinetic equation for catalytic reaction (3-1) is given by Eq. (8-7). It follows from this relationship that the dependence of $\ln([R]_1/[R]_2)$ or $\ln\{[R]_0/([R]_0 - x)\}$ on time is linear and its slope is proportional to catalyst concentration $[C]_0$. By plotting the $\ln\{[R]_0/([R]_0 - x)\}$ vs time dependence for various catalyst concentrations, several straight lines are obtained passing through a single point, with slopes proportional to the catalyst concentration. The plot of these slopes vs the catalyst concentration is

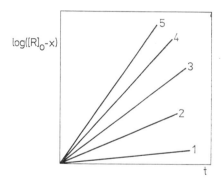

Fig. 14. Method of tangents. Dependence of $\ln([R]_0 - x)$ on time for various catalyst concentrations C.

another straight line which can be used as a calibration curve for the determination of the catalyst. Fig. 14 depicts the dependence of $\ln([R]_0 - x)$ on time for three concentrations of the catalyst, curve *1* corresponding to the reaction in the absence of the catalyst ($[C]_0 = 0$). Fig. 15 shows the calibration curve — the dependence of the slopes of the straight lines from

Fig. 14 on the catalyst concentration. Because concentration $[R]_0$ is constant in all measurements, it is sufficient to plot only $\ln([R]_0 - x)$ against time; the dependence $\ln\{[R]_0/([R]_0 - x)\}$ vs time need not be plotted, as only the slopes of the resultant straight lines are analytically important.

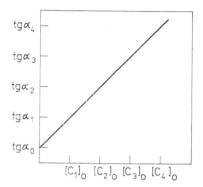

Fig. 15. Method of tangents. Calibration curve.

The method of tangents can also be used in a differential modification, i.e. under the conditions described in Section 8.2.1.1, on the basis of Eq. (8-2) according to which the dependence of Δx on t is linear and its slope is proportional to $[C]_0$. Similar to the previous method, the dependence of the slope on the catalyst concentration is a calibration curve for the determination of the catalyst.

Because indicator reactions mostly take place at a certain rate even in the absence of a catalyst, the slope of the straight line corresponding to the kinetic equation of the indicator reaction without the catalyst must also be considered in constructing the calibration curve.

When using the method of tangents, it is unnecessary to plot straight lines as a dependence of a function of concentration, e.g. $\log([R]_0 - x)$, on time; it is sufficient to plot a measured value that is proportional to the concentration (limiting current, absorbance). The method is thus considerably simplified and all its advantages are preserved.

It is now possible to obtain the slope of a straight line automatically, by using electronic circuits, by means of which the analytical procedures are substantially shortened and simplified. Such devices are described in Section 9.3. Here a technical simplification of the method of tangents described by

121

Yatsimirskii [2] will be mentioned, which can be used when the product of the indicator reaction exhibits an absorption maximum which can be measured spectrophotometrically or colorimetrically. The measurement is based on the fact that double-beam instruments measure the difference in the absorbances of two solutions placed in two cuvettes. If the two cuvettes contain the same reaction mixture prepared under identical conditions, with the only difference being that the reaction in one cuvette begins time interval Δt earlier than that in the other cuvette, then the solution absorbances are A and $(A + \Delta A)$. The instrument signal then corresponds to the ΔA value for time Δt, proportional to the derivative of the A vs t dependence. Therefore, the recorded value is also proportional to the slope of the linear dependence of Δx vs t (Eq. (8-2)). To obtain the calibration curve it is sufficient to measure the dependence of ΔA on the catalyst concentration. This procedure is very simple, no special instrumentation is required and the whole analytical procedure is simplified.

The method of tangents has very wide use and fails only when the indicator reaction is complex, so that the integration of the kinetic equation is difficult or impossible. It can even be used for reactions with an induction period; the data necessary for construction of the plot given in Fig. 14 are measured after the induction period. As a greater number of measurements is required for the determination by the method of tangents, the method is more precise than the constant time and constant concentration methods.

8.2.1.4 The Method Based on the Measurement of the Induction Period

As mentioned in Section 3.3, some catalytic reactions are characterized by an induction period. For most of them, the empirical relationship [218]

$$[C]_0 = \text{const.}/t_i \qquad (8-9)$$

is valid, where t_i is the induction period.

Analysis based on Eq. (8-9) is very simple. The reagents and the test or standard solution are thermostatted and known volumes are transferred into cuvettes in a manner suitable for exact determination of the beginning of the reaction (e.g. by injection of the sample into the reaction mixture). Time t_i is mostly determined visually; photometric determination is more precise, because t_i is found by extrapolation of the absorbance to zero value.

A linear calibration curve is obtained by plotting $1/t_i$ against the catalyst concentration. This method is not universal and is not based on exact knowledge of the reaction kinetics, but it is useful in some determinations of inorganic substances [219−221].

The comparative method of Bognar [250] is a simple method involving preparation of a series of solutions containing the same amounts of one of the reactants and known increasing amounts of the catalyst; one of the solutions contains the sample instead of the catalyst. The indicator reaction must yield a coloured product or the colour of the original solution must fade during the reaction. Using a "double starting pipette" (Section 9.2), the indicator reaction is begun by adding the other reactant and the sample solution is compared with the coloration of the standard solutions.

8.2.1.5 Catalytic End-Point Titrations

Recently titrations have been studied in which the titrant catalyzes the indicator reaction [222−224]. Inhibitors can be titrated in this way, i.e. substances that reduce the catalyst activity through formation of complexes or precipitates. The inhibitor is consumed during the titration and traces of free catalyst appear at the equivalence point, immediately accelerating the indicator reaction. This process can be monitored visually, photometrically, potentiometrically, thermometrically, etc. One of the most frequent indicator reactions is that of ceric ions with arsenic (III),

$$2\,Ce^{4+} + As^{3+} \rightleftharpoons 2\,Ce^{3+} + As^{5+} \tag{8-10}$$

which is slow in acidic media and is catalyzed by iodide ions. Table 10 lists some other indicator reactions and titrants suitable for catalytic end-point titrations.

An advantage of catalytic end-point titrations is mostly their high sensitivity and smaller error than common titrations. For use in titrations, the catalytic mechanism of the indicator reaction need not be rigorously known.

8.2.1.6 Catalytic Methods Employing Open Systems

The above methods use closed systems in which catalytic reactions take place without any external interference. However, it is also possible to add or remove some reaction component during the reaction. Such reaction systems are termed open [251, 252] and considerably broaden the analytical applicability of catalytic methods.

In the stat method [253], reactant A is added to a mixture of reactant B and catalyst C at a rate equal to the rate of its consumption. The addition rate is automatically regulated as follows: A small amount of substance A is added to the system. The state of the system is then characterized by a certain value of a measurable quantity (pH, absorbance, electrode potential, etc.) proportional to the concentration of a reactant. Any deviation from this state as a result of the catalytic reaction is then compensated for by automatic addition of substance A using a pH-stat, an absorption-stat, etc., maintaining the preset quantity at a constant value. A detailed description of the instrumentation can be found in Chapter 9.

The rate of addition of substance A is proportional to the catalyst concentration, as follows from the following reasoning. In the beginning of reaction $A + B \xrightarrow{C} P$, substance B is present in a great excess over the other components and its concentration can be considered constant. The reaction rate is then given by

$$- \frac{d[A]}{dt} = k[A]_0 [C]_0 \qquad (8\text{-}11a)$$

If substance A is added at the same rate as the rate of its consumption, it then holds that

$$- \frac{d[A]}{dt} = k'[C]_0 \qquad (8\text{-}11b)$$

The reaction thus assumes quasi-zero order, its rate is constant for a relatively long time period and the rate of addition of substance A is proportional to the concentration of catalyst C.

The stat method can be modified and the reaction product can be removed instead of adding substance A; an auxiliary reagent is used for this purpose and rapidly reacts with the product.

The Steady-State Method

In this method reactant A is added at a constant rate to a mixture of reactant B, present in a large concentration, and catalyst C [254]. The catalytic reaction of A with B is very slow in the beginning, because the concentration of substance A in the system is very small. During addition of substance A the reaction rate increases and the consumption of substance A in the catalytic reaction also increases. A stage is finally attained at which the consumed and added amounts are equal for a certain time interval. The system thus assumes steady-state in which it remains for a relatively long time and the concentration of substance A in the system is constant. The concentration of substance A is monitored continuously and a motor-driven burette is used for addition at a constant rate. A stationary concentration of substance A corresponds to a constant value of the measured quantity, which stabilizes after several minutes. The test substance concentration (catalyst C) is found from a calibration curve constructed by plotting the reciprocal of the stationary concentration of substance A against the catalyst concentration.

Flow-Through Methods

In flow-through methods all the reaction components (the reactants and the catalyst) are transported at a certain velocity to a point where they are mixed and are pumped out of the compartment at the same velocity. At a certain point in the flow-through system a physico-chemical quantity proportional to the reactant concentration is monitored. The measurement can be carried out in two ways.

First, the continuous flow method is used [255] (Fig. 16a). Reactants A and B and the test solution of catalyst C are transported into mixing

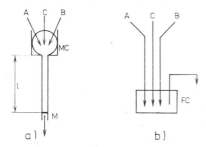

a)   b)

Fig. 16. (a) Flow-through cell — continuous flow method. A, B — reactants, C — catalyst, *MC* — mixing chamber, *M* — point where the measurement is done, *l* — distance between mixing chamber *MC* and point *M*.
(b) Flow-through cell — stationary concentration method. *FC* — flow-through cell, A, B — reactants, C — catalyst.

chamber MC. At point M in the exit tube a suitable quantity proportional to the concentration of a reactant or product is measured. At a constant flow-rate the distance, l, between point M and mixing chamber MC corresponds to a certain time interval; thus the continuous flow method is a flow-through modification of the constant time method. The method can be modified by measuring at two places, the difference between the two signals being proportional to the catalyst concentration.

Second, the concentration of the reactants or products can be measured in a flow-through cell [256]. The effect of mixing the initial solutions is used. As shown in Fig. 16b, all the components of the catalytic reaction are transported into flow-through cell FC at the same velocity and are rapidly and completely mixed. Stationary state is established in the cell and spectrophotometric measurement of the stationary concentration of a reactant or a product yields a signal proportional to the catalyst concentration. For details, see Chapter 9.

Flow-through methods are very important, because they enable continuous monitoring of changes in the test substance concentration in a stream of liquid. Another great advantage lies in the possibility of using these methods for fast reactions. They are ideally suited for automation (see Chapter 9).

8.2.2 METHODS BASED ON UNCATALYZED REACTIONS

Chapter 2 surveys the kinetic equations for various types of uncatalyzed reactions. Knowledge of the pertinent kinetic equation is necessary for determination of the initial concentration of a reactant, for which changes of the reactant concentration with time must be found. These two conditions thus must be met when an analytical method is to be developed. However, these analytical methods are rather disadvantageous compared with equilibrium methods, as can be seen from the following discussion.

Substance A reacts at a certain rate with reagent R. If substance A is to be determined, its concentration during the measurement must be constant, which can be attained only for large concentrations of A. If a smaller amount of substance A is to be determined, the amount of R must be decreased accordingly, which leads to a decrease in the reaction rate and to difficulties in determination of the reacted amount at a certain time, x. The sensitivity of the determination also decreases.

The reaction rate is given by

$$\frac{dx}{dt} = k_A([R]_0 - x)[A]_0 \tag{8-12}$$

where k_A is the appropriate second-order rate constant. Concentration $[A]_0$ can be determined in several ways. The change in this concentration at the beginning of the reaction can be measured or the constant time and constant concentration methods can be used; these methods are based on the equation

$$[A]_0 = \frac{dx/dt_{init}}{k_A[R]_0} \tag{8-13}$$

In measurements at the beginning of the reaction, concentrations $[A]_0$ and $[R]_0$ can be comparable.

Methods based on the equation

$$[A]_0 = \frac{\ln([R]_1/[R]_2)}{k_A(t_2 - t_1)} \tag{8-14}$$

where concentrations $[R]_1$ and $[R]_2$ correspond to times t_1 and t_2, can be used only when the concentration of substance A is at least 50 times that of reagent R.

The method of tangents is also applicable to uncatalyzed reactions, because the integrated forms of the kinetic equations are equations of straight lines whose slopes are proportional to the reagent concentration. These methods are of practical importance only when suitable equilibrium methods are not available.

In determinations of a single substance, the use of uncatalyzed reactions can be important when another substance is present that also reacts with the reagent, but at a different rate. If it is assumed, for the sake of simplicity, that substances A and B react with a single reagent to form common product P, both the reactions being first order, then the sum of the instantaneous concentrations of substances A and B at time t, i.e. ($[A]_t +$ $+ [B]_t$), is given by

$$[A]_t + [B]_t = [A]_0 e^{-k_A t} + [B]_0 e^{-k_B t} = [P]_\infty - [P]_t \tag{8-15}$$

where k_A and k_B are the rate constants for the reactions of substances A and B, respectively, and $[P]_\infty$ is the concentration of the product on com-

127

pletion of the reaction. If substance A reacts much faster than substance B, then term $[A]_0 e^{-k_A t}$ is very small compared with $[B]_0 e^{-k_B t}$ and can be neglected in Eq. (8-15). Then it can be derived from Eq. (8-15) that

$$[B]_0 = \frac{[P]_\infty - [P]_t}{e^{-k_B t}} \qquad (8\text{-}16)$$

Therefore, the concentration of slowly reacting substance B can be determined on condition that the product concentration is experimentally determinable at time t and rate constant k_B is known. The $[P]_\infty$ value is determined by an independent method or is found after completion of the reaction. For a successful determination at least 10% of unreacted B must be present in the reaction system after complete reaction of substance A. The error in this determination of substance B can be found from Eq. (8-15); the error expressed in terms of % B is

$$\frac{[A]_0 e^{-k_A t}}{[B]_0 e^{-k_B t}} \cdot 100$$

Hence, in analysis of mixtures of two substances the faster reaction can be neglected and the more slowly reacting substance can be relatively simply determined.

8.3 Methods of Analysis of Mixtures of Substances

Methods permitting determination of several substances in mixtures without prior separation are undoubtedly one of the greatest successes of kinetic analysis. These include methods that employ different rates of the reactions of the test substances with a single reagent, the most important being kinetic differential methods.

8.3.1 METHODS OF DETERMINATION OF TWO SUBSTANCES THAT REACT AT DIFFERENT RATES WITH A SINGLE REAGENT

In a system of substances A and B reacting with reagent R, when, for example, substance A reacts completely within 10 minutes, whereas the reaction of substance B takes 500 times longer, substance A can be determined by a kinetic method in the presence of substance **B**. On the other hand, the slow rate of the reaction of substance B renders the determination of this substance impossible. However, if the reaction conditions are changed on completion of the reaction of substance A, the reaction of substance B can be hastened and substance B can then be determined. The following procedures can be used for this purpose: a) a catalyst specific for substance B can be added; b) the temperature of the system can be increased; c) the concentration of reagent R can be increased; d) the composition of the mixture of solvents can be changed. Fructose and glucose have been determined in this way [225].

8.3.2 DIFFERENTIAL KINETIC METHODS

Some compounds, chiefly organic, react at similar rates with a single reagent. Therefore, they cannot be determined by thermodynamic methods or by the above kinetic methods without prior separation, but they can be analyzed using differential kinetic methods.

Consider a mixture of substances A and B that react with reagent R,

$$A + R \xrightarrow{k_A} P \tag{8-17a}$$

$$B + R \xrightarrow{k_B} P \tag{8-17b}$$

These reactions are bimolecular and irreversible. On examining the concentration conditions in the system described by Eqs. (8-17a) and (8-17b) it is found that three limiting cases are possible. First, reagent R is present in a large excess over the sum of the concentrations of A and B. The kinetic dependences are then described by pseudofirst order reactions with respect to $([A]_0 + [B]_0)$. Second, the sum of the concentrations of A and B is much greater than the concentration of reagent R. The kinetics of the system are again described by pseudofirst order reactions, but with respect to reagent R. Third, the concentration of reagent R equals the sum of the concentrations of substances A and B; second-order kinetics are then involved. A survey of the differential methods is given in Table 8 and the individual methods are described below.

8.3.2.1 Methods Based on First or Pseudofirst Order Reactions with Respect to the Reactants

These differential kinetic methods can be used for analyses of mixtures in which the concentration of reagent R is at least 50 times greater than the sum of the concentrations of substances A and B, i.e. $[R]_0 \gg ([A]_0 + [B]_0)$.

TABLE 8

Classification of differential kinetic methods

Concentration conditions in the reaction system	Reaction order	Differential kinetic methods applicable under these conditions
$[R]_0 \gg ([A]_0 + [B]_0)$	Pseudofirst order reaction with respect to $([A]_0 + [B]_0)$	Logarithmic extrapolation method Single point method Method of tangents for analysis of binary mixtures Method of proportional equations Linear graphical method
$[R]_0 > ([A]_0 + [B]_0)$ max. 50fold excess of $[R]_0$ over $([A]_0 + [B]_0)$	Second order reaction	Logarithmic extrapolation method for second order reactions
$[R]_0 = ([A]_0 + [B]_0)$	Second order reaction	Linear extrapolation method Logarithmic extrapolation method Single point method
$[R]_0 < ([A]_0 + [B]_0)$ max. 50fold excess of $([A]_0 + [B]_0)$ over $[R]_0$	Second order reaction	Logarithmic extrapolation method
$[R]_0 \ll ([A]_0 + [B]_0)$	Pseudofirst order reaction with respect to $[R]_0$	Method of proportional equations Single point method according to Roberts and Regan
$[R]_0 \ll ([A]_0 + [B]_0)$ $[R]_0 = ([A]_0 + [B]_0)$ $[R]_0 \gg ([A]_0 + [B]_0)$	Pseudozero order reaction	Single point method Method of proportional equations

If substances A and B form common product P with reagent R (e.g. glucose and fructose react with MoO_4^{2-} ions to form molybdenum blue), the system can be defined by the irreversible reactions,

$$A \xrightarrow{k_A} P \tag{8-18}$$

$$B \xrightarrow{k_B} P \tag{8-19}$$

that are first or pseudofirst order with respect to A and B. The concentration of product P at time t is then given by

$$[P]_\infty - [P]_t = [A]_t + [B]_t = [A]_0\, e^{-k_A t} + [B]_0\, e^{-k_B t} \tag{8-20}$$

where $[P]_\infty$ is the concentration of product P on completion of reactions (8-18) and (8-19). If substance A reacts faster than substance B, the term $[A]_0\, e^{-k_A t}$ in Eq. (8-20) can be neglected provided that $[A]_t = 0$ (i.e. when

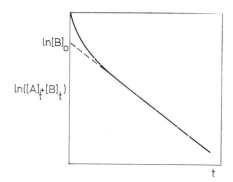

Fig. 17. Logarithmic extrapolation method.

the reaction of substance A is virtually completed). By taking logarithms, Eq. (8-20) is converted into

$$\ln\left([P]_\infty - [P]_t\right) = \ln\left([A]_t - [B]_t\right) = -k_B t + \ln[B]_0 \tag{8-21}$$

from which it follows that plots of $\ln\left([P]_\infty - [P]_t\right)$ or $\ln\left([A]_t - [B]_t\right)$ vs time are straight lines with slopes equal to $-k_B$ and with an intercept on the vertical axis $(t = 0)$ equal to $\ln[B]_0$ (see Fig. 17). The concentration of substance B at time $t = 0$ is thus obtained. The sum of the concentrations

$[A]_0 + [B]_0$ is usually determined by a common method or is calculated from the $[P]_\infty$ value (of course, on condition that the reaction mechanism is not affected by side reactions).

An example of the logarithmic extrapolation method is the determination of cobalt and nickel employing the reaction of the EGTA complexes of these metals with PAR [226] (EGTA is ethyleneglycolbis(2-aminoethylether)--N,N,N',N'-tetraacetic acid and PAR is 4-(2-pyridylazo)resorcinol). In an alkaline medium the nickel complex reacts with PAR more slowly than the cobalt complex. Instead of the logarithm of the concentration difference, the logarithm of the difference in the Ni-PAR absorbances is plotted, because the absorbance is proportional to the complex concentration.

As already pointed out, the logarithmic extrapolation method is applicable to systems with $[A]_0\, e^{-k_A t} \ll [B]_0\, e^{-k_B t}$, as it is necessary that substance A react relatively rapidly in order to measure the $[B]_t$ values. The applicability and precision of the method thus depend on ratios k_A/k_B and $[A]_0/[B]_0$. It can be derived theoretically that a value of k_A/k_B exceeding 4 suffices with $[A]_0/[B]_0$ equal to unity; the error of the method is then smaller than that of the measurement of the parameter, e.g. of the absorbance. The precision of the method deteriorates with increasing ratio $[A]_0/[B]_0$, independently of the rate constant ratio. If, e.g., $k_A/k_B = 7.5$ and the concentration of substance B is so small that $[A]_0/[B]_0 = 4$, then the method can no longer be used. It also cannot be used for reversible reactions and in the presence of side reactions, but it belongs among the most common differential kinetic methods for reactions with rate constant ratios greater than 6.

Linear Graphical Method

This method can be used for analyses of mixtures of substances A and B that react with formation of a common product according to Eqs. (8-18) and (8-19). The concentration of product P at time t is given by Eq. (8-20). This equation can be formulated as

$$([P]_\infty - [P]_t)\, e^{k_A t} = [A]_0 + [B]_0\, e^{(k_A - k_B)t} \tag{8-22}$$

from which it follows that the dependence of $([P]_\infty - [P]_t)\, e^{k_A t}$ vs $e^{(k_A - k_B)t}$ is linear with slope equal to $[B]_0$. At $t = 0$, $\exp(k_A - k_B) = 1$ and thus the intercept on the ordinate yields the sum of the initial concentrations

$([A]_0 + [B]_0)$. If concentration changes of the product are monitored spectrophotometrically during reactions (8-18) and (8-19), the absorbance can be plotted instead of concentrations $[P]_\infty$ and $[P]_t$, the slope and the intercept also being in absorbance units.

A mixture of three substances A, B and D can also be analyzed by this method, provided that the rate constants decrease in the series, $k_A > k_B > > k_D$. The concentration data are then plotted in the $([P]_\infty - [P]_t) e^{k_B t}$ vs $e^{(k_B - k_D)t}$ dependence only after completion of the reaction of substance A. The dependence is linear, its slope corresponds to the initial concentration of substance D and the intercept to the sum of concentrations $[B]_0 + [D]_0$. As $[P]_\infty = [A]_0 + [B]_0 + [D]_0$, the concentrations of all three substances are thus determined. If the measurement of the time dependence of the product concentration is performed before completion of reaction of A, the plot deviates from linearity for small values of t. The intercept can then be found by extrapolation.

An advantage of the linear graphical method over the logarithmic extrapolation method is the possibility of analyzing substances with similar rate constants (a small value of the ratio k_A/k_B) and mixtures with a large excess of one substance (ratios $[A]_0/[B]_0$ up to 20), because the data obtained at the beginning of the reaction are also used. Of course, the logarithmic extrapolation method is more advantageous for systems with very different reaction rates. The linear graphical method was verified on analyses of mixtures of esters, using their alkaline hydrolysis [257].

The Extrapolation Graphical Method

This method can be used for the determination of two substances that react with the formation of a common product according to Eqs. (8-17a) and (8-17b) which are first order with respect to A and B and zero order with respect to R. If only the reaction of substance A is considered, then the product concentration at time t is given by

$$[P]_t = [A]_0 - [A]_t = [A]_0 (1 - e^{-k_A t}) \tag{8-23}$$

Hence the $[P]_t$ vs $e^{-k_A t}$ dependence is a straight line whose intercept yields concentration $[A]_0$ at the end of the reaction when $e^{-k_A t} \to 0$. If both substances A and B react according to Eqs. (8-17a) and (8-17b), the concentration of the product at time t is given by

$$[P]_t = [A]_0 (1 - e^{-k_A t}) + [B]_0 (1 - e^{-k_B t}) \qquad (8\text{-}24)$$

When the rate constants differ substantially $(k_A \gg k_B)$, then $(1 - e^{-k_B t})$ approaches zero at the beginning of the reaction and Eq. (8-24) can be expressed by Eq. (8-23). The $[P]_t$ vs $e^{-k_A t}$ dependence is then linear and the initial concentration of substance A that reacts faster can be found by linear extrapolation. In a further time interval the $e^{-k_A t}$ value becomes almost zero, whereas $e^{-k_B t}$ changes from unity to zero. In this time interval, the concentration of the product is given by

$$[P]_t = [A]_0 + [B]_0 (1 - e^{-k_B t}) \qquad (8\text{-}25)$$

The plot of $[P]_t$ against $e^{-k_B t}$ is a straight line whose intercepts yield $[A_0]$ (at time $t = 0$) and $[A]_0 + [B]_0$ (at $t \to \infty$). Mixtures of two substances can readily be analyzed in this way [246], but the precision greatly depends on ratios k_A/k_B and $[B]_0/[A]_0$. For example, with $k_A/k_B \simeq 10$, substance A virtually does not affect the reaction of substance B and it is thus more advantageous to use the data obtained in the later stage of the reaction, extrapolating the $[P]_t$ vs $e^{-k_B t}$ dependence. The method readily lends itself to automation.

The Single Point Method

This method is used for analysis of a mixture of substances A and B reacting with reagent R under conditions described in the previous section (Eqs. (8-18) and (8-19)). If Eq. (8-20) is divided by the sum of the initial concentrations, $[A]_0 + [B]_0$, and is rearranged, the relationship

$$\frac{[A]_t + [B]_t}{[A]_0 + [B]_0} = \frac{[P]_\infty - [P]_t}{[P]_\infty} = (e^{-k_A t} - e^{-k_B t}) \frac{[A]_0}{[A]_0 + [B]_0} + e^{-k_B t}$$
$$(8\text{-}26)$$

is obtained. Eq. (8-26) expresses a linear dependence between $([A]_t + [B]_t)/([A]_0 + [B]_0)$ or $([P]_\infty - [P]_t)/[P]_\infty$ and the mole fraction of substance $A(y_A)$, with a slope of $(e^{-k_A t} - e^{-k_B t})$ and intercepts on vertical axes equal to $e^{-k_B t}(y_A = 0)$ and $e^{-k_A t}(y_A = 1)$, as can be seen in Fig. 18. If rate constants k_A and k_B are known, the dependence given in Fig. 18 can be obtained by calculating $e^{-k_B t}$ and $e^{-k_A t}$ for time t. When the rate constants

134

are not known, substances A and B are allowed to react separately with reagent R for time t. The values obtained determine intercepts $e^{-k_A t}$ and $e^{-k_B t}$ and their combination yields the calibration curve.

To analyze a mixture, a single measurement taken at a suitable time interval after addition of the reagent is sufficient; the y_A value is then found from the calibration curve. The sum of the initial concentrations of A

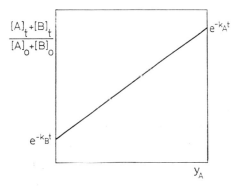

Fig. 18. Single-point method for first order reactions.

and B is determined by an equilibrium method. The precision of analysis depends on the slope of the calibration curve. The single point method generally has an absolute error of 2% when the ratio of the rate constants for substances A and B is greater than 4.

Method of Proportional Equations

The method of proportional equations [227] has been developed for mixtures of substances in which the determination of the sum of the initial concentrations of reactants A and B is difficult or impossible. For a first-order or pseudofirst order reversible reaction,

$$A \xrightarrow{k_A} n_A P \qquad (8\text{-}27)$$

where n_A is the stoichiometric coefficient, the concentrations of reactant A and product P at time t are given by

$$[A]_t = [A]_0 \, e^{-k_A t} \qquad (8\text{-}28)$$

$$[P]_t = n_A([A]_0 - [A]_t) = n_A[A]_0 (1 - e^{-k_A t}) = K_{A,t}[A]_0 \qquad (8\text{-}29)$$

135

where $K_{A,t}$ is a proportionality constant defined by

$$K_{A,t} = n_A(1 - e^{-k_A t}) \tag{8-30}$$

and dependent on the reaction stoichiometry, time and rate constant k_A.

Eq. (8-29) indicates that the concentration of the product at any time t different from zero is proportional to the initial concentration of substance A. When the concentration of the product is monitored continuously and is proportional to the physico-chemical parameter π,

$$\pi = \lambda[P] \tag{8-31}$$

then the relationship

$$\pi = \lambda n_A[A]_0 (1 - e^{-k_A t}) = K'_{A,t}[A]_0 \tag{8-32}$$

is valid. It follows from Eqs. (8-29) and (8-31) that the initial concentration of starting substance A at a certain time t is proportional to the concentration of the product or to measured parameter π. This proportionality holds not only for simple first and pseudofirst order reactions, but also for complex first order reactions, so that the method of proportional equations has broad application.

If substances A and B react to form common product P, as indicated by reactions (8-27) and (8-33),

$$B \xrightarrow{k_B} n_B P \tag{8-33}$$

(n_B is the stoichiometric coefficient in reaction (8-33)), which are first or pseudofirst order, then the reaction of substance A alone is decreased by Eq. (8-29) and that of substance B alone by the equation

$$[P]_t = K_{B,t}[B]_0 \tag{8-34}$$

where proportionality constant $K_{B,t}$ is defined by

$$K_{B,t} = n_B(1 - e^{-k_B t}) \tag{8-35}$$

If the reaction of substance A is not affected by the presence of substance B and that of substance B by the presence of substance A, then the concentration of product P at time t_1 is determined by the equation,

$$[P]_t = K_{A,t}[A]_0 + K_{B,t}[B]_0 \tag{8-36}$$

and that at time $t_2 (t_2 > t_1)$ by the equation,

$$[P]_{t_2} = K_{A,t_2}[A]_0 + K_{B,t_2}[B]_0 \tag{8-37}$$

The solution of Eqs. (8-36) and (8-37) for $[A]_0$ and $[B]_0$ yields the initial concentrations of substances A and B, provided that constants K_A and K_B at times t_1 and t_2 and product concentrations $[P]_{t_1}$ and $[P]_{t_2}$ are known. Constants K_A and K_B can be calculated from Eqs. (8-30) and (8-35). It is advantageous to determine the amount of product P formed by the reaction of a known amount of substance A alone at times t_1 and t_2 and by that of substance B alone, because in this way all factors affecting the reaction system can be studied. The method of proportional equations was used for analysis of a mixture of sugars that react with ammonium molybdate with formation of molybdenum blue [228]. Another advantage of this method is the fact that the sum of the initial concentrations of the reactants need not be known and that it can also be applied to systems in which more than two substances react with a single reagent and to mixtures of substances that form various reaction products with a common reagent.

Tangent Method for Analysis of Binary Mixtures

The tangent method is based on the determination of the rate of changes in the concentrations of the reactants or of the products. It can be used for a mixture of substances A and B that form product P on condition that reactions (8-18) and (8-19) are irreversible and first order. Eq. (8-20) then holds and its differentiation with respect to time yields the equation

$$-\frac{d([P]_\infty - [P]_t)}{dt} = k_A[A]_t + k_B[B]_t \tag{8-38}$$

If substance A reacts faster than substance B, then at the instant when $[A]$ approaches zero the expression, $-d([P]_\infty - [P]_t)/dt$, approaches the value $k_B[B]_t$.

The plot of Eq. (8-38) is the curve obtained by plotting $([P]_\infty - [P]_t)$ against time (see Fig. 19). At any point on this curve, corresponding to time t_n, a tangent can be constructed whose slope determines the rate of the change in the concentration of product P. If substance A reacts faster than substance B, then the ratio of the slope to $([P]_\infty - [P]_t)$ approaches

137

a constant value of k_B with decreasing $[A]_t$, as indicated by the relationship

$$-\frac{d([P]_\infty - [P]_t)}{dt}\left(\frac{1}{[P]_\infty - [P]_t}\right)_{[A]_{t\to0}} = k_B \qquad (8\text{-}39)$$

When k_B is determined, the initial concentration of substance B is calculated from

$$\ln[B]_0 = \ln[B]_t + k_B t = \ln([P]_\infty - [P]_t) + k_B t \qquad (8\text{-}40)$$

Concentration $[A]_0$ must, however, be determined from the independently

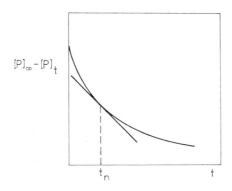

Fig. 19. The tangent method for analysis of binary mixtures.
The determination of the rate of the product concentration change using the tangent to the curve $([P]_\infty - [P]_t)$ vs time.

measured value of $[A]_0 + [B]_0$. The tangent method is used for systems for which the time dependence of $([P]_\infty - [P]_t)$ can be monitored continuously. The precision of the method depends on the precision of the graphical determination of the slope of the tangent to the dependence studied.

8.3.2.2 Methods Based on Second Order Reactions

The following differential kinetic methods are suitable for analysis of mixtures of two substances, A and B, when the concentration of reagent R equals the sum of the concentrations of substances A and B or differs only slightly from it.

The validity of the logarithmic extrapolation method has been extended to second order reactions where the reagent concentration does not equal the sum of the initial concentrations of the reactants [178], but it holds that $[R]_0 > ([A]_0 + [B]_0)$ or $[R]_0 < ([A]_0 + [B]_0)$. A second order irreversible reaction can be generally formulated as

$$M + R \xrightarrow{k_M} P \tag{8-41}$$

The reaction rate is given by

$$\frac{dx_M}{dt} = k_M([R]_0 - x_M)([M]_0 - x_M) \tag{8-42}$$

where x_M is the amount of substance M or R that reacts in time t and k_M is the rate constant. If $x_M = 0$ at time t_0 and $x_M = x_M$ at time t, then integration of Eq. (8-42) yields

$$k_M = \frac{1}{t([R]_0 - [M]_0)} \ln \frac{[R]_0 - x_M}{[M]_0 - x_M} - K \tag{8-43}$$

If M denotes a mixture of substances A and B that react irreversibly with reagent R and the appropriate rate constants are k_A and k_B, then $[M]_0 = [A]_0 + [B]_0$ and $x = x_A + x_B$, where x_A and x_B are the concentrations of substances A and B that react in time t. If substance A reacts with reagent R faster than substance B, then $[A]_0 = x_A$ at the time of complete reaction of A. It then holds that

$$[R]_0 - x = [R]_0 - x_A - x_B = [R]_0 - [A]_0 - x_B \tag{8-44}$$

$$[M]_0 - x = [A]_0 + [B]_0 - x_A - x_B = [B]_0 - x_B \tag{8-45}$$

On substitution of Eqs. (8-44) and (8-45) into integrated Eq. (8-43) and rearrangement, the relationship

$$\ln \frac{[R]_0 - [A]_0 - x_B}{[B]_0 - x_B} = k_B t\{[R]_0 - ([A]_0 + [B]_0)\} + K \tag{8-46}$$

139

is obtained. If follows from Eqs. (8-44) and (8-45) that the time dependence of $\ln\left(([R]_0 - x)/([M]_0 - x)\right)$ is linear for values of t for which $[A]_t = 0$ and the extrapolated value of intercept K for $t = 0$ is given by

$$K = \ln \frac{[R]_0 - [A]_0}{[M]_0 - [A]_0} \qquad (8\text{-}47)$$

The initial concentration of substance A can be determined from the plot of $\ln\left(([R]_0 - x)/([M]_0 - x)\right)$ vs time, and from the linear part of this dependence that corresponds to the reaction of the more slowly reacting substance B, the intercept for $t = 0$ (point 0) can be extrapolated (see Fig. 20). A parallel to the t-axis is constructed through point 0, obtained

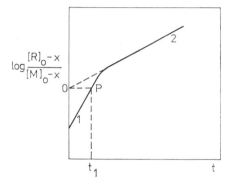

Fig. 20. Logarithmic extrapolation method for second order reactions.

by extrapolating straight line 2, intersecting straight line 1 (the reaction of faster reacting substance A) at point P, and time t_1 corresponding to intercept P is found. The concentration of faster reacting substance A corresponds to the concentration of mixture A + B that reacted in time t_1.

In addition to this graphical metod, the concentration of substance A can also be calculated [230]. During the reaction of substances A and B with reagent R, the amounts of the reagent that reacted in times t_1 and t_2, x_1 and x_2, are determined; times t_1 and t_2 correspond to the times during which the reaction of substance A is virtually complete. If the values of $[R]_0$ and $[A]_0 + [B]_0$ are known, it is possible using Eq. (8-43) to formulate two equations for x_1 and t_1 and x_2 and t_2, whose solution for $[A]_0$ yields

140

the relationship

$$[A]_0 = \frac{[R]_0 \left\{ \left(\dfrac{[M]_0 - x_1}{[R]_0 - x_1} \right)^{t_2} \dfrac{[M]_0 - x_2}{[R]_0 - x_2} t_1 \right\}^{1/(t_1 - t_2)} - [M]_0}{\left\{ \left(\dfrac{[M]_0 - x_1}{[R]_0 - x_1} \right)^{t_2} \dfrac{[M]_0 - x_2}{[R]_0 - x_2} t_1 \right\}^{1/(t_1 - t_2)} - 1} \tag{8-48}$$

from which the concentration of substance A can be calculated.

The determination of substance A by this calculation is simple, because two measurements during the reaction suffice; however, this method is less precise than the graphical method.

The logarithmic extrapolation method was used for analysis of a mixture of organic substances containing the same characteristic group that reacts with a common reagent [231]. For the method to be applicable, the reaction rates of the components of the test mixture must be sufficiently different. The resultant dependence is then expressed by two straight lines with unequal slopes, as shown in Fig. 20. The method cannot be used when the initial concentration of the reagent equals the sum of the initial concentrations of the test substances, as Eq. (8-43) then has no meaning. It is necessary that a method for determination of the sum of the initial concentrations of the test substances be available, because this value is required for construction of the calibration curve.

The Linear Extrapolation Method for Second Order Reactions

The linear extrapolation method can be used for analysis of mixtures of two substances A and B which react with common reagent R on condition that $[R]_0 = [A]_0 + [B]_0$ and that the reactions are irreversible [232]. The rate of the consumption of reagent R is then expressed by the equation

$$- \frac{d[R]}{dt} = k_A [R] [A] + k_B [R] [B] \tag{8-49}$$

where rate constants k_A and k_B correspond to the reactions of reagent R with substances A and B, respectively. If $-d[R]/dt = dx/dt$, x_A is the amount of substance A consumed and x_B is the amount of substance B consumed, then $x = x_A + x_B$ is the amount of reagent R consumed in

141

time t. On substituting these expressions into Eq. (8-49), the relationship

$$dx/dt = k_A([R]_0 - x)([A]_0 - x_A) + k_B([R]_0 - x)([B]_0 - x_B) =$$
$$= ([A]_0 + [B]_0 - x_A - x_B)\{k_A([A]_0 - x_A) + k_B([B]_0 - x_B)\}$$

$$(8\text{-}50)$$

is obtained. If substance A reacts faster than substance B, then, at the instant of completion of the reaction of substance A, it holds that

$$[A]_0 = x_A \qquad (8\text{-}51)$$

This expression is substituted into Eq. (8-50), obtaining

$$dx/dt = k_B([B]_0 - x_B)^2 \qquad (8\text{-}52)$$

Integration of Eq. (8-52) yields

$$\frac{1}{[B]_0 - x_B} = k_B t + K \qquad (8\text{-}53)$$

Because $K = 1/[B]_0$ at time $t = 0$, Eq. (8-53) can be rewritten to yield

$$\frac{1}{[B]_0 - x_B} - \frac{1}{[B]_0} = k_B t = \frac{x}{[B]_0([B]_0 - x_B)} \qquad (8\text{-}54)$$

Eq. (8-54) holds only when the reaction of substance A is complete. Because $[R]_0 = [A]_0 + [B]_0$ and $[A]_0 = x_A$, it holds that $[B]_0 = [R]_0 - x_A$ and $x - [A]_0 = x_B$; thus Eq. (8-52) can be formulated as

$$x = k_B[B]_0([R]_0 - x)t + [A]_0 \qquad (8\text{-}55)$$

It follows from Eq. (8-55) that the dependence of x on $([R]_0 - x)t$ is linear with a slope of $k_B[B]_0$ and the intercept on the vertical axis $(t = 0)$ corresponds to the value $[A]_0$. A typical dependence of x on $([R]_0 - x)t$ is given in Fig. 21. This dependence is linear from the time instant corresponding to complete reaction of substance A. Extrapolation of the linear part to $t = 0$ yields an intercept corresponding to the initial concentration of substance A.

The precision of reading of the $[A]_0$ value depends on the validity of the relationship $[R]_0 = [A]_0 + [B]_0$. If, for example, the reagent is added in a 2% excess, the $[A]_0$ value obtained from the graph is 2% less.

If more slowly reacting substance B is to be determined, the appropriate relationship can be analogously derived from Eq. (8-54),

$$\frac{1}{[R]_0 - x} = k_B t + \frac{1}{[B]_0} \qquad (8-56)$$

Eq. (8-56), which holds on condition that $[A]_t = 0$, confirms the linear dependence of $1/([R]_0 - x)$. A typical plot of this dependence is shown in

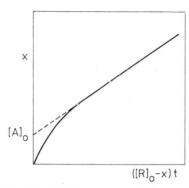

Fig. 21. The linear extrapolation method for second order reactions. Determination of the faster reacting substance.

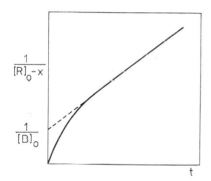

Fig. 22. The linear extrapolation method for second order reactions. Determination of the more slowly reacting substance.

Fig. 22; the linear part of the dependence for $[A]_t = 0$ has a slope equal to k_B and the intercept obtained by extrapolation to $t = 0$ equals $1/[B]_0$.

The linear extrapolation method can readily be simplified and the analysis can be carried out by determining the x value one or two times after

143

completion of the reaction of the faster reacting substance, but during the period of the reaction of the more slowly reacting substance. Because $[B]_0 = [R]_0 - [A]_0$, this expression can be substituted into Eq. (8-55), giving

$$[A]_0 = \frac{x - k_B([R]_0 - x)[R]_0\, t}{1 - k_B([R]_0 - x)\, t} \tag{8-57}$$

Using Eq. (8-57) $[A]_0$ can be calculated on the basis of a single measurement of x at time t under the above conditions. This determination, of course, requires knowledge of rate constant k_B.

If the amounts of reagent R consumed at time instants t_1 and t_2 are determined, the x_1 and x_2 values are obtained. It holds for these values, according to Eq. (8-55), that

$$k_B = \frac{x_1 - [A]_0}{[B]_0\,([R]_0 - x_1)\, t_1} = \frac{x_2 - [A]_0}{[B]_0([R]_0 - x_2)\, t_2} \tag{8-58}$$

By solving Eq. (8-58) for $[A]_0$, the expression,

$$[A]_0 = \frac{x_1 - \{([R]_0 - x_1)\, t_1/([R]_0 - x_2)\, t_2\}\, x_2}{1 - ([R]_0 - x_1)\, t_1/([R]_0 - x_2)\, t_2} \tag{8-59}$$

is obtained, that enables calculation of the concentration of substance A from two measurements, without knowledge of rate constant k_B.

Relationships corresponding to Eqs. (8-57) and (8-58) determining the concentration of more slowly reacting substance B can be derived analogously. It holds for measurement at time t that

$$[B]_0 = \frac{1 - k_B t([R]_0 - x)}{[R]_0 - x} \tag{8-60}$$

and for measurement at two times,

$$[B]_0 = \frac{([R]_0 - x_1)([R]_0 - x_2)(t_2 - t_1)}{([R]_0 - x_2)\, t_2 - ([R]_0 - x_1)\, t_1} \tag{8-61}$$

The Single Point Method for Second Order Reactions

A single point method has been developed for the determination of two substances in a mixture when the concentration of reagent R equals the sum of the concentrations of the two reactants, $[R]_0 = [A]_0 + [B]_0$

[233]. It is then sufficient to determine the sum of the concentrations of substances A and B and the decrease in the amounts of these substances at a certain optimum time after the beginning of the reaction. The calibration curve is the dependence of the unreacted amount of substances A and B (in percent) in the reaction mixture at time t_{opt} on the amount of substance A (in percent), i.e. $\{[A]_0/([A]_0 + [B]_0)\} \cdot 100$. Time t_{opt} that corresponds to the most suitable time for the determination of the amount (%) of substances A and B that remain unreacted is given by

$$t_{opt} = \frac{(k_B/k_A)^{1/2}}{k_B[R]_0} \tag{8-62}$$

The determination itself is very simple. The $[A]_0 + [B]_0$ value is determined by some equilibrium method. The amount of substances A and B is determined at time t_{opt} and the percentage of substance A in the mixture is obtained directly from the calibration curve.

A single point method for second order reactions is suitable for repeated analyses of mixtures of two substances with varying composition. The calibration curve is then constructed on the basis of analyses of mixtures of substances A and B with a known composition, as the dependence of the unreacted amounts of substances A and B (%) on the amount of substance A (%) in the mixture. The single point method requires knowledge of rate constants k_A and k_B corresponding to the reactions of substances A and B with reagent R.

The values necessary for the construction of the calibration curve can also be calculated using the kinetic equations for the reactions of substances A and B with reagent R. The solution is rather complicated and requires combination of numerical and graphical techniques [132].

The Graphical and Integration Method of Schmalz and Geisler [234]

The graphical differential and integration method can be used when the pertinent reactions are second order. The simultaneous reactions of substances A and B with reagent R are described by the kinetic equations

$$\frac{dx_A}{dt} = k_A([A]_0 - x_A)([R]_0 - x_A - x_B) \tag{8-63}$$

$$\frac{dx_B}{dt} = k_B([B]_0 - x_B)([R]_0 - x_A - x_B) \tag{8-64}$$

145

where x_A and x_B are the amounts of substances A and B that react in time t. The amount of reagent R that reacts in time t is given by the sum $x_A + x_B$. Eqs. (8-63) and (8-64) cannot be solved for values $[A]_0$ or $[B]_0$, because the solution leads to an equation that cannot be integrated, Schmalz and Geisler [234] therefore proposed a general solution for the determination of $[B]_0$. This consists of the following steps: first the concentration of reagent R that reacted in time t is determined experimentally; this value equals the sum of x_A and x_B. Now a tangent is constructed at time t to the experimental concentration (of reagent R) vs time curve. The slope of this tangent has the value q, defined by

$$q = \frac{dx}{dt} = \frac{dx_A}{dt} + \frac{dx_B}{dt} =$$

$$= \{k_A([A]_0 - x_A) + k_B([B]_0 - x_B)\} ([R]_0 - x_A - x_B) \qquad (8\text{-}65)$$

Using the relationship $[R]_0 = [A]_0 + [B]_0$, the terms containing $[A]_0$ can be eliminated from Eq. (8-65) and the equation for $[B]_0$ can be derived. The graphical interpretation of this equation is rather complicated because a high degree equation is involved; the exponent of the $[B]_0$ term has a value of k_A/k_B.

The authors of the method also proposed a graphical determination of constants k_A and k_B, so that the graphical differential method is mathematically well developed. In the calculation, the slope of the tangent to the experimental kinetic curve (the x vs t dependence) is used. The precision of the determination of the slope depends on the precision of the construction of the kinetic curve and on that of the construction of the tangent to it at time t. Hence the graphical treatment of the experimental data is subject to a certain error. The complexity and tediousness of the calculation result in the method having a rather limited practical importance.

The graphical integration method is also based on kinetic equations (8-63) and (8-64) that are formulated as follows:

$$\frac{dx_A}{dt} = k_A([A]_0 - x_A) f(t) \qquad (8\text{-}63)$$

$$\frac{dx_B}{dt} = k_B([B]_0 - x_B) f(t) \qquad (8\text{-}64)$$

146

because relationship (8-65) is valid, $[R]_0 - x_A - x_B = f(t)$. This dependence can be found from the experimental kinetic curve, the time dependence of x, as $x = x_A + x_B$. Integration of Eqs. (8-63) and (8-64) is carried out graphically and the applicability of the method depends on knowledge of the rate constants. As the solution is rather complicated, it has not yet been used analytically.

8.3.2.3 Methods Using First and Pseudofirst Order Reactions with Respect to the Reagent

This section contains differential kinetic methods for analysis of a mixture of two substances, A and B, when the concentration of reagent R is very small compared with the sum of the concentrations of substances A and B.

The Roberts and Regan [235] Single Point Method

In this method, the changes in the concentration of reagent R are monitored during the reaction of substances A and B with reagent R; hence sensitive determination of reagent R is required.

If reagent R reacts with substances A and B present in an at least 50-fold excess and if the reactions are irreversible, then the loss of the reagent in dependence on time can be expressed by

$$- \frac{d[R]}{dt} = k_A[A]_0 [R] + k_B[B]_0 [R] = K^*[R] \qquad (8\text{-}66)$$

where

$$K^* = k_A[A]_0 + k_B[B]_0 \qquad (8\text{-}67)$$

In Eqs. (8-66) and (8-67), k_A and k_B are the rate constants for the reactions of substances A and B with reagent R. The amounts of substances A and B consumed in reaction with reagent R can be neglected, because these substances are present in excess. The sum of the initial concentrations of substances A and B is denoted as $[M]$:

$$[M]_0 = [A]_0 + [B]_0 \qquad (8\text{-}68)$$

The solution of Eqs. (8-67) and (8-68) then makes it possible to calculate the initial concentrations of substances A and B. The values of constants K^*,

k_A and k_B and the sum of the initial concentrations, $[M]_0$, must be known. The $[M]_0$ value is determined by an equilibrium method. Constant K^* is the overall rate constant for a pseudofirst order reaction, defined by

$$K^* = \frac{\ln [R]_0/[R]_1}{t_1 - t_0} \qquad (8\text{-}69)$$

where the reagent concentration, $[R]_1$, corresponds to time t_1. The values of rate constants k_A and k_B are determined in the same way as the rate constant for the second order reaction of substance A or B with reagent R.

The original Roberts and Regan method was later modified [217] so that a single measurement using a calibration curve is sufficient for the determination. Integration of Eq. (8-66) yields the relationship,

$$\ln \frac{[R]_t}{[R]_0} = -(k_A[A]_0 + k_B[B]_0)\, t = -K^* t \qquad (8\text{-}70)$$

which, on substitution for $[B]_0$ from Eq. (8-68) and rearrangement, assumes the form

$$\frac{[A]_0}{[M]_0} = \frac{\ln ([R]_t/[R]_0)}{t[M]_0 (k_B - k_A)} + \frac{k_B}{k_B - k_A} \qquad (8\text{-}71)$$

It follows from Eq. (8-71) that, at a certain reagent ratio, $[R]_t/[R]_0$, the

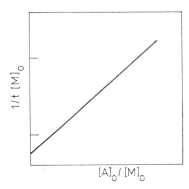

Fig. 23. The single point method of Roberts and Regan.

$1/t[M]_0$ vs $[A]_0/[M]_0$ dependence is linear and can be plotted as a calibration curve for the analysis (Fig. 23) ($[A]_0/[M]_0$ is the mole fraction of substance A). To construct the calibration curve and carry out the measure-

148

ment itself, the optimum time interval, t_e, in which the determination error is minimal must be found. This interval lies in the part of the experimental $[R]_t$ vs t dependence where the change in the concentration of R per time unit is maximal. It can be derived from Eq. (8-70) that time interval t_e is limited by the values of concentration R that satisfy the ratio equality

$$\frac{[R]_t}{[R]_0} = \frac{1}{e} \qquad (8\text{-}72a)$$

where e is the base of natural logarithms.

When mixing the reactants takes e.g. 2 or 3 seconds, Eq. (8-72a) can be used in the form

$$\frac{[R]_{t_2}}{[R]_{t_1}} = \frac{1}{e} \qquad (8\text{-}72b)$$

where $[R]_{t_1}$ is the concentration of R at time t_1, shortly after the beginning

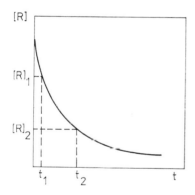

Fig. 24. The single point method.
Determination of time interval t_e.

of the reaction. The determination of t_e is evident from Fig. 24. The calibration curve itself is readily constructed; provided that substances A and B are available in pure form, the $1/t_e[A]_0$ and $1/t_e[B]_0$ values are determined, recorded on the $1/t[M]_0$ vs $[A]_0/[M]_0$ plot (Fig. 23) and connected by a straight line.

The Method of Proportional Equations for Systems with a Small Reagent Concentration

The method of proportional equations has also been modified for systems in which two substances react with a common reagent, the corresponding reactions being pseudofirst order with respect to the reagent [236]. Then, according to Eq. (8-67), the relationship

$$K_1^* = k_{A1}[A]_0 + k_{B1}[B]_0 \qquad (8\text{-}73)$$

is valid. On a change in the reaction conditions an analogous expression is valid,

$$K_2^* = k_{A2}[A]_0 + k_{B2}[B]_0 \qquad (8\text{-}74)$$

Most often, the water/solvent ratio or the composition of a mixture of solvents is changed. If the rate constants satisfy the inequality, $k_{A1}k_{B1} \neq k_{A2}k_{B2}$, then the solution of Eqs. (8-73) and (8-74) yields the initial concentrations of substances A and B. It should be emphasized that it is necessary to record experimental data before the change in the reaction conditions and after it; if, e.g., only the initial reagent concentration is changed, it is impossible to obtain two equations solvable for $[A]_0$ and $[B]_0$. From the experimental data obtained from measurements on substance B, substance A and for a mixture of A with B under two different sets of experimental conditions, constants k_{A1}, k_{B1}, k_{A2}, k_{B2}, K_1^* and K_2^* are obtained and used for solution of Eqs. (8-73) and (8-74). Mixtures of aldehydes and ketones were analyzed in this way [236].

The applicability and precision of the method depends on the rate of reaction of substance A or B with the reagent. Precise results are obtained when the rate constant ratio, k_A/k_B, is at least 2.0. Under optimal conditions the rate constant ratio is reversed on a change in the reaction conditions, e.g. $k_A/k_B > 1$ becomes $k_A/k_B < 1$.

Graphical Interpolation Method

A graphical interpolation method has been proposed [258] for analysis of three-component mixtures, on condition that the test substances form a common product with reagent R by first or pseudofirst order reactions.

The curve of the time dependence of the product formation is recorded in a solution of the test mixture. A reference mixture with known concentrations of the test substances is then prepared and the same dependence is recorded under identical conditions. If the compositions of the test and reference mixtures are the same, the curves are also the same. A suitable composition of the reference mixture is found graphically, on the basis of data obtained for defined mixtures of substances A, B and C. After addition of reagent R, the product concentration is measured at two time intervals. The graphical interpolation method is very simple and requires no numerical calculations. However, as programmable pocket calculators are now commonly available, it is probably faster to treat the experimental data numerically, using the method of proportional equations, a geometrical version of which is the graphical interpolation method.

8.3.2.4 Methods Based on Pseudozero Order Reactions

When substances A and B react with common reagent R, the reactions are pseudozero order if the measurement is carried out over a short interval at the beginning of the reaction when it can be assumed that the concentrations of all the reactants are virtually constant, without respect to their ratio.

The Single-Point Method for Pseudozero Order Reactions

The initial rate of reaction of A with R, v', is proportional to the initial concentration of substance A,

$$v' = K_A[A]_0 \tag{8-75}$$

When substances A and B react with a common reagent, the overall initial rate is given by

$$v'_{(A+B)} = K_A[A]_0 + K_B[B]_0 \tag{8-76}$$

If the sum of the initial concentrations $[A]_0 + [B]_0$ is known, then the initial concentrations of A and B can be determined from the measured $v'_{(A+B)}$ value similarly as in the Roberts and Regan method [237]. An advantage of this method lies in the possibility of using any ratio of reactants

151

in the absence of side reactions, rapidity, because the initial rate is measured, and in simplicity of the calculations. The method was applied to analysis of mixtures of sugars.

Method of Proportional Equations for Pseudozero Order Reactions

The method is entirely analogous to the method of proportional equations for systems with a low reagent concentration (Section 8.3.2.3), but the initial reaction rates are measured.

According to Eq. (8-76) it holds for reaction of substances A and B with reagent R that

$$v'_{(A+B)1} = K_{A1}[A]_0 + K_{B1}[B]_0 \qquad (8\text{-}77)$$

and, after a change in the reaction conditions, that

$$v'_{(A+B)2} = K_{A2}[A]_0 + K_{B2}[B]_0 \qquad (8\text{-}78)$$

Solution of equations (8-77) and (8-78) yields the $[A]_0$ and $[B]_0$ values in the same way as in Section 8.3.2.3. Mixtures of alcohols were analyzed by this method [238].

8.4 The Accuracy, Precision and Sensitivity of Kinetic Methods

It follows from the very principle of the kinetic methods that their accuracy is ensured only when the particular reaction obeys the assumed mechanism, i.e. no side reactions of the reactants and products are involved. Naturally, the analytical procedure employed for monitoring time changes in the concentration of a given component of the reaction system must also be accurate. The accuracy is further affected by some external factors that can be influenced by modification of the experimental conditions. The main factors are the temperature and the ionic strength (see Chapter 2). The temperature must be maintained constant and it is recommended that all solutions be thermostatted before mixing; the ionic strength must also be kept constant.

The relative standard deviation of kinetic determinations is about $2-10$ per cent.

Applications of modern instrumentation and data treatment methods have made the rapidity, precision and simplicity of kinetic methods comparable with those of equilibrium methods. However, kinetic methods are more strongly dependent on the experimental conditions, chiefly on the pH and temperature. Small differences in the experimental data obtained during kinetic measurements on samples and standards are ascribed to variations in the rate constant values. It would therefore be advantageous to find kinetic methods that would suppress the dependence of the analytical result on these variations. Methods based on first order reactions are very precise, because of the possibility of determining the rate constant without knowledge of the concentration of the rate-determining species. The rate constant can thus be determined independently for each sample. A multiple linear regression program has been developed for calculation of the rate constant and the initial and final absorbances, permitting the fit of the experimental data to the model of the first order reaction [260] and hence a further improvement in the precision.

8.4.1 METHODS EMPLOYING CATALYTIC REACTIONS

Analytically, the sensitivity of these methods is most important. The minimum determinable catalyst content can be theoretically derived from the equation for the rate of catalytic reaction (3-13), that can be simplified to give

$$dx/dt = N[C]_0 \qquad (8\text{-}79)$$

where coefficient N is the product of all concentration terms in Eq. (3-10), except for $[C]_0$, and of the appropriate rate constant. Eq. (8-79) is integrated for $[C]_0$ and rearrangement yields

$$[C]_0 = \frac{\Delta x}{\Delta t} \frac{1}{N} \qquad (8\text{-}80)$$

The lowest determinable catalyst concentration, $([C]_0)_{min}$, is given by the lowest concentration change, Δx_{min}, that can be measured analytically and time Δt_{max} during which the concentration changes can be found,

$$([C]_0)_{min} = \frac{\Delta x_{min}}{\Delta t_{max}} \frac{1}{N} \qquad (8\text{-}80a)$$

It follows from Eq. (8-80a) that, for a substance determinable at a concentration of 10^{-7} mol.l^{-1}, an indicator reaction time of 10 min and initial reactant concentrations of about 1 mol . l^{-1}, the $([C]_0)_{min}$ value equals 10^{-16} mol . l^{-1}. The theoretical sensitivity is thus very high. However, the actual determination limit is from 10^{-8} to 10^{-12} mol . l^{-1}, because the indicator reaction also takes place in the absence of the catalyst. The overall rate of a catalytic reaction is given by Eq. (3-13), i.e. by the sum of the rates of the indicator reaction alone and of the catalyzed reaction. In addition to the unfavourable effect of the rate of the uncatalyzed indicator reaction, effects from impurities in the solution and on the walls of reaction vessels must also be considered. These negative effects from the kinetic background must be suppressed as much as possible. For this reason, constant temperature, pH and ionic strength must be maintained, the purest chemicals must be used and, if possible, work must always be carried out with the same vessels. The addition method is also advantageous for this reason.

The error of kinetic determinations based on catalytic reactions is generally smaller, the greater is the difference between the rates of the catalytic and indicator reactions and the more the background is suppressed. In choosing among the kinetic methods listed in Section 8.2.1 for determining a catalyst, the following procedure is adopted. The rate of a catalytic reaction is generally described by

$$\frac{d[P]}{dt} = F([S_1], [S_2], ..., [S_i]) [C_0] \tag{8-81}$$

where F is a certain function of the rate constants and the reactants. Eq. (8-81) can be formulated as

$$\int_{[P]_1}^{[P]_2} \frac{d[P]}{F([S_1], [S_2], ..., [S_i])} = \int_{t_1}^{t_2} [C]_0 \, dt \tag{8-82}$$

and, on substituting

$$G([P]) = \int \frac{d[P]}{F[S_i]}$$

and integrating, it is simplified to yield

$$[C]_0 = \frac{G[P]_2 - G[P]_1}{t_2 - t_1} \tag{8-83}$$

Eq. (8-83) is generally valid, even for very complex functions F. If the numerator of the fraction in Eq. (8-83) is constant, then $[C]_0$ is proportional to $1/(t_2 - t_1)$. This condition is met in the constant concentration method, which is optimal for determination of catalysts; methods based on the measurement of the initial rate are unnecessary. The same holds for the determination of enzymes, where the results obtained by the constant time or concentration method are subject to a larger error if the concentration changes are measured for a small reaction degree [259].

8.4.2 METHODS EMPLOYING UNCATALYZED REACTIONS

Among these methods, when applied to determination of a single substance, those based on first and pseudofirst order reactions are most important. The appropriate integrated kinetic equation can be rearranged to give

$$-\Delta[A] = [A]_0 \left(e^{-kt_1} - e^{-kt_2}\right) \tag{8-84}$$

from which it holds for $[A]_0$ that

$$[A]_0 - \frac{-\Delta[A]}{e^{-kt_1}(1 - e^{-k\Delta t})} \tag{8-85}$$

where $\Delta t = t_2 - t_1$. Eq. (8-85) determines the test substance concentration as a function of the measured quantities and of parameters that can be influenced experimentally.

If the constant concentration method is used for determining A, the $\Delta[A]$ term in Eq. (8-85) is constant and the Δt value is measured. Because Δt and t_1 vary in dependence on $[A]_0$, the $1/\Delta t$ vs $[A]_0$ dependence is nonlinear. A virtually linear dependence is obtained only when $\Delta[A]$ is very small compared with $[A]_0$, which is the condition for precise determination of A. However, the $\Delta[A]$ value must not be too small in order to preserve a sufficient precision of the measuring technique used.

The situation is quite different with the constant time method, where t_1 and Δt in Eq. (8-85) are kept constant and thus the $[A]_0$ vs $\Delta[A]$ dependence is strictly linear. Hence the constant time method is more advantageous than the constant concentration method for first and pseudofirst order reactions.

When evaluating the methods based on measurement at the beginning of the reaction, Eq. (8-85) is used and it is assumed that the dependence of $[A]$ on t is linear. To find the time interval from the beginning of the reaction, within which the reaction exhibits pseudozero order behaviour, the $e^{-k\Delta t}$ term in Eq.(8-85) is expanded into the series

$$e^{-k\Delta t} = 1 - k\,\Delta t + (k\,\Delta t)^2/2! + \ldots \tag{8-86}$$

Eq. (8-85) can then be expressed in the form

$$[A]_0 = \frac{-\Delta[A]}{e^{-kt_1}[k\,\Delta t - (k\,\Delta t)^2/2! + \ldots]} \tag{8-87}$$

If $k\,\Delta t$ is very small, Eq. (8-87) is simplified to give

$$[A]_0 = \frac{-\Delta[A]}{k\,\Delta t\,e^{-kt_1}} \tag{8-88}$$

and thus the $[A]_0$ vs $-\Delta[A]$ and $[A]_0$ vs $1/\Delta t$ dependences are linear. Time interval Δt, in which these dependences are linear, is found from Eq. (8-87).

TABLE 9

Error of a determination by the constant concentration method when the measurement is carried out at the beginning of the reaction (according to ref. [248])

$k\,\Delta t$	$[A]_2/[A]_1$	Relative change of [A], %	Error %
0.002	0.9980	0.20	0.10
0.005	0.9950	0.50	0.25
0.010	0.9900	1.00	0.50
0.020	0.9802	1.98	1.00
0.050	0.9512	4.88	2.52

Table 9 shows the increase in the error of a determination by the constant concentration method in dependence on time interval Δt. The relative change in $[A]$ during Δt is expressed by the $[A]_2/[A]_1$ ratio that can be calculated from

$$\ln \frac{[A]_2}{[A]_1} = -k\,\Delta t \tag{8-89}$$

156

where $[A]_2$ and $[A]_1$ correspond to times t_2 and t_1, respectively, for which $\Delta t = t_2 - t_1$. It follows from Table 9 that, for the constant concentration method at the beginning of the reaction, the error is about 1% when the relative change in $[A]$ is 2%. This limitation holds only for the constant concentration method; application of the constant time method does not depend on Δt, as the $[A]_0$ vs $\Delta[A]$ dependence is generally linear.

8.4.3 DIFFERENTIAL KINETIC METHODS

The differential kinetic methods described in Section 8.3 are not suitable for any mixture. Many of the authors who contributed to the development of differential methods [178, 217, 232, 233, 239, 240] also dealt with the sources of error and factors that affect the applicability of the methods.

The basis for the use of both the extrapolation methods is the kinetic character of the reactions of both test substances with the reagent. These reactions must be irreversible and side reactions must not be present. Further factors determining the accuracy and precision of the logarithmic extrapolation method are the ratio of the rate constants, k_A/k_B, for the reactions of the test substances with the reagent and the concentration ratio $[A]_0/[B]_0$.

Conditions for the application of the logarithmic extrapolation method are given by the equations

$$\ln \frac{[A]_t}{[A]_0} - -k_A t \tag{8-90}$$

and

$$\ln \frac{[B]_t}{[B]_0} = -k_B t \tag{8-91}$$

describing the kinetics of reactions (8-18) and (8-19). The relationship

$$\ln \frac{[A]_t}{[A]_0} = \frac{k_A}{k_B} \ln \frac{[B]_t}{[B]_0} \tag{8-92}$$

is derived from Eqs. (8-90) and (8-91). A linear relationship between $\ln [A]_t/[A]_0$ and $\ln [B]_t/[B]_0$ follows from Eq. (8-92) (see Fig. 25 for various values of k_A/k_B). Fig. 25 yields a value of k_A/k_B for which a sufficient amount of the more slowly reacting substance remains in the solution on virtual completion of the faster reaction. It follows that, for $k_A/k_B = 5$, a value of $[B]_t/[B]_0 = 0.4$ corresponds to $[A]_t/[A]_0 = 0.01$, i.e. about 60% of substance B remains unreacted after reaction of 99% of substance A and the

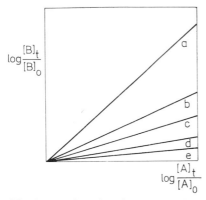

Fig. 25. The logarithmic extrapolation method.
The log $[B]_t/[B]_0$ vs log $[A]_t/[A]_0$ dependence.
The k_A/k_B ratio equals: $a - 1$, $b - 2$, $c - 3$, $d - 5$, $e - 10$.

time of further reaction of B is sufficient to yield enough values to construct a linear dependence allowing precise extrapolation.

The error of the logarithmic extrapolation method can be assessed from Fig. 26, where the dependence of the deviation of the experimental from the theoretical result on the k_A/k_B ratio is given for various ratios of the initial concentrations of the test substances. The curves in Fig. 26 correspond to various $[A]_0/[B]_0$ ratios, where $[B]_0$ is always 5×10^{-4} mol . 1^{-1}. Deviation $\Delta[B]$ from the theoretical value of $[B]_0$ is expressed in mol . 1^{-1}. If the maximum deviation for substance B does not exceed 5×10^{-5} mol . 1^{-1}, it follows from Fig. 26 that, e.g. with a ratio of $[A]_0/[B]_0 = 1$, the rate constant ratio $k_A/k_B = 13$, whereas for $[A]_0/[B]_0 = 0.25$ it suffices when $k_A/k_B = 5$.

The applicability and precision of the method of proportional equations are less limited by the $[A]_0/[B]_0$ and k_A/k_B values than those of the experimental methods. The precision depends on the rate constant ratio and on time t_2 at which the second measurement is carried out, required for formulation of the second of the two equations necessary for calculation of the

158

initial concentrations of substances A and B. The error decreases with increasing k_A/k_B; the lowest permissible value of this ratio is 3 to 4. For the error to be small, time t_2 must be as long as possible. The method of proportional equations has a minimum error for $t_2 = t_\infty$, i.e. for complete

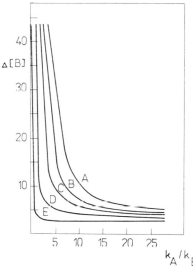

Fig. 26. The error of the graphical extrapolation method.
The ΔB vs k_A/k_B dependence was calculated for $[A]_0/[B]_0$ ratios $A - 9, B - 3, C - 1,$ $D - 0.25, E - 0.1.$

reaction of substances A and B. A detailed study [240] has shown that the optimal conditions involve $k_A/k_B = 4$ (or less) and t_2 approaching t_∞. If k_A/k_B lies between 4 and 20, t_2 must correspond to 97 to 99% reaction of substance A; at $k_A/k_B > 20$, the reaction of substance B must also be considered and t_2 must correspond to at least 30% reaction of substance B. Fig. 27 shows the dependence of the relative error of the determination of A on the k_A/k_B ratio, for a mixture containing 25% substance A and $k_A - 0.03$ min^{-1}, for various t_2 values.

The error of the single point method also depends on ratios $[A]_0/[B]_0$ and k_A/k_B. The error generally increases with increasing $[A]_0/[B]_0$ and decreasing k_A/k_B. It is unnecessary that the difference in the rates of the reactions of A and B be large. For example, the error is acceptable with

159

mixtures containing 90, 50 and 10% of substance A, if k_A/k_B equals 2, 1.7 and 1.5, respectively. The main advantage of the single point method is the possibility of determining very small quantities of the faster reacting substance. As shown in Fig. 28, the sensitivity of the method increases with increasing k_A/k_B; for $k_A/k_B = 50$, faster reacting substance A can be determined precisely even if its content in the mixture with B is only 0.1%. The validity of the theoretical dependence in Fig. 28 was experimentally verified

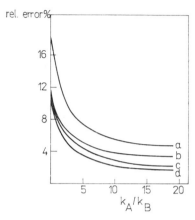

Fig. 27. Dependence of the relative error of the method of proportional equations on the k_A/k_B value.
Error expressed in % $[A]_0$; the amount of reacted A: $a - 97.6\%$, $b - 99.2\%$, $c - 99.9\%$, $d - 100\%$.

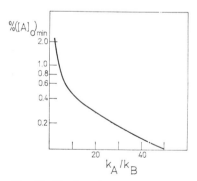

Fig. 28. Single point method.
Dependence of $([A]_0)_{min}$ (the lowest permissible concentration of the faster reacting substance) on k_A/k_B.

160

by determining 0.001% diethylamine in the presence of methylaniline [241]; k_A/k_B then equals approximately 160.

The applicability of differential kinetic methods is limited by a number of conditions. The methods thus must be compared and the types of analysis for which they are best suited must be found. For a large number of samples of the same mixture, the method of proportional equations or the single point method are evidently advantageous. Then it is worthwhile to determine rate constants k_A and k_B, which must be known if these methods are to be used. The single point method has broader use, in view of the values of ratios $[A]_0/[B]_0$ and k_A/k_B. Another advantage is the possibility of repeating the analysis with the same sample, as the initial reactants are present in a large excess over the added reagent. The method of proportional equations has an advantage in that the sum of the initial concentrations of the test substances, $([A]_0 + [B]_0)$, need not be determined by an independent method.

Graphical extrapolation methods are, on the other hand, preferable for analyses of smaller numbers of samples, because the appropriate rate constants need not be known; however, the method is more tedious, as it requires a greater amount of experimental data.

In evaluating differential methods, the kinetic character of the reactions involved must be considered. In the final stages of the reaction, side and reversible reactions may occur, etc., that may affect the precision, because the measurements in extrapolation methods are carried out practically during the whole reaction. This limitation applies only partially to the method of proportional equations, because time t_2 can be chosen so that the interfering reactions do not yet play a role. The complexity of the final stages of the reaction cannot affect the single point method, as the measurement is performed in a time interval within which the reactant concentrations change very little.

In addition to negative kinetic effects occurring in the final stage of the reaction, the rate constants may change in mixtures of some chemically similar substances (see Chapter 4). These changes are caused by synergistic effects [217], stemming from interactions of the reactants with the reaction medium, from changes in the activity coefficients of the reactants and from catalytic effects. Synergistic effects do not influence the graphical methods. If the method of proportional equations must be used, then the effect of any synergistic phenomenon can be eliminated by determining k_A and k_B

experimentally for mixtures of the test substances with known compositions.

The best precision is obtained using the method of proportional equations, as the decisive factor determining the error of the determination is the choice of time t_2. The extrapolation methods are considered less precise (provided that the measuring methods have the same precision), because the precision depends only on that of the graphical extrapolation. The precision of the single point method is comparable with that of the method of proportional equations [242, 243].

Finally, two important facts should be pointed out. First, the differential kinetic methods can also be automated and the apparatuses required for the method of proportional equations or the single point method are simpler than those for the extrapolation methods [244–246]. Second, it is theoretically possible to use any differential method even for analyses of multicomponent mixtures. However, only the graphical extrapolation method and the method of proportional equations are practically suitable and the differences in the test substance rate constants must be larger than in analyses of binary mixtures. It follows from the published works that the method of proportional equations is most suitable for analyses of multicomponent mixtures [247]; with sufficient differences in the rate constants, 3 to 4 components can be determined with an acceptable error.

9 Instrumentation for Kinetic Methods

The apparatus for kinetic analytical methods can vary in complexity, from a simple titration apparatus to fully automated instruments capable of collecting samples, carrying out the measurement and treating the results. Any analytical procedure based on measurement of the reaction rate consists of the following steps:

. sample $\searrow$
reagent $\nearrow$ reaction vessel $\rightarrow$ measurement of time changes in the concentration of a reaction component $\rightarrow$ handling of the experimental data.

In all kinetic methods the sample must be mixed completely and rapidly with the reagent at a constant temperature, an analytical method suitable for monitoring the concentration of a reactant or a product at constant temperature must be chosen and the results must be appropriately handled. Each of these steps can evidently be automated. Instrumental analytical methods readily enable automation of the monitoring step. Spectrophotometric and electroanalytical methods are mostly used, recording a certain physico-chemical parameter, proportional to the concentration as a function of time. Automation of the mixing of the sample with the reagent requires special mechanical devices, especially for reactions with half-times of less than 10 seconds. As slower reactions are mostly used for analytical purposes, it is only necessary to mix the reactants directly in the reaction vessel for sufficiently precise determination of the beginning of the reaction, i.e. of time t_0 (e.g. in a spectrophotometer cuvette, in a vessel with electrodes, etc.). The reagent is usually injected with a syringe into the prepared reaction mixture in a suitable vessel with a stirrer. The time is measured from the instant of injection. The reagent is injected against the solution flow to ensure the fastest possible mixing of the reactants. For fast reactions

163

(with half-times shorter than 10 seconds), special mixing techniques, used in the study of fast reactions, must be employed.

As the reaction rate depends on the temperature, the measurement must be performed at a constant temperature (see Section 2.5). A temperature change by 1 °C leads to a change in the reaction rate by about 10%. Thus, the reaction vessel must be thermostatted. Commercial thermostats with a precision of ± 0.01 °C suffice for analytical purposes.

9.1 Monitoring of the Reaction Course

Chemical and instrumental methods are available for this purpose. Most important among chemical methods are titration techniques. Samples are taken from the reaction mixture at certain time intervals and a component is titrated. For precise determination, the reaction must be stopped or slowed down in the sample immediately after collection, either by complexation of a reactant or by adding an inhibitor. Reactions dependent on the pH can be slowed down by a sudden change in the pH. The mixture can also be cooled suddenly to slow the reaction. Generally, monitoring by titration is simple, but tedious and often imprecise. Instrumental methods are therefore preferred and can be partially or fully automated.

9.1.1 INSTRUMENTAL METHODS

It is evident that the most suitable monitoring method is one that involves no interference with the reaction system (e.g. sample collection for titration). In this respect, instrumental methods allowing monitoring of a physico-chemical parameter proportional to the concentration of a component of the reaction system are most suitable.

9.1.1.1 Optical Methods

A suitable and frequently used parameter for kinetic measurements is the absorbance. If one or more components of the reaction system absorb visible or ultraviolet radiation, their concentrations can be monitored spectrophotometrically. The temperature of the solution in the cuvette must be constant within ± 0.2 °C and thus the cuvettes must be thermostatted. For kinetic analysis, spectrophotometers permitting the measurement of absorbance at a constant wavelength are advantageous. Double-beam instruments are suitable, in which measurements against a blank can be carried out. These requirements are met by many commercial spectrophotometers and thus kinetic measurements are easy to carry out in a well equipped laboratory. Not all double-beam spectrophotometers are suitable for monitoring in dependence on time, chiefly because their response may be delayed, affecting the precision of the measurement, especially with faster reactions. Therefore, some authors prefer single-beam instruments. It is also important that the light flux of the source does not fluctuate during the measurement. For highly precise measurements, regulating feedback circuits have therefore been proposed [261] that maintain a constant radiation flux even during prolonged measurements.

Measurement of fluorescence is closely related to spectrophotometry and is chiefly used in kinetic determinations of enzymes [262]. An advantage of fluorescence measurement is a greater sensitivity; the measuring instrument is similar to a spectrophotometer. It is also analytically important that the fluorescence intensity is proportional to the concentration in dilute solutions.

Optical methods are not used only for simple monitoring of the concentration in dependence on time. Modern spectrophotometers can substantially simplify some chemical kinetic methods. For example, methods based on the Landolt effect (Section 3.3) can be accelerated and partially automated by placing the same solutions with different catalyst concentrations in the two cuvettes of a double-beam photometer. The recorder then records the difference in the absorbances of the contents of the two cuvettes and thus the recorded absorbance is zero at the beginning of the reaction. On completion of the reaction in the cuvette with a greater catalyst content, the difference recorded changes sharply and remains constant until the reaction is completed in the other cuvette with the lower catalyst content; the absorbance then sharply decreases to zero. If a series of measurements is carried out, maintaining a constant catalyst concentration in one cuvette and changing it in the other, the time interval limited by the increase and decrease in the absorbance is proportional to the catalyst concentration and can be used for construction of a calibration curve. For precise measurements, the solutions must be mixed in the two cuvettes at exactly the same instant. For this purpose, a "double starting pipette" [249] is employed, which enables e.g. hydrogen peroxide to be simultaneously injected into two cuvettes containing ascorbic acid and potassium iodide with different concentrations of the catalyst (Mo(VI)). Then a calibration curve for the determination of molybdenum can be constructed [263].

Measurements with fast-scanning spectrophotometers [291] have recently become important. These modern spectrophotometers are provided with a Vidicon (television camera pick up tube) array detector system and a computer controlling the operation [292]. A great advantage of these instruments is the fact that they yield time resolved spectra for fast reactions and their analytical use in combination with stopped-flow methods produces more precise results [293]. Fast scanning has also been successfully used in fluorescence measurements of reaction rates [289].

9.1.1.2 Electrochemical Methods

Another large group of measuring techniques applicable to kinetic analytical methods comprises the electrochemical methods. Similar to optical methods, these methods do not affect the reaction system. The test solution can readily be thermostatted and the measured quantity (current or potential) continuously recorded.

The dropping mercury electrode is often used, either measuring the polarographic curves of a polarographically active component in the system at certain intervals or, preferably, monitoring the limiting current of the component at a constant potential. The use of solid electrodes is sometimes advantageous, especially rotating electrodes that ensure thorough stirring of the reactants.

If electrochemical methods are used to monitor the kinetics of a homogeneous reaction following the charge-transfer reaction at the electrode surface, the reaction mechanism may become complicated, e.g. through disproportionation or other second order reactions involving electrogenerated intermediates. The main cause of this phenomenon is inhomogeneity of the solution close to the electrode, owing to diffusion. The resultant complex concentration profile considerably complicates the dependence of the concentration on time obtained from the electrochemical measurement. These complications can often be overcome by electrogenerating reactive species under conditions where the solution remains homogeneous during the kinetic run, e.g. in thin-layer electrochemical cells [310] or in a thin-layer optically transparent cell [311] allowing selective spectrophotometric monitoring of some reactions.

166

Among potentiometric methods, measurements with ion-selective electrodes are advantageous, as these electrodes are well suited for monitoring time changes in the concentration of ions and are also often used for study of reaction mechanisms [265]. An example is the use of the iodide-selective electrode in the study of the perbromate reaction with iodide and in the determination of ferrous iron which induces this reaction [308]. A perchlorate-selective electrode was used for an automated determination of various glycols by the constant concentration method [309]. Glass pH-electrodes are often used in kinetic determinations of enzymes and substrates, as the reactions of these substances usually involve liberation or consumption of hydrogen ions.

Oxygen electrodes [267] are also used in enzymatic analyses, especially for systems that consume oxygen, such as in the determination of glucose in blood. These electrodes are mostly amperometric sensors; the Clark membrane electrode is noteworthy. Bare electrode systems are less important [290]. An ammonia membrane ion-selective electrode can also be used in enzyme analysis, e.g. for the determination of creatin [268].

9.1.1.2.1 Enzyme Electrodes

The use of enzyme electrodes for determination of organic and biologically important substances (amino acids, urea, glucose, etc.) has recently become very extensive, as the high selectivity of enzyme reactions is combined with a simple electrochemical method [294]. The extensive use of this method has been made possible only by mastering the preparation of immobilized enzymes. The simplest enzyme electrodes have an artificial enzyme-containing membrane fixed directly on the transducer [268]. The substrate diffuses through the thin layer of the catalyst, producing an electroactive substance that is detected by the sensor, either potentiometrically or amperometrically.

An advantage of potentiometric detection is the easy preparation of the electrode and low cost; however, a suitable ion-selective electrode must be available. Most potentiometric enzyme electrodes have a slow response, because the diffusion of the substrate and the product in the enzyme layer is rather slow, especially in artificial membranes about 10^{-2} cm thick. Quantitative relationships describing the processes in membranes are given in Section 4.4.1. Potentiometric enzyme electrodes have been used for the determination of amino acids [295], urea [296, 297], amygdalin, glucose, penicillin, etc.

Protolytic gases are formed in many enzymatic reactions and can be used for the determination of substrates or enzymes, employing potentiometric gas-sensing electrodes that are highly selective and respond rapidly. On the other hand, the practical application of these electrodes in biological fluids is often limited by blocking of the membrane pores. This limitation has been removed by the introduction of air-gap electrodes [297], in which the gas-permeable membrane is replaced by an air gap separating the surface of the indicator electrode from the sample solution. Air-gap electrodes have been successfully used in kinetic measurements based on the initial rate [300].

Amperometric enzyme electrodes are employed with enzyme reactions involving consumption or production of oxygen or hydrogen peroxide. A platinum electrode is covered with a plastic membrane to which an enzyme gel layer adheres. The measured current is proportional to the amount of electroactive substance generated in the membrane. For example, in determination of glucose the measured current is proportional to the loss of oxygen inside the immobilized glucose oxidase layer as a result of penetration of glucose into the layer from the solution.

An advantage of amperometric electrodes is their wide linear dynamic range that is about one order of magnitude greater than for potentiometric sensors. Mell and Malloy [298] have shown that the current is limited either by the rate of the enzyme reaction or by the diffusion rate. In the former case, for a small concentration of the substrate in the membrane, i.e. $\bar{S} \ll K_M$, steady-state current i is given by

$$i = \frac{1}{2} \frac{nFA\bar{X}\bar{V}}{K_M} [S] \qquad (9\text{-}1)$$

where n is the number of electrons involved in the electrode reaction, F is Faraday constant $(964\,870 \times 10^4 \text{ C . mol}^{-1})$, A is the electrode surface area, $\bar{X}$ is the thickness of the enzyme layer and $\bar{V}$ is the enzyme concentration per gel volume unit.

With large enzyme concentrations and low substrate concentrations, i.e. for $(\bar{V}/K_M D_S)\bar{X}^2 > 1$, stationary current i is limited by diffusion, is independent of the enzyme concentration and is given by

$$i = \frac{nFAD_s[S]}{X} \tag{9-2}$$

At high substrate concentrations the current becomes independent of the substrate concentration for any amount of enzyme. According to the authors [298] the stationary current is attained in a time shorter than $1.5\bar{X}^2/\bar{D}$; hence the theoretical response time of an amperometric enzyme electrode can be determined. These electrodes have been used for determination of glucose, some amino acids, phosphates, ureates and other substances.

An effort to combine enzyme reactions most effectively with electrochemical detection has led to construction of enzyme reactor electrodes. The enzyme is then completely separated from the electrode, in order that the catalytic reaction and the electrochemical detection take place under optimal acidity conditions that are generally different for the two processes. The enzyme, immobilized e.g. on glass beads, is placed in a small reactor (e.g. a liquid chromatographic column) through which the sample solution with a suitable buffer passes. The substrate is thus converted into the product with 100% efficiency. The effluent from the reactor is mixed with a suitable solution to adjust its pH to an optimal value for measurement in a flow-through gas electrode, which the effluent enters after passage through a heat exchanger coil. Enzyme reactor electrodes are micro-flow systems and have been used for determination of urea and amino acids [301, 302]. Obviously, the enzyme reactor principle can also be combined with other instrumental methods.

9.1.1.2.2 Bipotentiometry and Differential Potentiometry

In bipotentiometry, the potential difference is measured between two platinum electrodes polarized by a small constant current, using a voltmeter with a high input impedance or a voltage follower and recording the time dependence of the signal. The method is advantageous for catalytic reactions and has been widely used for determination of enzymes. A necessary condition is that the reactant and the product have different redox potentials. In the absence of an enzymatic catalyst, the reaction rate is zero and the potential difference is constant. On addition of a catalyst the potential difference begins to vary. The slope of its time dependence, $\Delta E/\Delta t$, at the beginning of the reaction is proportional to the catalyst concentration and can be used for construction of a calibration curve.

Differential potentiometry with constant current was recently applied in kinetic analysis [270]. Analogous to bipotentiometry, two platinum electrodes are polarized by constant current with different intensities and the potential difference between the electrodes is recorded as a function of time.

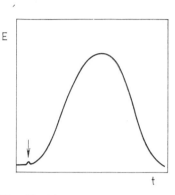

Fig. 29. The E vs t dependence obtained in differential potentiometry at constant current.
↓ denotes the beginning of the reaction.

Organic and inorganic substances have been determined in this way on the basis of their oxidation with an electroactive reagent. Stirred solutions are employed and thus electrode potentials are given by the usual voltammetric relationships. When the electroactive oxidant concentration decreases during the reaction, the E vs t plot is a curve that increases to a maximum and then decreases (Fig. 29). The width of the peak, Δt, is proportional to the reaction rate.

The potential difference is measured by a voltage follower and recorded in dependence on time (Fig. 30). Currents I_1 and I_2 polarizing the electrodes must be found experimentally. If the reaction is first or pseudofirst order, then the rate of the reaction of A with B is given by

$$\ln \frac{[A]_0}{[A]_t} = k[B]_0 \Delta t \tag{9-3}$$

Expressing concentrations $[A]_0$ and $[A]_t$ in terms of currents I_1 and I_2

170

leads to conversion of Eq. (9-3) into

$$[B]_0 = \frac{1}{\Delta t} \left(\frac{1}{k} \ln \frac{I_1}{I_2} K \right) \tag{9-4}$$

Constant K in Eq. (9-4) includes the other quantities defining the current in voltammetry with a stationary electrode in a stirred solution that are constant under particular conditions (electrode dimensions, vessel shape, rate of stirring).

Among other electrochemical methods, coulometry should be mentioned. Coulometrically generated reagents can be used for titration of products of

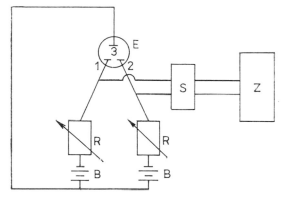

Fig. 30. Block scheme of the apparatus for differential potentiometry at constant current.
E — cell, $1, 2$ working platinum electrodes, 3 — auxiliary platinum electrode, R — resistor, B — voltage source, 90 V, S — voltage follower, Z — recorder.

some enzyme reactions [272]. Conductometric methods are used less often, e.g. in the differential kinetic determination of some organic substances [217, 273]. The bipolar pulse modification of conductometric measurements seems to be promising, as its sensitivity is much higher than that of normal conductometry [274].

9.1.1.3 Other Methods

For example, optically active isomers can be determined polarimetrically. Radiochemical methods can be used for systems containing radioisotopes, applying tracer techniques. Enthalpimetric methods have been recommended

171

for clinical kinetic determinations. Direct injection enthalpimetry [303] is most important. Excess reagent is injected into the test solution under virtually adiabatic conditions. The resultant heat pulse is proportional to the amount of test substance and is measured with a sensitive thermistor forming one arm of a Wheatstone bridge. The method was used for determination of some substrates and enzymes [304]. It is advantageous that whole blood analyses can sometimes be carried out by this method without prior deproteinating.

9.2 Instrumentation for Differential Methods

Section 8.2.1.1 describes the principles of a differential kinetic method for determining catalysts. The method was originally developed for photometric measurements, in which the two cuvettes of a double-beam photometer contain the same reaction mixtures differing only in the catalyst concentration; the photometer output signal is then proportional to the difference in the solution absorbances. A recent development in chemical instrumentation has also permitted measurement of differences in other physico-chemical quantities, such as the conductance or temperature, in dependence on time.

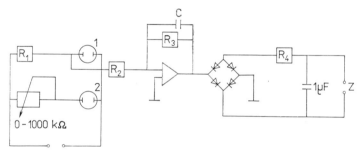

Fig. 31. Scheme of the apparatus for the differential kinetic method with conductance measurement (from ref. [275]).
1, 2 — conductivity vessels with electrodes, $E = 2$ V with a frequency of 50 Hz, $R_1 = 500\ \Omega$, $R_2 = 100\ k\Omega$, $R_3 = 1\ M\Omega$, $R_4 = 10\ k\Omega$, $C = 0.01\ \mu F$, Z — recorder.

An apparatus for recording the time dependence of the difference between the conductances of two solutions is depicted in Fig. 31 and is advantageous when the indicator reaction is accompanied by a large change in the conductance, such as in oxidation of thiosulphate to sulphate. The two thermostatted cells are filled with the initial reactant of the indicator reaction and a buffer. A constant amount of the catalyst is added to one vessel and to the other vessel are added various known amounts of the catalyst (for construction of the calibration curve) or of the sample. After temperature equilibration the same amounts of the other reactant (mostly an oxidant) are added to the vessel. The signal recorded is first constant; on addition of the oxidant its value begins to increase and the slope of the straight line obtained

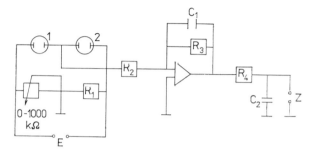

Fig. 32. Apparatus for the differential kinetic method with temperature measurement (from ref. [275]).
$E = 1.3$ V, $R_1 = 68$ kΩ, $R_2 = 10$ kΩ, $R_3 = 1$MΩ, $R_4 = 10$ kΩ, $C_1 = 0.01$ μF, $C_2 = 100$ μF, $1, 2$ — Dewar vessels with thermistors (100 kΩ at 20 °C), Z — recorder.

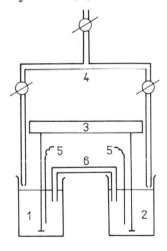

Fig. 33. Double starting pipette.
$1, 2$ — vessels, 3 — mechanical stirrer, 4 — double starting pipette, 5 — electrodes, 6 — liquid bridge.

173

is proportional to the catalyst concentration in the second vessel. The calibration curve obtained is linear.

If the test substance catalyzes an exothermic reaction, thermometric indication can be used. The procedure is the same as above (the apparatus is shown in Fig. 32) and variation of temperature with time is similar to that of conductance with time; its slope is proportional to the catalyst concentration in the second vessel.

A similar apparatus can also be used for the biamperometric differential method [275].

For a differential method to be sufficiently precise, the second reactant (oxidant) must be added instantaneously and simultaneously to both vessels. For this purpose, a "double starting pipette" (Fig. 33) was proposed by Weisz and Ludwig [249].

9.3 Stat Methods

The principle of the stat method is described in Section 8.2.1.6. The apparatus automatically controls addition of the reagent (usually one of the reactants) to the reaction system at a rate equal to the rate of its consumption in the catalytic reaction. Any measurable property of the solution that is proportional to the concentration of a particular substance and that is

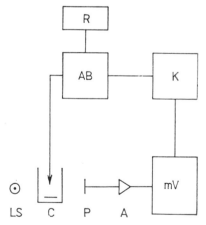

Fig. 34. Absorption-stat according to Weisz and Rothmaier [305].
LS — light source, *C* — cuvette with stirrer, *P* — photocell, *A* — amplifier, *mV* — mV-meter, *K* — comparator, *AB* — automatic burette, *R* — recorder.

174

maintained constant can be used to regulate the reagent addition. pH-static measurements have found wide use [266], especially in enzyme analysis. The H^+ ion concentration is maintained constant in the reaction system by adding an acid or an alkali. Coulometric generation of the reagent improves the precision [304].

The stat method also employs other instrumental methods. For example, an absorption-stat [305] can be used when the measured species absorb light. A scheme of this instrument is shown in Fig. 34. The photocell signal is amplified and measured by a mV-meter and is compared with a preset working potential corresponding to a particular absorbance, and thus to a defined reagent concentration that must be kept constant. A difference between the actual and working potential activates a device opening an automatic burette. The time dependence of the reagent consumption is recorded and the rate of reagent addition is found graphically.

In the determination of Mo on the basis of its catalytic effect on the oxidation of iodide by hydrogen peroxide [271, 305], the rate of the catalytic reaction is given by

$$\frac{d[I_2]}{dt} = -\frac{1}{2}\frac{d[I^-]}{dt} = k[I^-][H_2O_2][Mo] \tag{9-5}$$

In the stat method, a certain absorbance due to the presence of iodine is kept constant by removing excess iodine formed by addition of ascorbic acid.

$$I_2 + \text{ascorbic acid} \rightarrow 2\,I^- + 2\,H^+ + \text{dehydroascorbic acid} \tag{9-6}$$

The concentrations of I^- and I_2 in the system are constant, the change in the concentration of H_2O_2 can be neglected because of the high peroxide concentration in the system and thus the rate of addition of ascorbic acid is proportional to the concentration of molybdenum,

$$\frac{d[I_2]}{dt} = -\frac{1}{2}\frac{d[I^-]}{dt} = -\frac{d[\text{ascorbic acid}]}{dt} = k^*[Mo] \tag{9-7}$$

From the recorded consumption of KI in time, the rate of addition of KI is found, equal to cotan α (Fig. 35). According to Eq. (9-7), this rate is proportional to the Mo concentration. The cotan α vs [Mo] dependence is used as the calibration curve.

The stat methods have been thoroughly elaborated by Weisz and co-workers. The authors proposed a "biamperostat" [306], consisting of an amperometric polarized system and an autotitrator combined with a current-voltage transducer to maintain a constant titrant concentration.

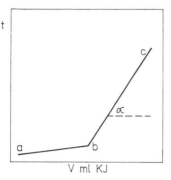

Fig. 35. Stat method — a determination of molybdenum.
Dependence of the consumption of a KI solution on time; *ab* — amount of KI necessary for attaining the working potential, *bc* — the actual measurement.

Pantel and Weisz [307] proposed a "luminostat" for the catalytic reaction of luminol with hydrogen peroxide, based on a similar principle to the absorption-stat, but monitoring the luminescence intensity.

9.4 Flow-through Methods

Flow-through methods, extensively developed over the last decade, are important in practical analysis and in automation of kinetic methods.

9.4.1 FLOW-THROUGH MEASUREMENTS ON PAPER

A very simple version of a flow-through apparatus employs filter paper as a capillary support for chemical reactions and transport of the reaction mixture. A scheme of the apparatus described by Weisz [312] is depicted in Fig. 36. It is very simple and requires no pump; transport of the solution at a rate of about 3 ml per 12 h is ensured by capillary elevation in a column of circular filter papers *F* (about 130 papers with a diameter of 6 cm). In determination of molybdenum using the catalytic effect of Mo(VI) on the reaction of hydrogen peroxide with potassium iodide, the solutions of the reactants are brought from vessels *A* and *B* through filter paper strips *a* and *b*

(7 mm wide) to main filter paper strip c (20 mm wide). One end of the strip is immersed in flow-through cell C, through which the sample solution flows, and the other is placed at the bottom of column F. The maximum sample flow-rate through cell C is $0.2-0.3$ ml . min^{-1}. Strips a, b and c

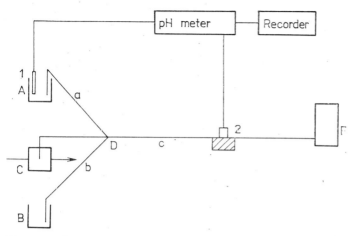

Fig. 36. Scheme of the apparatus for flow-through measurements on paper. (For description see the text.)

are connected by plastic clamps. About 3 cm from point of connection D, iodide selective electrode 2 is applied to strip c at a constant pressure. Reference calomel electrode 1 is immersed in a vessel containing hydrogen peroxide and potassium sulphate base electrolyte. The potential of the iodide-selective electrode is monitored against time. In the beginning of the measurement, water is passed through the flow-through cell for about 1 h; the selective electrode potential has a value corresponding to zero Mo(VI) concentration. Standard Mo(VI) solutions (from 0.2 to 10 µg . ml^{-1}) are used for the calibration. The apparatus can operate for 24 h; then reagent solutions and papers in column F must be replaced.

9.4.2 FLOW-THROUGH CELL

Section 8.2.1.6 describes the principles of measurement with a flow-through cell. In the cell proposed by Weisz and Ludwig [256], reactants A and B and catalyst C flow at a constant rate into a flow-through cuvette,

177

where they are mixed instantaneously and completely and the reaction mixture is then led to waste. Stationary state characterized by a constant composition is established in the cuvette and in time interval dt the number of moles of the reactants entering the cuvette equals that leaving the cuvette and consumed in the reaction. Then

$$Q_V[A]_0 \, dt = Q_V[A]_s \, dt + rV \, dt \qquad (9\text{-}8)$$

where Q_V is the flow-rate, V is the cuvette volume, r is the reaction rate under stationary conditions, $[A]_0$ is the concentration of A entering the cuvette and $[A]_s$ is the concentration of A under stationary conditions. If substance B is present in a large excess, the reaction rate is given by the relationship

$$-\frac{d[A]}{dt} = k'[C][A] \qquad (9\text{-}9)$$

and at stationary state it holds that

$$r = k'[C][A]_s \qquad (9\text{-}10)$$

Substitution of Eq. (9-10) into (9-8) yields

$$\frac{1}{[A]_s} = \frac{1}{[A]_0} + \frac{k'V}{[A]_0 \, Q_V}[C] \qquad (9\text{-}11)$$

defining a linear dependence of $1/[A]_s$ on $[C]$. According to Eq. (9-11), the reciprocal absorbance measured during the flow at a stationary state is proportional to the catalyst concentration.

The flow-through cell enables continuous determination of catalysts. The method was used for determination of traces of I^- (indicator reaction of Ce^{4+} with As^{3+}), Mn^{2+} (indicator reaction of alizarin S with H_2O_2) and Hg^{2+} that decrease the catalytic activity of I^- [256]. Traces of silver that catalyze the oxidation of sulphanilic acid by $K_2S_2O_8$ in the presence of 2,2'-bipyridine were also determined in this way [264].

9.4.3 CLOSED FLOW-THROUGH SYSTEMS WITH SAMPLE INJECTION

Dutt and Mottola [313] used the sample injection technique common in flow-through methods [314] for kinetic determinations in flow-through systems with reagent recirculation. All the necessary reagents are placed in a single container and circulate at a constant rate through a flow-through cell in which a component of the reaction mixture is monitored. An aliquot portion of the sample solution is rapidly injected into the flowing reaction mixture and takes part in a rapid reaction with the reagent, leading to change in the concentration of the species monitored. A subsequent slower reaction removes the reaction product from the system and a new determination can be carried out.

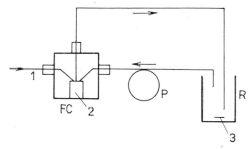

Fig. 37. Scheme of a closed flow-through system with sample injection.
1 — teflon needle for sample injection, *2* — spectrophotometer cell spin-bar or an electrochemical detection system, *3* — magnetic stirring bar, *P* — peristaltic pump, *R* — reagent solution reservoir, *FC* — flow-through cell, → flow direction.

A scheme of the apparatus for repeated determinations is shown in Fig. 37, employing a simple circulation device and suitable reactions. For example, a product is formed by the first reaction, yielding a signal in the measuring circuit. The subsequent reaction removes the product and the signal returns to the original value. Isonicotinic acid hydrazide was determined by Dutt and Mottola [315] using the reaction

$$\text{isoniazide} + V(V) \rightleftharpoons \text{coloured complex} \tag{9-12}$$

followed by a slower redox process,

$$\text{coloured complex} + V(V) \rightleftharpoons \text{isonicotinic acid} + N_2 + V(IV) \tag{9-13}$$

If the rate of reaction (9-13) is less than that of reaction (9-12), but sufficient for destruction of the coloured complex within a short time (about 10 s), the signal profile for repeated determinations has the character depicted in Fig. 38. Because the product is coloured, spectrophotometric

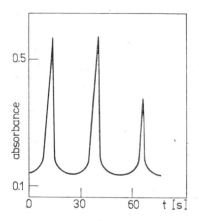

Fig. 38. Typical signal profile on repeated determinations in a closed flow-through system with sample injection.

measurement can be used. Reaction (9-13) has an optimum rate when the circulating reaction mixture contains a high concentration of VO_3^- (10^{-2} M) and phosphoric acid to maintain a suitable pH. The complex decomposition is then sufficiently fast, enabling 3 determinations to be carried out per min.

Spectrophotometric measurement is also suitable for repeated determinations of iron [316] based on formation of the Fe(III) thiocyanate complex. The complex is decomposed either by a ligand substitution reaction with NTA, EDTA, citrate or oxalate or by oxidation catalyzed by iodide ions. The latter reaction is more suitable, because it is sufficiently fast and leads to complete dissociation of the Fe(III)-SCN complex. The signal profile then corresponds to Fig. 38, whereas the use of a ligand substitution reaction does not result in return of the signal to zero value.

Electrochemical detection in a closed flow-through system was used for repeated determination of glucose in serum [317], on the basis of the reaction

$$\text{D-glucose} + O_2 + H_2O \xrightarrow{\text{GLOD}} \text{D-gluconic acid} + H_2O_2 \qquad (9\text{-}14)$$

where GLOD is glucose oxidase. The oxygen concentration is indicated

180

amperometrically in a flow-through three-electrode cell with a platinum working electrode. The baseline (zero glucose concentration) corresponds to the current in a solution saturated with oxygen. Reaction (9-14) causes a sudden current drop. Restoration of the baseline is attained by continuous feeding of air into the circulating mixture, in the direction of the flow, with air bubbles leaving in an open reservoir.

In repeated determinations in closed flow-through systems, the monitored substance can also be removed by washing out from the detection zone by flow of the medium [318]. If X is the test substance injected into the circulating system and Y is the reagent present in the circulating mixture, the reaction

$$(X + Y)_D \underset{k_{-1}}{\overset{k_1}{\rightleftarrows}} Z_D \tag{9-15}$$

takes place on injection of X into the detection zone, where Z is the substance monitored and subscripts D denote the presence of the substances in the detection cell. Removal of substance Z from the detection cell is given by

$$Z_D \xrightarrow{k_2} Z' \tag{9-16}$$

where Z′ is substance Z after passage through the detection cell. If reaction (9-15) is fast, then equilibrium is formed before removal of substance Z_D from the detection cell. Rapid equilibration is also facilitated by a great excess of reagent Y over substance X in the circulating system. The signal profile is thus given by

$$\frac{d[Z]}{dt} = v(Z)_f - v(Z)_r \tag{9-17}$$

where $v(Z)_f$ is the rate of formation of the substance monitored and $v(Z)_r$ is the rate of removal of this substance from the detection cell. If $k_1 \gg k_{-1}$ and $k_2 \gg k_{-1}$, Eq. (9-17) can be formulated as

$$\frac{d[Z]}{dt} = k_1[X]_0 e^{-k_1} - k_2[Z] \tag{9-18}$$

This equation indicates that the signal profile corresponds to Fig. 38, provided that k_1/k_2 is at least 1000. This method was practically verified on determination of iron (II) with ferrozine [318].

The advantages of closed flow-through systems with sample injection are obvious. The apparatus is simple, air bubbles separating the samples are avoided and the analyses are very rapid. For example, in the above determination of iron [316], 120 analyses per h. can be performed. In enzyme analyses it is important that the reagent consumption be small; 10,000 samples for the determination of glucose in serum can be analyzed at a rate of 700 analyses per hour without container replenishment.

9.4.4 CONTINUOUS FLOW AND STOPPED-FLOW METHODS

These methods were treated in Sections 5.3 and 8.2.1.6 and are important in automation. The stopped-flow method is very important in the study of fast reactions and in their application in kinetic analysis, especially when the apparatus is coupled with a computer. Rapid scanning spectrophotometers and vidicon-based spectrophotometric systems [293, 319] are suitable for monitoring rapid concentration changes, because they suppress errors stemming from e.g. stray light, poor resolution, vidicon lag time and random errors in absorbance measurement. Fast reactions are readily monitored using a rapid scanning spectrophotometer with a silicon vidicon detector controlled by a computer.

In continuous flow and stopped-flow methods, the most advantageous detection technique is absorbance measurement. However, electrochemical quantities can also be measured and the apparatus can thus be simplified. Porterfield and Olson [320] have developed a differential redox potentiometric method with tubular electrodes applicable in both continuous flow and stopped-flow modes. A flow-through concentration cell contains two tubular carbon electrodes with the cell compartments separated by a thin membrane. The electrode potentials, E_1 and E_2, are given by the Nernst equation and the cell voltage E equals $E_1 - E_2$, because the interfacial potential is virtually zero, as the same solutions are present on both sides of the membrane. A continuous flow apparatus has been constructed (Fig. 39a). Steady-state conditions are established at the two electrodes and the difference in the concentrations represent an integrated concentration change that occurs as the solution element moves through the delay line between the electrodes. The delay times for the movement of the solution from the mixing point to the first electrode (t_1) and to the second electrode (t_2) can be measured. The cell voltage is given by

$$\exp \frac{nFE}{RT} = \frac{[Ox]_2}{[Ox]_1} \tag{9-19}$$

subscripts *1* and *2* correspond to electrodes *1* and *2*, respectively. Because

$$[Ox]_1 = [Ox]_0 + kt_1 \tag{9-20}$$

$$[Ox]_2 = [Ox]_0 + kt_2 \tag{9-21}$$

the reaction rate is given by

$$k = \frac{[Ox]_0 \left(\exp \dfrac{nFE}{RT} - 1 \right)}{(t_2 - t_1) \exp \dfrac{nFE}{RT}} \tag{9-22}$$

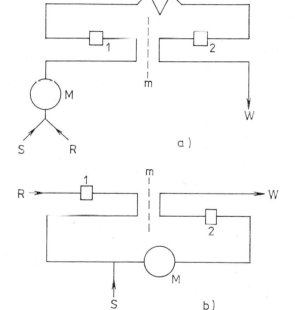

Fig. 39. Scheme of an apparatus for differential redox potentiometry in (a) continuous-flow and (b) stopped-flow mode.
R — reagent, *S* — sample, *M* — mixer, *W* — waste, *DL* — delay line, *m* — membrane, *1, 2* — electrodes.

When the differential redox potential is used in the stopped-flow method, the reagent is mixed with the sample between the electrodes (Fig. 39b); electrode *1* then has a constant potential corresponding to the reagent concentration. On stopping the flow, the cell voltage variations reflect the concentration changes occurring at the second electrode. This voltage is measured at times t_1 and t_2 after initiation of the reaction. Measured voltages E_{t_1} and E_{t_2} are defined by

$$E_{t_1} = \frac{RT}{nF} \ln \frac{[Ox]_0 + kt_1}{[Ox]_1} \tag{9-23}$$

$$E_{t_2} = \frac{RT}{nF} \ln \frac{[Ox]_0 + kt_2}{[Ox]_2} \tag{9-24}$$

where $[Ox]_0$ is the concentration at the mixing point when the reaction starts and $[Ox]_1$ is the stable concentration at the upstream electrode. Subtraction of Eqs. (9-23) and (9-24) leads to

$$E_{t_1} - E_{t_2} = \Delta E = \frac{RT}{nF} \ln \frac{[Ox]_0 + kt_2}{[Ox]_0 + kt_1} \tag{9-25}$$

from which it follows for k that

$$k = \frac{[Ox]_0 \left(\exp \dfrac{nF \, \Delta E}{RT} - 1 \right)}{(t_2 - t_1) \exp \dfrac{nF \, \Delta E}{RT}} \tag{9-26}$$

The validity of Eqs. (9-22) and (9-26) was verified by measuring known rates of enzyme reactions. It has been shown that, in electrochemically suitable systems, differential redox potentiometry can replace more costly spectrophotometric measurements in flow-through systems.

9.5 Automation of Kinetic Methods

In the introduction to this chapter, the operations required for kinetic analyses are summarized. Except for sample preparation involving dissolution, pH adjustment, buffer addition, etc., all the other stages can be automated. The present level of instrumentation permits acceleration of the

184

originally tedious kinetic measurements and thus contributes to rapid development of this analytical field.

The first part of an automated kinetic analyzer is a sample collecting device that transports the samples to the mixing chamber into which the reagent is fed. The reaction mixture is then pumped into a measuring cell from which the solution is transported to waste. The apparatus is then washed and the cycle is repeated with another sample.

The measurement itself is carried out spectrophotometrically or electro-chemically. The meter output signal is an electric quantity proportional to the test substance concentration, which is further treated using a constant time or constant concentration technique, by the tangent or the stat method [276, 277].

9.5.1 AUTOMATION OF METHODS FOR DETERMINATION OF A SINGLE SUBSTANCE

When substance A reacts with substance B in the presence of catalyst C,

$$A + B \xrightarrow{\quad C \quad} P \qquad\qquad (9\text{-}27)$$

then meter signal S is related to the concentrations of A and P by the relationships

$$S = f([P]) \qquad\qquad (9\text{-}28)$$

$$S = f'([A]) \qquad\qquad (9\text{-}29)$$

or

$$[P] = g(S) \qquad\qquad (9\text{-}30)$$

$$[A] = g'(S) \qquad\qquad (9\text{-}31)$$

Hence signal difference ΔS over time interval $t_2 - t_1 = \Delta t$ corresponds to a certain change in the concentration of P or A,

$$\Delta S = f([P]_2) - f([P]_1) \qquad\qquad (9\text{-}32)$$

$$\Delta S = f'([A]_2) - f'([A]_1) \qquad\qquad (9\text{-}33)$$

Similarly, the change in the product concentration, $\Delta[P]$, in time interval Δt

can be defined by

$$\Delta[P] = g(S_2) - g(S_1) \tag{9-34}$$

In the simplest case, i.e. when reaction (9-24) is zero or pseudozero order, the product concentration at time t is given by

$$[P]_t = k[A]_0[B]_0[C]t \tag{9-35}$$

The change in the product concentration, $\Delta[P]$, in time interval Δt is then

$$\frac{\Delta[P]}{\Delta t} = k[A]_0[B]_0[C] \tag{9-36}$$

Substituting from Eq. (9-34), Eq. (9-36) can be formulated as

$$\frac{g(S_2) - g(S_1)}{\Delta t} = k[A]_0[B]_0[C] \tag{9-37}$$

or

$$[C] = \frac{1}{k[A]_0[B]_0\,\Delta t}\,\{g(S_2) - g(S_1)\} \tag{9-38}$$

Because the first term on the right-hand side of Eq. (9-38) is constant, the constant time method can be used for automated analysis. An equation for first and pseudofirst order reactions can be derived analogously, obtaining

$$[C] = \frac{1}{k[B]_t\,\Delta t\{[A]_0 - g(S_1)\}}\,\{g(S_2) - g(S_1)\} \tag{9-39}$$

An equation similar to Eq. (9-39) can be derived for uncatalyzed reactions, to which the constant time method is also applicable.

Originally, instruments automatically handling experimental data used operational amplifier circuits [277, 278]. Cordos et al. [279] proposed a system that reads and integrates the meter signal, where the time and order of the computing operations is controlled by digital logic circuits. The function of this type of instrument was verified on analyses of glucose in serum [279]. This type of analyzer is also suitable for relatively fast reactions, when the signal from a stopped-flow apparatus spectrophotometer is fed into its input [121]. The high degree of automation permits recording of up to one value per second, thus improving the precision, as demonstrated

186

on analyses of phosphates by fast reaction of molybdate with phosphoric acid, in which the standard deviation was less than 1% [121]. A similar system was proposed by Chlapowski and Mottola [323].

Further development in computing techniques has shown the immense importance of digital systems. Ingle and Crouch [280] constructed the first digital ratemeter with two modes of operation: a) a single measurement mode in which one rate measurement is made and the results are held indefinitely and b) a continuous measurement mode in which the results are held for one measurement period, then cleared and the measurement and hold process is repeated indefinitely.

Another digital ratemeter reported by Irachi and Malmstadt [324] is interfaced to a minicomputer. This instrument has single and continuous measurement modes. The interfaced minicomputer enables reduction of the delay between the measurement periods to only 5% of the measurement time, while the instrument described by Ingle and Crouch [280] has a minimum delay time proportional to the measurement time.

A ratemeter that need not be connected to a computer, but which is also suitable for measurement of higher reaction rates, was described by Wilson and Ingle [322]. It is comparable with the previous digital fixed-time systems, but has a differential voltage-frequency converter and a multi-integration mode. Compared with the Irachi and Malmstadt ratemeter [321], this instrument exhibits better linearity and a larger dynamic range.

The main advantage of digital systems in automatic kinetic analyzers include the following; 1, all non-linear effects in the computing unit are eliminated; 2, the computing unit contains no mechanical or analog switches, so that integration times can be very short and consequently the calculation can be repeated several times, improving the precision of the analysis.

Fully automated instruments have also been constructed for the constant concentration method with analog [281] or digital [282] computing units. Many designers of analog analyzers prefer the tangent method, in which the tangent to the kinetic curve is calculated for times close to t_0. For example, a spectrophotometer unit with an analog differentiating circuit for direct differentiation of the signal was constructed [280].

9.5.2 AUTOMATIC ANALYSIS OF MIXTURES OF SUBSTANCES

Automation of kinetic methods suitable for analyses of mixtures of substances is rather demanding and requires small laboratory computers. An apparatus for automated analysis of binary mixtures that does not need a commercial computer was described by Pinkel and Mark [283]. It contains relatively simple analog circuits solving two equations with two unknowns, permitting the use of the method of proportional equations (Section 8.3.2.1). The authors did not propose using the apparatus for this purpose, but rather for simulation of kinetic curves. Now, with the ready availability of small laboratory computers, the construction of analog based circuits is losing its importance.

In analyses of multicomponent mixtures computers are necessary. Graphical and numerical methods employed in differential kinetic analysis yield sufficiently precise results for binary mixtures. Even here, a computerized system enabling handling of hundreds of pieces of data during the measurement substantially improves the precision. In analyses of multicomponent mixtures, the entire response signal must be analyzed (mostly absorbance vs time), which is virtually impossible without a computer. An on-line computer processes the spectrophotometer data and performs linear least-squares fit of the regression equation [287, 324]. This procedure was successful in analyses of mixtures containing components reacting with common reagent(s) to form a common product. Ridder and Margerum have extended this procedure to mixtures with different reactants and products, all with measurable absorbances. The measuring system consists of an on-line minicomputer and a stopped-flow spectrophotometer. The data handling technique includes variable rates of data acquisition, linear least squares regression analysis, centering of data and reparametrization of matrices [293].

9.6 The Use of Minicomputers in Kinetic Analysis

Originally computers were used mainly in the final stage of analysis, in calculation of the test component concentration. For example, a computer was used to calculate the initial concentrations of the test substances by the method of proportional equations, provided that the rate constants of the individual reactions are known [287]. Only later was wider use of computers suggested. The calculation should be carried out during the measurement, so that experimental parameters can be optimized during the measurement. For example, James and Pardue [284] designed a system for kinetic analyses by the constant time or constant concentration method. The program is very general and the signal vs time dependence is stored in the memory, so that the calculation is continuously refined with increasing experimental data.

Another approach in applying computers is based on the computer not carrying out the calculation, but controlling the measurement, analyzing the data and generally controlling all the analytical operations, such as the start of the reaction, selection of a time scale, the function of a generator of a calibration signal, etc. [285]. The operator can also modify the operation through the computer during the reaction. The instrument is rather costly, but enables the study of reaction kinetics in addition to analytical use.

Laboratory computers are very advantageous for kinetic determinations involving fast reactions. In the stopped-flow method the time dependence of the concentration is usually monitored on an oscilloscope, the curve is photographed and only then can the data be handled. A computer can be coupled with the spectrophotometer output directly, and the data handling and calculation can be carried out directly, using the value of the slope of the signal vs time curve at the beginning of the reaction [286].

In computer-controlled instruments, the minicomputer must handle all the instrument control functions in addition to data acquisition, storage, processing, display and decision-making operations. Hence there have been attempts to relieve the minicomputer of some time-consuming functions. A fully automated stopped-flow instrument was proposed [325], with a hierarchial computer system using a microcomputer for control and minicomputer for data acquisition and processing.

The method of analog simulation has a somewhat different basis [288]. The experimental kinetic curve is compared with a curve calculated by a computer. The data introduced into the computer input are the electrical analogues of the initial concentrations of the reactants. When the simulated and experimental curves match perfectly, the initial concentrations can be obtained, provided that the reaction mechanism and rate constants are known.

189

10 Kinetic Analysis

This chapter surveys the methods of determining various substances. The classification of the methods given in Section 8.1 is basically observed. The principles of determinations are listed in Tables with appropriate literature references. Detailed procedures are given only for selected determinations.

10.1 Catalytic Methods

10.1.1 DETERMINATION OF INORGANIC SUBSTANCES

Catalytic methods with the test substance acting as a catalyst of an indicator reaction are especially advantageous for determination of very small amounts of inorganic substances. The methods are listed in Table 10, covering the literature published from 1967 to 1978; older works based on 65 redox reactions catalyzed by metal ions and anions of inorganic acids can be found in Yatsimirskii's monograph [2].

The catalytic methods given in Table 10 also include procedures based on chain reactions. Chain reactions of complexes are analytically important, as described in detail in Section 6.3.2. Several procedures of catalytic determinations that have found practical use will be given below.

Determination of NH_3 in Waters [452]

An amount of 1 ml of a solution containing 10^{-4} M Hg^{2+} salt and 10^{-2} M perchloric acid is transferred to a distillate collecting flask. To 25 ml of water in the distillation flask are added 10 ml 10^{-2} M EDTA and 25 ml sat. K_2CO_3. The mixture is slowly distilled at 25° for about 50 min with slow passage of air through the solution. The distillate is then diluted with water to 6 – 10 ml and 5 ml of this solution is transferred to a quartz cuvette. An amount of 2 ml of a buffer solution $\left(0.05 \text{ M } Na_2B_4O_7 + 0.1 \text{ N NaOH},\right.$

191

TABLE 10

Catalytic reactions used for determination of elements

Determinand	Indicator reaction	Sensitivity	Interferences	Ref.
Ag	$As^{3+} + Ce^{4+}$ (catalyzed by I^-)	$1\ \mu g\ ml^{-1}$	compds. reacting with I^-	330
	$Na_2HAsO_3 + Ce^{4+}$ (cat. by I^-)	$0.1\ \%$	appl. to Ag in ores after extraction of Ag with dithizone	345
	$H_3AsO_3 + Ce^{4+}$	$10^{-6}\ M$	$I^-, I_2, Hg(II)$	347
	lucigenine $+ H_2O_2$	$0.2\ \mu g\ ml^{-1}$	Co, Cr, Cu, Mn, Ni, Os, Pb	403
	$Fe(CN)_6^{4-} + 2,2'$-bipyridine	$0.1-80\ \mu g\ ml^{-1}$	Fe(II), Fe(III), Hg(II)	404
	$Fe(CN)_6^{4-} + $ 4-nitrosophenol-4-nitrosoresorcinol	$2\ \mu g\ ml^{-1}$	$C_2H_2O_4$, EDTA, thiourea, thiosemicarbazide	444
	$Mn^{2+} + S_2O_8^{2-}$	$10\ \mu g\ ml^{-1}$	Pd	349, 408, 442
	sulphanilic acid $+ K_2S_2O_8$ (activated by $2,2'$-bipyridine)	$0.1\ \mu g\ ml^{-1}$	sulphide, Ir, Ru, Pb, Fe(III)	327, 329, 535
	bromopyrogallol red $+ K_2S_2O_8$ (activated by 1,10-phenanthroline)	$0.5\ \mu g\ ml^{-1}$	Ag extracted prior to detn.	67
	polyazo dyes $+ K_2S_2O_8$ (activated by $2,2'$-bipyridine)	$2 \times 10^{-4}\ \mu g\ ml^{-1}$		443
	tropaeolin O $+ K_2S_2O_8$ (activated by $2,2'$-bipyridine)	$6 \times 10^{-5}\ \mu g\ ml^{-1}$		472
	p-nitrophenyldiazoaminobenzene-p-azobenzene ("cadion") $+ K_2S_2O_8$ (activated by $2,2'$-bipyridine)	$0.1\ ng\ ml^{-1}$	Cl^-, Br^-, I^-	518
	oxine-5-sulphonic acid $+ S_2O_8^{2-}$	$6\ ppb$	interferants removed by extraction with dithizone	695
	luminol $+ S_2O_8^{2-}$			710

TABLE 10 — CONTD.

Determinand	Indicator reaction	Sensitivity	Interferences	Ref.
Ag	$Ce^{4+} + Cl^-$	2 ppb		723
Al	$Cr(III) + EDTA$ (activated by Na acetate)	0.25 mg ml^{-1}	In	468
As(III)	ferroin + $Cr(VI)$	1 μg ml^{-1}	$V(IV)$, $Cr(VI)$, $Mo(VI)$, oxalate	576
Au	$Fe^{2+} + Ag^+$	3×10^{-6} μg ml^{-1}	Pt-metals, W, Zn, F$^-$	430
	N-methylthiourea + Fe^{3+}	10 μg ml^{-1}	Cu	477
	$Hg^+ + Ce^{4+}$	3 ng ml^{-1}	Ir, Pt(IV), Pd, Ru, Ga	542
	$NaN_3 + I_2$	0.1 μg ml^{-1}	Pb, Hg, Pt-metals	754, 815
Bi	$PbO_2^{2-} + SnO_2^{2-}$	0.5 μg ml^{-1}		350
	PO_4^{3-} hydrolysis (catalyzed by phosphatase)	0.3 μg ml^{-1}	Be	473
BO_3^{3-}	sulphonate hydrolysis	1 μg ml^{-1}	metal ions	835
Br$^-$	$BrO_3^- \rightarrow Br^-$	3 ng ml^{-1}	As, $S_2O_3^{2-}$, SO_3^{2-}, SCN$^-$, NO_2^-, Fe(II)	502
	4,4'-dihydroxybiphenyl + Tl(III)		Cl$^-$	807
Ca	Cu(II)-EGTA + 4(2-pyridylazo) resorcinol	$10^{-3} - 10^{-5}$ M	Mg does not interfere	460
Ce	H_2O_2 decomposition	10^{-5} M	WO_3^-, Fe(II), Mn(II)	447
	luminol + H_2O_2	0.02 μg ml^{-1}	Cu	474
	lucigenine + H_2O_2	2 μg ml^{-1}		702
CN$^-$	indolylacetate hydrolysis (in the pres. of hyaluronidase)	0.1 μg ml^{-1}	Fe, Cu	526

TABLE 10 — CONTD.

Determinand	Indicator reaction	Sensitivity	Interferences	Ref.
CN^-	H_2O_2 catalytic decomposition	$3\ \mu g\ ml^{-1}$	Cu	760
	4-nitrobenzaldehyde	0.01 ppm		805
CO_3^{2-}	Cr(III) + xylenol orange			806
Co	lucigenine + H_2O_2	$1\ \mu g\ ml^{-1}$	Tl, Fe(II), Ni, Ag, Cu	355, 409
	alizarin S, tiron + H_2O_2	$10^{-5}\ \mu g\ ml^{-1}$		342
	indigo carmine + H_2O_2	$1\ \mu g\ ml^{-1}$		353
	alizarin + H_2O_2	$5 \times 10^{-5}\ \mu g\ ml^{-1}$	Ni, Ba, Mg, Fe(III), Cu, Mn, Cd	354, 467
	diphenylcarbazone + H_2O_2	$2 \times 10^{-5}\ \mu g\ ml^{-1}$	Ni	417
	pyrocatechol violet + H_2O_2	$10^{-5}\ \mu g\ ml^{-1}$	Mn	433
	orange G + H_2O_2	$5\ ng\ ml^{-1}$		445
	bromopyrogallol red + H_2O_2	$0.1\ ng\ ml^{-1}$		446
	disodium salt of 3-(4-carboxy-3--hydroxyphenylazo)chromotropic acid + H_2O_2	$3\ \mu g\ ml^{-1}$	Fe(II), Fe(III), Os, Ti(IV)	440
	α,α-diantipyridyl-3,4-dimethoxytoluene + H_2O_2	$0.16\ pg\ ml^{-1}$	$KMnO_4$, strong oxidants	711
	H_2O_2 decomp. in alk. media, catalyzed by the Co complex with 2-aminopyridine	$10\ ng\ ml^{-1}$	Fe(III), Mn(II), Cu(II), Ni(II)	717
	gallic acid autooxidn. at pH 10.3	0.08 µmol	Ag, Ni, Fe(III), CN^-	725
	luminol + H_2O_2	$0.1\ ng\ ml^{-1}$	Fe	735, 736
Cr(VI)	I^- + H_2O_2	$40\ ng\ ml^{-1}$		356, 475
	tropaeoline OOO No. 1 + H_2O_2	$10\ ng\ ml^{-1}$		476

TABLE 10 — CONTD.

Determinand	Indicator reaction	Sensitivity	Interferences	Ref.
Cr(VI)	indigo carmine + H_2O_2 (sometimes activated by 2,2'-bi-pyridine)	2×10^{-7} M, 6×10^{-8} M		515
	o-aminophenol + H_2O_2	30 ng ml^{-1}	Mn	410
	o-dianisidine + H_2O_2 (activated by carboxylic acids)	10^{-5} µg ml^{-1}	Fe(III)	437, 472
	BrO_3^- + I^-	10^{-7} M	W, Mo	757
Cr(II, III)	lucigenine + H_2O_2		Cr(VI)	465
Cr(III, VI)	azomethine compd. + H_2O_2	8 ng ml^{-1} Cr(VI), 80 rg ml^{-1} Cr(III)		728
Cr(III)	luminol + H_2O_2	0.6 ng ml^{-1}	Fe, Co, Cu	730, 736
	o-dianisidine + H_2O_2	10^{-5} µg ml^{-1}		748
Cu	$S_2O_3^{2-}$ + H_2O_2	4 µg ml^{-1}	transition elements	275
	hydroquinone + H_2O_2	5×10^{-4} µg ml^{-1}	Fe, Ni	351, 466, 530
	indigo carmine + H_2O_2	0.18 ng ml^{-1}	10times Fe(III), Cr(VI), Ag, V 100times Sb(V), Mo(VI)	436
	amidol + H_2O_2	5×10^{-5} g ml^{-1}	EDTA	448, 449
	K-diphenylbisazonaphthionate + + H_2O_2	$1-2C$ ng ml^{-1}	Co, Ag, Ce, V, SCN^-, $Fe(CN)_6^{3-}$, EDTA, Mn	503
	I^- + O_2	6 µg ml^{-1}	Fe	334
	$TiCl_3$ + O_2	0.64 µg ml^{-1}		432
	Fe^{3+} + $S_2O_3^{2-}$	6 µg ml^{-1}		352
	Fe^{3+} + $CS(NH_2)_2$	1 µg ml^{-1}	Hg(II), Au, F^-, I^-, PO_3^-	534

TABLE 10 — CONTD.

Determinand	Indicator reaction	Sensitivity	Interferences	Ref.
Cu	Fe^{3+} + N-methylthiourea	$1 \mu g \, ml^{-1}$	Au	477
	$[Fe(CN)_5NOS]^{4-}$ decomp.	10^{-7} M		427
	$NaN_3 + KI + I_2$ (catal. by Cu-diethyldithiocarbamate)	0.06 ppm	Pb, Cd	519
	$S_2O_8^{2-} + I^-$ (in the pres. of $S_2O_3^{2-}$)	$1 \mu g \, ml^{-1}$	Fe	533
	phenol + Chloramine T	$50 \, ng \, ml^{-1}$	Fe	520
	indoxyl acetate hydrolysis (in the pres. of hyaluronidase)	$0.1 \, \mu g \, ml^{-1}$	Fe(II), CN^-	526
	ascorbic acid autooxidation	10^{-6} M	Fe(III), Mn(II), Pd(II)	544
	catalytic decomp. of H_2O_2 in alkaline media	$60 \, ng \, ml^{-6}$		704, 760
	$Fe(CN)_6^{3-} + CN^-$	10^{-7} M	Fe, Co, Cr	736
	luminol + H_2O_2	$0.5 \, ng \, ml^{-1}$	Fe, Ni do not interfere	741
	hydroquinone + H_2O_2 in mixed media	$0.1 \, ng \, ml^{-1}$		
	$I_2 + NaN_3$ (inhibiting effect of Cu-thioammeline complex)	$1 \mu g \, ml^{-1}$	S^{2-}	830
	sulphanilic acid + H_2O_2 (activated by pyridine)	$3 \mu g \, ml^{-1}$	Ni, Zn, Hg(II)	70
Fe(II)	$I^- + O_2$	$60 \, \mu g \, ml^{-1}$	Cu	334
	$I^- + S_2O_8^{2-}$	$0.3 \, \mu g \, ml^{-1}$		359
	p-phenetidine + H_2O_2 (in the pres. of 1,10-phenanthroline + Fe(III)	$1 \, ng \, ml^{-1}$		69, 478,700

TABLE 10 – CONTD.

Determinand	Indicator reaction	Sensitivity	Interferences	Ref.
Fe(III)	pyrocatechol violet + H_2O_2	$0.1\ \mu g\ ml^{-1}$		357
	methyl orange + H_2O_2	$10\ pg\ ml^{-1}$		358
	variamine blue B + H_2O_2	$0.1\ ppm$		479
	Chrome blue (C. I. Mordant Blue 13)	$2\ ng\ ml^{-1}$	Ti, Ga, Nb, Sb, Sn, Mo, W	504
	eriochrome black T + H_2O_2	$0.4\ \mu g\ ml^{-1}$		434
	photochem. oxidn. of methyl orange in the pres. of Fe(III) and oxalate	$15\ ng\ ml^{-1}$		343
	ascorbic acid + Br_2	$0.1\ \mu g\ ml^{-1}$		220
	o-toluidine + KIO_4 (activated by 2,2'-bipyridyl)	$0.5-50\ \mu g\ ml^{-1}$	Mn(II, VII), Cr(III)	412
	indoxyl acetate hydrolysis in the pres. of hyaluronidase	$0.2\ \mu g\ ml^{-1}$	Cu, CN^-	526
	4,4'-bis(dimethylamino)diphenyl-amine + H_2O_2	$1\ ng\ ml^{-1}$	Cu(II)	724
	luminol + H_2O_2	$3\ ng\ ml^{-1}$	Co	735, 736
	2,4-diaminophenol + H_2O_2	$2\ \mu g\ ml^{-1}$	Sn(II), Mn(II), MoO_4^{2-}, CrO_4^{2-}, oxalate	738
	tris(oxalato)cobaltate(III) + ascorbic acid	$10^{-5}\ M$	Co, Ni, V(IV), Mo(VI)	761
	p-phenetidine + H_2O_2 (activated by 1,10-phenanthroline)		detn. of Fe in blood serum	701
	α,α-diantipyridinyl-3,4-dimetho-xytoluene	$10\ ng\ ml^{-1}$	Ce(IV), V(V), Cr(VI), Mn(VII), NO_3^-	712
F^-	BO_3^- + I^- (cat. by Zr)	$1\ \mu g\ ml^{-1}$	oxalate, citrate	539, 763

197

TABLE 10 — CONTD.

Determinand	Indicator reaction	Sensitivity	Interferences	Ref.
Ge	$I^- + MoO_4^{2-}$	$1\ \mu g\ ml^{-1}$		360
	pyrocatechol $+ H_2O_2$	$3\ \mu g\ ml^{-1}$		470
	$KI + MoO_4^{2-}$	$30\ ng\ ml^{-1}$	Si, As(V)	220
	$I^- + MoO_3$	$8\ ng\ ml^{-1}$		505
	$Na_2MoO_4 +$ ascorbic acid	$1\ \mu g\ ml^{-1}$		204
	$MoO_4^{2-} + Sn^{2+}$	$10\ ng\ ml^{-1}$	Sn(IV), Hg(I, II), Sb(V), Pb(II), W(VI), V(V), PO_4^{3-}	
Hf	$I^- + H_2O_2$	$1 \times 10^{-6}\ M$	Zr	385, 706
Hg	$As^{3+} + Ce^{4+}$	$5\ \mu g\ ml^{-1}$	Ag, I^-	348, 737
	$As^{3+} + Ce^{4+}$ (catalyzed by I^-; inhibited by Ag^+)	$50\ ng\ ml^{-1}$	Br^-, Cr(III), Ni, Ag, Cu, Co, Fe(III), PO_4^{3-}, Hg	424
	$K_4Fe(CN)_6 + 2,2'$-bipyridine	$10^{-5}\ M$	Br^-, I^-, Pd, Ag	425, 426
	luminol $+ H_2O_2$	$0.2\ \mu g\ ml^{-1}$	Pb, Cu	450
	$Fe(CN)_6^{4-} + C_6H_5NO$	$0.25\ \mu g\ ml^{-1}$		481
	$Fe(CN)_6^{4-} + 1,10$-phenanthroline		detn. in biological materials	521
				814
In	Cr(III) $+$ EDTA (activated by Na acetate)	$0.3\ mg\ ml^{-1}$	Al	469
Ir	luminol $+ KIO_4$	$20\ ng\ ml^{-1}$	Pt, Pd, Os, Rh, Ru	326
	As^{3+} or $Sb^{3+} + Ce^{4+}$	$4\ ng\ ml^{-1}$	Os, Ru, Rh, Au, Ag, Pt, Co, Pd, Te, Fe, Bi, Cr, Se, Cu, SO_4^{2-}	335
	lucigenine $+$ hydrazine $+ O_2$	$20\ ng\ ml^{-1}$	Pt, Rh	339
	$Ag^+ + Fe^{2+}$	$0.02\ \mu g\ ml^{-1}$	Pt-metals	431
	$Hg^+ + Mn^{3+}$	$0.2\ \mu g\ ml^{-1}$	Os, Ru	482, 439

TABLE 10 — CONTD.

Determinand	Indicator reaction	Sensitivity	Interferences	Ref.
Ir	$Ce^{4+} + H_2O_2$	10 ng ml^{-1}	Pt, Pd, Rh	532
	$Ce^{4+} +$ diphenylamine	10 ng ml^{-1}	Re	545
	$IO_4^- + H_2PO_2$ (inhibited by NTA)	3 ppb		696
	$Cu(H_4TeO_6)_2^{2-} + BrO^-$	0.5 ng ml^{-1}	Rh, Cr	744
	lumirol $+ H_2O_2$	10^{-4} % in Pt salts	Pd, Pt do not interfere	447
I_2	$As^{3+} + Ce^{4+}$	0.4 ng ml^{-1} in serum	Hg(II), Ag, Cu, Cr, Ni, Co, Fe, Sn	341
		2.5 µg ml^{-1}		363, 525
	triphenylmethane dyes $+$ chloramine B	10 µg ml^{-1}	sulphur-containing anions	333
I^-	$As^{3+} + Ce^{4+}$	0.3 µg ml^{-1}		361, 362, 456, 737
	triphenylmethane dyes $+ H_2O_2$	2 ng ml^{-1}	100times Ag, Hg, Fe(III), Pb, Cu	411
	$Hg(II) - PAR + DCTA$	5×10^{-8} M	Br^-, most cations	461
	pyrocatechol violet $+ H_2O_2$	0.1 ppm	Cl^-, ClO_3^-, F^-, SO_4^{2-}, Co, Al, Ba, Li, Mg, Pb, Sr	483, 715
	$As^{3+} + Ce^{4+}$		Ag, Hg, Os	484, 256, 718, 759
	p-nitrophenyldiazoaminobenzene-p-azobenzene ("cadion") $+ K_2S_2O_8$	10^{-7} M	10times Br^-	485
	$BrO_3^- \rightarrow Br^-$	10 ng ml^{-1}	$S_2O_3^{2-}$, SO_3^{2-}, SCN^-, NO_2^-, Fe(II)	502
	$AsO_3^{3-} + IO_3^-$	5 ng ml^{-1}	F^-	506, 553
	$NaNO_2 + Fe(CSN)^{2+}$	20 ng ml^{-1}		513
	benzidine $+ H_2O_2$	2 ng ml^{-1}	I^- in soils	716
	o-tolidine			

TABLE 10 — CONTD.

Determinand	Indicator reaction	Sensitivity	Interferences	Ref.
I^-	o-phenylenediamine + H_2O_2 diphenylcarbazide + H_2O_2	2 ng ml^{-1}	I^- in mineral waters	740
Mn	$I^- + IO_4^-$	0.5 ng ml^{-1}	Fe(III), Co	336
	diethylaniline + IO_4^-	0.5 ng ml^{-1}	Fe(III)	364
	malachite green + KIO_4	0.2 ng ml^{-1}	Fe, Al	414, 462
	N,N,N',N'-tetraethyl-o-dianisidine + IO_4^-	80 ng ml^{-1}	Fe(II), citrate, oxalate, tartrate	429
	malachite green + KIO_4 (in the pres. of NTA)	10^{-7} M		547, 731
	effect of Mn on the fluorescence of the Be-morine complex	5 ng ml^{-1}		451
	o-anisidine + KIO_3	0.1 ng ml^{-1}		458
	oxalic acid + MnO_4^-	0.1 µg ml^{-1}		365
	xylenol orange + $ZrOCl_2$	20 ng ml^{-1}		537
	indigo carmine + H_2O_2 (activated by 1,10-phenanthroline)	20 ng ml^{-1}	Fe	486
	5-(2-hydroxy-3-sulphochlorophenyl-azo)barbituric acid + H_2O_2	6 ng ml^{-1}	$P_2O_7^{2-}$	487
	Na-carminate + H_2O_2 (in the pres. of Pb^{2+} and ethylenediamine)	0.1 ng ml^{-1}		508
	alizarin S + H_2O_2	0.3 ng ml^{-1}	Cu, borate	536
	α-aminoalcohol + IO_4^- (inhibited by NTA)	0.3 µg ml^{-1}	detn. in non-ferrous alloys	697, 762

TABLE 10 – CONTD.

Determinand	Indicator reaction	Sensitivity	Interferences	Ref.
Mn	$Cr(III) + EDTA$ (77°C)		$Sn(II)$, Ag, AsO_2^-, $P_2O_7^{4-}$, I^-, $S_2O_3^{2-}$	705
	$IO_4^- + Sb(III)$		NTA, EDTA, DTPA	834
	Acid blue 45 + H_2O_2	4 ng ml^{-1}	Cu, $Fe(III)$	714
	H_2O_2 decomp. catalyzed by $Mn(II)$-1,10-phenanthroline	5 ng ml^{-1}	$Fe(III)$	720
	lumomagnesone + H_2O_2	0.1 ng ml^{-1}	highly selective	776
	1,4-dihydroxyphthaldimine autooxidation	10 ng ml^{-1}		732
	o-dianisidine + H_2O_2	0.1 ng ml^{-1}	$Fe(III)$, $Cr(III)$	743
	o-dianisidine + KIO_4	0.1 ng ml^{-1}	Mn electrochem. separated	831
	Azorubine S + H_2O_2 (C.I. Acid Red 14)	1 ng ml^{-1}	$Mo(VI)$, $W(VI)$	765
Mo	$KI + H_2O_2$	60 ng ml^{-1}	W	344
	$I^- + H_2O_2$	2×10^{-5} %		367, 366, 542
	$I^- + H_2O_2$	50 ng ml^{-1}	W	383
	$I^- + H_2O_2$ (induction period)	3×10^{-7} %		463
	$I^- + H_2O_2$	2 ng ml^{-1}	W, Mo, Zr, Ta, Fe	507, 766
	rubeanic acid + H_2O_2	0.01 µmol l^{-1}		488
	$I^- + KBrO_3$	50 ng l^{-1}		255, 327
	$BrO_3^- + 9 I^- + 6 H^+$	10^{-8} M	W, Cr(III)	327, 756, 757
	1-amino-2-naphthol-4-sulphonic acid + BrO_3^-			368
	methylene blue + hydrazine	20 µg ml^{-1}	W, Fe(III)	441

TABLE 10 — CONTD.

Determinand	Indicator reaction	Sensitivity	Interferences	Ref.
Mo	Landolt reaction	$10\ \mu g\ ml^{-1}$	F^-, Th, W	443, 548, 549, 550
	$SeO_4^{2-} + Sn^{2+}$	$10\ ng\ ml^{-1}$		511
	$S_2O_3^{2-} + H_2O_2$	$3\ \mu g\ l^{-1}$	V(V), Cu(II)	713
	1-naphthylamine + $KBrO_3$	$2 \times 10^{-7}\ M$		66
	2-aminophenol + H_2O_2	$10\ ng\ ml^{-1}$	W(VI)	739
	azorubine S + H_2O_2	$1.5\ \mu g\ ml^{-1}$	Mn	765
NH_3	Hg-o-cresolphthaleincomplexone + DCTA	$1\ \mu mol\ l^{-1}$	cations	452
NO_2^-	Mn-EDTA + H_2O_2	$10\ ng\ ml^{-1}$	nitrogen oxides	464
NO_3^-	continuous-flow immobilized enzyme reaction	ppb		804
Nb	$S_2O_3^{2-} + H_2O_2$			405
Nb + (Ta)	$I^- + H_2O_2$	$Nb - 8 \times 10^{-5}\ M$ $Ta - 5 \times 10^{-5}\ M$		413
Ni	o-diphenylcarbazone + H_2O_2	0.1–0.7 ppm	Cu	421
	MnO_4^- decomp.		Ag, Co, Cu, Fe	512
	1-naphthylamine + percaprinic acid	$2\ ng\ ml^{-1}$	Cu, Co	742
	$NaN_3 + I_2$ (Na-diethyldithiocarbamate induction)	$30\ ng\ ml^{-1}$	Ni, Co	758
Os(IV, VI, VIII)	$Ce^{4+} + As^{3+}$	$1\ ng\ ml^{-1}$	I^-, Ni, Ti	61, 386, 387
	$Fe(CN)_6^{3-} + BH_4^-$	$1\ ng\ ml^{-1}$	Re, Ru, Rh, Ir do not interfere	726
Os(VI, VIII)	p-phenetidine + BrO_3^-	$1\ ng\ ml^{-1}$	Ru, Fe(II, III), Cu, V, Mo, W	392, 393

TABLE 10 — CONTD.

Determinand	Indicator reaction	Sensitivity	Interferences	Ref.
Os(VIII)	As^{3+} + $KBrO_3$	0.1 ng ml^{-1}	Pt-metals (except for Ru)	328
	1-naphthylamine + NO_3^-	2 ng ml^{-1}	Ru	50,719
	2-naphthylamine + HNO_3	2 ng ml^{-1}	Pt does not interfere	428
	I^- + $NaClO_2$	19 ng ml^{-1}	Rh, Pd, Pt	407
	I^- + ClO_3^-	2 ng ml^{-1}	Cu, Cr(III), V, Fe(III), oxalic acid, thiourea	370
	I^- + BrO_3^-	0.5 ng ml^{-1}	Ru	389, 390
	MnO_4^- + As^{3+}	2 μg ml^{-1}	Cu, Ni, Mn(II), Mo	388
	ascorbic acid + ClO_3^-	100 ng ml^{-1}		220
	p-phenylenediamine + H_2O_2	40 pg ml^{-1}	Pt, Rh	332
	1-naphthylamine + sulphanilic acid	3 ng ml^{-1}	Mo, Fe(III), oxidants	391
	2-aminophenol-4-sulphonic acid + $K_2S_2O_8$	0.3 μg ml^{-1}	oxidants, reductants, Fe(III)	394
	ClO_3^- + org. subst. (benzidine, phenolphthalein, malachite green, p-phenylenediamine, 3,3-dimethylnaphtidine)	~ 50 ng ml^{-1}	oxidants, Ru	395
	hydroquinone + H_2O_2	0.2 ng ml^{-1}	Ag, Co, Bi, Cu,	396
	resorcinol + H_2O_2	0.2 ng ml^{-1}	Zn, Cr(III), Ni, Cu, Ti(IV)	396
	1,3-naphthalenediol + H_2O_2	0.2 ng ml^{-1}	Ag, Co, Bi, Cu	396
	lucigenine + H_2O_2	5 μg ml^{-1}	Cu, Cd, Pb, Hg, Co, Ni(III), Ca	369
	chloramine T + As(III)	10 μg ml^{-1}	I^-	750
PO_4^{3-}	MoO_4^{2-} + I^-	100 ng ml^{-1}		371
	MoO_4^{2-} + ascorbic acid	10 ng ml^{-1}		538

TABLE 10 — CONTD.

Determinand	Indicator reaction	Sensitivity	Interferences	Ref.
Pb	gallic acid + $K_2S_2O_8$	20 ng ml^{-1}	Cu, Co, Fe(III), Cr(III)	490
	tiron + $K_2S_2O_8$	20 ng ml^{-1}	Cu, Co, Fe(III)	490
	stilbazo + $K_2S_2O_8$	10 ng ml^{-1}		491
	3,4-dihydroxyazobenzene and pyro-gallol red + $K_2S_2O_8$	30 ng ml^{-1}	BO_3^{3-}	492
	pyrocatechol violet + $K_2S_2O_8$	10 ng ml^{-1}	Ni, Co, Cd, Ag, Cu, Mn(II)	509
Pd	$Ag^+ + Fe^{2+}$	20 pg ml^{-1}	Pt-metals	431
	$SnCl_4 + HAsO_2$	200 ng ml^{-1}	Pt, Ir, Re, Rh	489
	AgBr + metol-hydroquinone	1 µg ml^{-1}	Rh	434
	$Mn^{3+} + Cl^-$	4 ng ml^{-1}	Ru, Rh, Ir, Pt, Au do not interfere	729
	Dye + $H_2PO_2^-$	10^{-7} M	I^-, CN^-, S^{2-}, Ru, Pt	753
Pt	hydrazine and lucigenine + O_2	1 ng ml^{-1}	Rh, Ir	339
	$I_2 + NaNO_3$ (inhibited by the 6-mer-captopurineplatinum complex)		for pure Pt solutions	257
	luminol + H_2O_2	10 pg l^{-1}	for pure Pt solutions	775
Re	$TeO_4^{2-} + Sn^{2+}$	2 ng ml^{-1}	Pt, Pd, Ir	372
	$Fe^{3+} + SnCl_2$	10 ng ml^{-1}		438
Rh	hydrazine and lucigenine + O_2	5 ng ml^{-1}	Pt, Ir	339
	$Mn^{2+} + NaBrO_3$	1 ng ml^{-1}	Pt-metals do not interfere	493
	AgBr + metol-hydroquinone	50 ng ml^{-1}	Pd	494
	$Cu^{2+} + KIO_4$		Pt-metals, Au do not interfere	722
	$Cu(OH)_4^{2-} + H_4TeO_6^{2-}$	1 ng ml^{-1}	Fe, Cr, Bi, Pt-metals	751
	$Ag^+ + Fe^{2+}$	0.1 ng ml^{-1}	Pd, Ir, Ru, Os	752

TABLE 10 — CONTD.

Determinand	Indicator reaction	Sensitivity	Interferences	Ref.
Ru(III)	1-naphthylamine + NO_3^-	5 ng ml^{-1}	Os	50, 719
	Cu^{2+} + KIO_4	4×10^{-9} M		331
	p-benzoquinone + $Mn_4(P_2C_7)_3$ + (Cl-compounds of Pd-catalyst)	10—50 ng ml^{-1}	Os, Pd	64
	I^- + H_2O_2	0.2 rg ml^{-1}	Pt-metals, Au(III), S^{2-}	340, 527
	Ag^+ + Fe^{2+}	0.08 ng ml^{-1}		420
	p-anisidine + $MnSO_4$	60 ng ml^{-1}	Pt-metals do not interfere	457
	Fe^{3+} + $SnCl_2$	4×10^{-7} M	Cu, Mo, F^-, oxalate, $C_4H_4O_6^{2-}$	510
	IO_4^- + ferroin	10^{-8} M	Ir, Re, Os, Fe(III), V(IV, V), CN^-, EDTA	402, 528
	I^- + IO_4^-	1 ng ml^{-1}	Os, Pd, Ir	397
	$Fe(SCN)_3$ + $SnCl_2$	10 ng ml^{-1}	Cu, Mo, Pd, Al, V, F^-, oxalic acid	398
	benzidine + H_2O_2	100 ng ml^{-1}	Os, SO_4^{2-}	400
	Fe^{3+} + $Na_2S_2O_3$ (in the pres. of KSCN)		Cu(II), Pt(IV), W(VI)	708
	o-dianisidine + IO_4^-	10 ng ml^{-1}	Ir, Pd	401, 418
Ru(III, IV)	benzidine + IO_4^-	20 ng ml^{-1}	Ir, Pd	401, 418
	I^- + ClO_4^-	5 ppb	Os, Ir	734
Ru(IV)	methyl red + IO_4^-	10 ng ml^{-1}	Os, Pt, Pd, Re do not interfere in detn. in ores	727
	I^- + H_2O_2	100 ng ml^{-1}	Os, Fe(II)	399
	Hg^+ + Ce^{4+}	200 ng ml^{-1}	Mo, Fe(III), oxidants	422
	2-naphthylamine + sulphanilic acid	5 ng ml^{-1}	Os, I^-	391
Ru(III,IV,VIII)	Ce^{4+} + As^{3+}	5 ng ml^{-1}		61, 386, 387

TABLE 10 — CONTD.

Determinand	Indicator reaction	Sensitivity	Interferences	Ref.
S^{2-}	$N_3^- + I_2$	10^{-3} ng ml^{-1}	larger conc. of Cl^-	373, 374
	$Ag^+ + Fe^{2+}$	1 ng ml^{-1}	S_2O_3, CSN^-, SO_3^{2-}	514
	$KBrO_3$ decomp. in acid media	6 ng ml^{-1}		746
$S_2O_3^{2-}$	$N_3^- + I_2$	3 ng ml^{-1}	CNS^-, S^{2-}	376, 377
	$2\,NaN_3 + I_2$	10^{-2} ppm	Fe(III), Cu	423
	$KBrO_3$ decomp. in acid media	9 ng ml^{-1}	S^{2-}, SCN^-, SO_3^{2-}	746
SCN^-	$KBrO_3$ decomp. in acid media	5 ng ml^{-1}	S^{2-}, SO_3^{2-}, $S_2O_3^{2-}$	746
	$I_2 + NaN_3$	50 ng ml^{-1}		803
	$N_3^- + I_2$	4 ng ml^{-1}	S^{2-}. $S_2O_3^{2-}$	378, 379
Se^{2-}	AgBr + metol-hydroquinone	0.05 mg ml^{-1}	Au, Ir, Pd, Rh	415
Se(IV)	2,3-diaminophenazine + glyoxal	5 ng ml^{-1}	I^-, greater conc. Te(IV)	375
	1,6-dihydro-1,4,11-tetrazo-naphthacene + glyoxalic and hypophosphic acids			529
	phenylhydrazine-p-sulphonic acid + ClO_3^-	500 ng ml^{-1}	Fe(II, III), Cu, Ce, V, Cr(VI)	532
	Fe(II)-EDTA + NO_3^-	5 ng ml^{-1}	Rh, NO_2^-	813
	Co(III)-EDTA + Sn^{2+}	2 ng ml^{-1}	Mo(VI), Re(VII), Hg(II)	816
	p-hydrazinobenzenesulphonic acid oxidn.	10^{-7} M	Cu, Ce(IV), Mo(VI), I^-, Fe(III), Cr(VI), Te(IV)	764
Si	$I^- + (NH_4)_2\,MoO_4$	3 ng ml^{-1}	P, As, Ge	338
	$Si(OH)_4$ + molybdenic acid dimer	0.3 ppm	Fe(III), F^-, As(V), PO_4^{3-}	531

TABLE 10 – CONTD.

Determinand	Indicator reaction	Sensitivity	Interferences	Ref.
Sn	reduction of Na-isopolymolybdate by ascorbic acid	5 ng ml^{-1}	Cu, Bi(III), EDTA	721
Ta	oxalic acid + H_2O_2	14 µg ml^{-1}		380
	$S_2O_3^{2-}$ + H_2O_2	200 ng ml^{-1}	V, Ti, W, Th, F$^-$	381
Te(IV)	tartrate + Cl$^-$	traces		817
Ti	I$^-$ + H_2O_2	50 ng ml^{-1}	Mo, W, V, EDTA, F$^-$, oxalate	495
	$S_2O_8^{2-}$ + H_2O_2	5 µg ml^{-1}	Fe, Mo, W, V	516
	o-phenylenediamine + H_2O_2	10 ng ml^{-1}	Cu, WO_4^{2-}, CrO_4^{2-}, Zr(IV)	745
UO$_2$	photodegradation of carotenoid bixine	10^{-2} ppm		416
U(VI)	reduction of Victoria blue with Ti(III)	10^{-2} µg ml^{-1}		703
V	gallic acid + $(NH_4)_2S_2O_8$	5 ng ml^{-1}	Cu, Fe(II, III)	346, 524
	Landolt reaction	20 ng ml^{-1}	Os, Se, Cu, T, Fe	220, 221 551, 552
	benzidine + $KClO_3$			382
	meturine + KBrO$_3$		Cr(VI)	337
	chromotropic acid + BrO$_3^-$	5 ng ml^{-1}	Fe(III), Cu, W	406
	bromopyrogallol red + KBrO$_3$	3 ng ml^{-1}	Ti, Fe(II, III)	419
	p-phenetidine + BrO$_3^-$	$0.2-1.8$ ng ml^{-1}	Mo, W, Cr(VI)	435, 766
	gallic acid + KBrO$_3$	3 ng ml^{-1}	Fe(III), Ti(IV)	454, 733

TABLE 10 — CONTD.

Determinand	Indicator reaction	Sensitivity	Interferences	Ref.
V	1,5-diphenylaminopyrrazoline + KBrO$_3$	0.1 µg ml^{-1}		455
	eriochrome blue P + KBrO$_3$	0.04 ng ml^{-1}	Ti	459
	o-chloroaniline + KBrO$_3$ (activated by 2-quinoline)		EDTA, oxalic acid	496
	I$^-$ + BrO$_3^-$ (activated by tartaric or succinic acid)	10 pg ml^{-1}	Ta, Mo	497
	4-aminodiphenylamine-2-sulphonic acid + BrO$_3^-$ (activated by loretine, citric, tartaric acid, tiron)	8 × 10^{-8} M	EDTA, oxalic acid, F$^-$, 8-quinolinol	498
	p-phenetidine + BrO$_3^-$ (activated by sulphosalicylic acid)	20 pg ml^{-1}		499
	Solochrome violet RS + BrO$_3^-$	30 ng ml^{-1}	Ni, Fe(II), Mn(II), Th	500
	SCN$^-$ + BrO$_3^-$	50 ng ml^{-1}	Fe(III)	517
	bordeaux + BrO$_3^-$	5 ng ml^{-1}	Hg, CNS$^-$, Br$^-$, I$^-$, oxalate	546
	phenylhydrazine-p-sulphonic acid + NaClO$_3$	10 ng ml^{-1}	interferants removed by extr.	554
	α-aminophenol + NaClO$_3$	4 ng ml^{-1}	Mo(VI), W(VI), Mn(VII), Cr(VI), Al(VIII), Fe(III), Mg, thiosulphate	698
	gallic acid + KBrO$_3$	0.03 ppb	detn. in natural waters after ion-exchanger separation	698
	ClO$_3^-$ + I$^-$	0.5 µmol	Fe(III)	707
W	KI + H$_2$O$_2$		Mo	344, 541

TABLE 10 — CONTD.

Determinand	Indicator reaction	Sensitivity	Interferences	Ref.
W	rubeanic acid + H_2O_2	0.4 ng ml^{-1}	Mo	384
	dithiooxamide + H_2O_2	2.5 ng ml^{-1}	Mo(VI), V(V), Cu(II)	713
	2-aminophenol + H_2O_2	10 ng ml^{-1}	Mo(VI) does not interfere	739
	$BrO_3^- + I^-$	10^{-5} M	Cr	756, 757
	Azorubin S + H_2O_2 (C.I. Acid red 14)	6 μg ml^{-1}	Mn, Mo(VI)	765
	triarylmethane dyes + Ti^{3+} (e.g. malachite green)	1 μmol		776
Zn	oxalacetic acid decarboxylation	5 μg ml^{-1}		540
Zr	$I^- + H_2O_2$	10^{-6} M	Hf	385
	$S_2O_3^{2-} + H_2O_2$	100 ng ml^{-1}		501

209

pH 9.5) and 2 ml 10^{-3} M *o*-cresolphthalein complexone (CPC) are added to the solution in the cuvette. Then 1 ml of 4×10^{-3} M DCTA is mixed with this solution, initiating the reaction Hg-CFC + DCTA → Hg-DCTA + + CFC. The reaction is monitored spectrophotometrically at 583 nm (the absorbance maximum of the Hg-CFC complex). The constant concentration method is used to calculate the concentration.

Determination of Gold in Iron Ores [543]

A sample is dissolved in aqua regia, the solution is evaporated to dryness and the residue is dissolved in 0.6 N HCl. Gold is then separated by extraction into ethyl acetate. The organic phase is evaporated and the residue is dried, dissolved in aqua regia and evaporated with nitric acid until all the HCl is removed. The residue is dissolved in 0.6 N H_2SO_4, the solution is transferred to a cuvette and thermostatted at 30 °C. A solution containing 6×10^{-3} M $Hg(NO_3)_2$ and 3.7×10^{-3} M $Cd(SO_4)_2$ is added and the absorbance is monitored at 420 nm. The tangent method is used for the calculation.

Determination of Copper in Serum [669]

An amount of 0.1 to 0.2 ml of blood is burned at 500 °C. The ashes are dissolved in 5 drops of HNO_3 (1 : 4) and transferred to a test tube. The solution is neutralized and phenolphthalein, 1 ml of 4×10^{-3} M pyridine and 5 ml of pH 7.4 phosphate buffer are added. The test tube is placed in a water bath at 25 $\pm$ 0.5 °C for about 10 min, 1 ml of 0.05 M hydroquinone is then added, the solution is transferred to a photometer cuvette and the absorbance is measured at 1 min. intervals using a blue filter. The tangent method is used for the calculation.

Determination of Vanadium in Biological Materials [346]

A sample containing 0.02 to 0.25 µg V/ml is decomposed in a 100 ml Kjeldahl flask by boiling for 15 min. with 10 ml HNO_3, 2 ml 70% $HClO_4$ are added and the mixture is boiled until the appearance of white fumes. If the sample contains larger amounts of fats, it must be boiled with HNO_3 to remove them. The contents of the flask is then transferred to a 125 ml separating funnel and the pH is adjusted to 4 with ammonia. Then 10 ml

8-quinolinol (a 5% solution in chloroform) are added and extraction is carried out for 5 min. The extraction is repeated once more with a new portion of 8-quinolinol and the combined organic phase is extracted for 30 min with 25 ml of an ammonia buffer pH 9.4−9.5. The pH of the aqueous phase is adjusted to 9.4−9.5 and the extraction is repeated for another 90 min. A 10 ml aliquot of the aqueous phase is transferred to a 50 ml cuvette, 1 ml of 0.30% $Hg(NO_3)_2$ is added and the cuvette is placed in a water bath at $30 \pm 0.5\,°C$ for 30 min. Then 1 ml of thermostatted 70% $HClO_4$ and 1 ml of a 2% solution of gallic acid are added. The cuvette is left in the water bath for another 60 min and the absorbance is then measured at 415 nm. The same procedure is carried out with the standards and the blank determination is performed. The sensitivity equals that of neutron activation analysis.

10.1.2 CATALYTIC END-POINT TITRATIONS

The principles of the method are described in Section 8.2.1.5 and the reactions employed are surveyed in Table 11. These also include titrations of ions that act as activators in some enzymatic reactions. For example, in the determination of manganese, cobalt and magnesium, the indicator reaction is the conversion of isocitric acid catalyzed by isocitric acid dehydrogenase (ICD), with the test cations acting as activators. The ions can be titrated with EDTA and the rate of the indicator reaction serves for end-point detection [578]. Many indicator reactions have been proposed for determining ions acting as activators of enzymic reactions [579]. The determinations are very simple, but they do not seem to be very important for analysis of inorganic substances. The use of catalysis by substances in common titrations, e.g. in cerimetry [580] is not included among kinetic methods. However, the use of catalysts is very important, especially in titrations of organic substances in non-aqueous solutions [581, 582].

10.1.3 INDIRECT CATALYTIC DETERMINATION OF ORGANIC SUBSTANCES

The modifying effect of various ligands on the rate of catalytic reactions (Section 6.4) has been used for determination of some organic substances; for a survey, see Table 12.

TABLE 11 — Catalytic end-point titrations

Indicator reaction	Titrant	Titrand	Note	Ref.
$Ce^{4+} + As^{3+}$	KI	Ag^+	Automatic titration	254, 330, 347, 555, 556, 557, 558
autooxidation of ascorbic acid	Cu^{2+}	CN^-	1 equiv. $Cu^{2+} \sim 2$ equiv. CN^- at the end point	559
methyl orange + $S_2O_8^{2-}$	Ag^+	Cl^-, Br^-, I^-		560
H_2O_2 + diphenols	Co^{2+}	EDTA		560, 561
malachite green + IO_4^-	Mn^{2+}	EDTA		562
$Ce^{4+} + As^{3+}$	KI	Pd(II)		563
$Ce^{4+} + Sb^{3+}$ + ferroin	Ag^+, Hg^{2+}	I^-		564
$KBrO_3$ + tiron	Co^{2+}	EDTA		560, 565
H_2O_2 + hydroquinone	Cu^{2+}	oxine		351
H_2O_2 + aromatic amines	EDTA	Mn^{2+}	back-titration of EDTA with Mn^{2+}	566
H_2O_2 + alizarin S	EDTA	Co^{2+}		567
$Ce^{4+} + As^{3+}$	KI	Ag^+, Hg^{2+}, Pd^{2+}		568
H_2O_2 + resorcinol	EDTA	Zn, Cd, Cu, Ni, Mn, Pb, Al, In, Ga, Th	back-titration of EDTA with Mn^{2+}	569
H_2O_2 + resorcinol	EDTA	Mg, Ca, Sr, Ba	back-titration of EDTA with Zn^{2+} (the Mn-EDTA catalyst is activated at the equivalence point)	570

212

TABLE 11 — CONTD.

Indicator reaction	Titrant	Titrand	Note	Ref.
$Ce^{4+} + As^{3+}$	KI	Ag^+	coulometrically generated I^-	571
$Ce^{4+} + As^{3+}$	KI	I^-, Br^-, CNS^-	back-titration of Ag^- by coulometrically generated I^-	571
$Ce^{4+} + As^{3+}$	KI	Au(III), Pd(II)	sens. 1 µg Pd	818
$Ce^{4+} + As^{3+}$	KI	Hg(II)		819
periodate + diethylaniline	Mn^{2+}	EDTA	sens. µg EDTA	820
alcohol + acetanhydride	$HClO_4$	tertiary amines, metal carboxylates		821
$KI + H_2O_2$	Th(IV)	F^-, SiF_6^{2-}	titrated in 50% ethanol	822

TABLE 12

Determination of organic substances based on their modifying effect on the rate of catalytic reactions

Indicator reaction	Determinable substances	Ref.
H_2O_2 + pyrocatechol violet (Cu catalyst)	Amino acids	572
H_2O_2 + pyrocatechol violet (Cu^{2+} catalyst)	Nitrogen-containing complexing agents (e.g. EDTA)	573
ferroin + Cr^{6+}	oxalic acid	574
Malachite green + IO_4^- (Mn^{2+} catalyst)	EDTA	575
Malachite green + IO_4^- (Mn^{2+} catalyst)	NTA	170
ferroin + Cr^{6+}	oxalic and citric acids	576
ferroin + Cr^{6+}	uric acid (pseudoinduction period measured)	577
I^- + ClO_3^- in the presence of V(V)	ascorbic acid	808
NaN_3 + I_2	rubeanic acid and its derivatives	767
NaN_3 + I^-	thioketones	803
Alizarin S + H_2O_2 in the presence of Co(II)	8-hydroxyquinoline	810, 811
Luminol + H_2O_2	vitamin B_{12}	812
Phosphinate + periodate catalyzed by Mn^{2+}	DCTA	616
IO_4^- + Sb(III) catalyzed by Mn^{2+}	NTA, DTPA, EDTA	834
n-phenetidine + KIO_4 in the presence of 2,2'-bipyridyl	Hg	838

10.1.4 DETERMINATIONS BASED ON MEASUREMENT OF CATALYTIC POLAROGRAPHIC WAVES

Table 13 surveys the analytical use of catalytic polarographic waves (Section 7.2). The sensitivity is always higher than with normal polarographic methods. Methods based on the catalytic effects of proteins (the Brdička reaction) are not given, as they are described in detail in monographs devoted to polarography and clinical analysis.

TABLE 13

Determinations based on measurement of polarographic catalytic waves

Determinand	Condition for formation of catalytic wave	Note	Ref.
CN^-	Ni-ethylenediamine	hydrogen wave	768
CN^-	Ni + NH_2OH	hydrogen wave	587
CN^-, Br^-, I^-	reduction of In^{3+}		583
Co(II), Fe(II)	thiourea + ClO_4^- excess of Mn(II)		769
Co	Co(II)-dithiocarbamate	hydrogen wave	770
CrO_4^{2-}	maximum at 1.7 V in the presence of NO_3^- ions		771
Cu(II), Ni(II), Co(II)	dimethylglyoxime	hydrogen wave	772
Cu(II)	2-aminophenol		773
Fe	Fe-bissalicylal		598
Ga	salicylic or p-amino-salicylic acid		774
Hf	reduction of H_2O_2		584
Ir, Rh	thiosemicarbazide	Ru(III, IV), Os(IV) interfere	775
La	Cr-TTHA + La	some lanthanoids interfere	585
Mo(VI)	MoO_4^{2-} + NO_3^-	sens. 5 ng ml^{-1}	586
Mo(VI)	glutaric acid + $KClO_3$		779
Mo(VI)	titration of Mo(VI) with Pb^{2+}		780
Mo(VI)	BrO_3^-	sens. 0.1 ng ml^{-1}	781
Mo(VI)	glycolic acid + $NaClO_3$		782
Mo(VI)	ClO_4^- or NO_3^- in acid media	sens. 0.04 μM	783
Mn(II), Fe(II)	Mn- and Fe-dithiocarbamates		784
Nb	Nb + NH_2OH	sens. 10^{-7} mol. l^{-1}	599
Nb	catalytic wave of ClO_3^- in citric acid medium		785
Ni	thiocyanate solution containing Co(II) ions	increase in Ni wave	786
Pt	Pt + EDTA	Rh, Ir, Pd, Ru do not interfere	600
Pt	Pt-ethylenediamine	Rh(III) interferes	787
Rh	thiosemicarbazide	hydrogen wave	588

215

TABLE 13 — CONTD.

Determinand	Condition for formation of catalytic wave	Note	Ref.
Ru	nitrosylchlorocompounds + + Ru(III)		589
Ru	rubeanic acid + Ru(III)		590
Sb(III)	hydrogen catalytic wave in the presence of Co-5-sulpho-8-mercaptoquinolate	sens. 6×10^{-5} mol l^{-1}	601
Ti	trihydroxyglutamic acid + + ClO$_3^-$	det. of Ti in steel	591
Ti	Ti(IV) + EDTA + BrO$_3^-$	Zr, V do not interfere	592, 593
Ti	H$_3$PO$_4$ + KClO$_3$	sens. 10^{-7} g l^{-1}	788
Ti	n-phenylbenzohydroxamic acid		789
Ti	2-dibutylphosphinyl-2--hydroxypropionic acid		790
U	UO$_2^{2+}$ + NO$_3^-$		594
U	oxidation of U(III) by ClO$_3^-$		791
U	UO$_2^{2+}$ + glyoxylic acid	sens. 5×10^{-6} mol. l^{-1}	792
V	V(V) + BrO$_3^-$	rot. graphite disk electrode	778
V	o-aminophenol + ClO$_3^-$	sens. 0.2 μg l^{-1}	793
W	oxalic acid + H$_2$O$_2$		595
W	catechol + NH$_2$OH	sens. 10^{-7} mol l^{-1}	794
W	ascorbic acid + NH$_2$OH	sens. 10^{-7} mol l^{-1}	795
W	hydrogen catalytic wave in H$_2$SO$_4$		796
catechol	Sn(IV) in the pres. of ClO$_3^-$ and catechol		797
catechol	Mo(VI) + ClO$_3^-$		798
catechol	Ge(IV) + ClO$_4^-$		799
citric acid	Ti(IV) + ClO$_4^-$		597
oxalic acid	In(III)-oxalate in chlorate medium		800
o-phenylene-diamine	Ni(II) + o-phenylenediamine	sens. 10^{-7} mol l^{-1}	597
phthalic acid	In(III)-phthalate in chlorate medium		802
thiamine	Co(II)-thiamine in borate buffer		803

10.2 Kinetic Methods

10.2.1 METHODS FOR DETERMINATION OF A SINGLE SUBSTANCE

As was mentioned in Section 8.2.2, these kinetic methods are not as important as those for analysis of mixtures. Some recently described determinations are listed in Table 14.

TABLE 14

Some chronometric determinations of inorganic and organic substances

Determinand	Reagent or reaction conditions	Ref.
Al^{3+}, Zn^{2+}	the effect on the rate of discoloration of crystal violet (indicator reaction with OH^- ions)	602
Al^{3+}	Al^{3+} + oxine-5-sulphonic acid	836
AsO_3^{3-}	IO_3^-	603
AsO_3^{3-}	MoO_2^{4-}	607
Br^-	$I^- + IO_3^-$	832
	$I^- + ClO_3$	
Br^-	methyl orange + BrO_3^- (induction period)	844
Ca^{2+} in the presence of Mg^{2+}	the effect of Ca^{2+} on the rate of reaction of Cu-EGTA + PAR	605
Cr^{3+}	EDTA	606
Fe(II)	Fe(II) + ClO_4^-	840
$KBrO_4$	$BrO_4^- + I^-$	837
Nb^{5+}	tiron	607
NO_2^-	sulphanilamide + N-(1-naphthyl)ethylenediamine	608
PO_4^{3-}	MoO_4^{2-}	609
PO_4^{3-}	MoO_4^{2-} + ascorbic acid	610
hypophosphite	$IO_3^- + H_2PO_2^-$	839
SCN^- in serum	$Br_2 + SCN^-$	833
prim., sec. alcohols	XeO_3	611
methanol, ethanol	Ce^{4+}	621
2-propanol	oxidn. by chromic acid	841
amino acids	ninhydrine	612
aniline, antipyrine, oxine	Br_2	613
cyanamide	cyanamide + pentacyanoammineferrate(II); stopped-flow method	843

TABLE 14 — CONTD.

Determinand	Reagent or reaction conditions	Ref.
cystein	Ag^+	615
cystein	effect on the reaction, $I_2 + N_3^-$	616
o-, m-, p-dihydroxybenzene	Br_2	617
phenols	Br_2	618
phloroglucinol	Br_2	617
furfural	HSO_3^-	619
hydrocortisone	reduction of blue tetrazolium; stopped-flow method	842
cholesterol	H_2SO_4	620
creatinin	picric acid	614
nitric esters	hydrolysis in alkaline methanolic solution	845
acetic acid, acetanhydride	aromatic amines	623
oxalic acid	XeO_3	622

10.2.2 ANALYSIS OF MIXTURES

Table 15 surveys differential kinetic methods for organic and inorganic substances, specifying the method and the measuring technique. It is evident from Table 15 that computers are required for the calculations involved.

10.3 Enzyme Analysis Methods

These methods include determinations of enzymes, substrates, activators and inhibitors and are at present the most important kinetic methods in analytical practice, undergoing constant development.

10.3.1 DETERMINATION OF ENZYMES

Determination of enzymes is one of the main tasks of enzymatic analysis, because it forms the basis for all other determinations, i.e. those of substrates, activators and inhibitors. Because of their catalytic activity, enzymes can be determined by any catalytic method. Obviously, physico-chemical methods are now most often used for monitoring of the indicator reaction (the reaction of the substrate), the most important being spectrophotometric, fluorescence and electrochemical methods.

TABLE 15

Analysis of mixtures of substances (differential kinetic methods)

Mixture	Method	Measuring technique	Ref.
unsaturated organic substances in synth. rubber	single-point method (1st order)	titration	624
unsaturated organic compounds	logarithmic extrapolation (1st, 2nd order)	spectrophotometry	625
fructose, glucose	single-point method	spectrophotometry	626
prim., sec. butanol	logarithmic extrapolation (2nd order)	titration	627
prim., sec. alcohols	logarithmic extrapolation (2nd order)	titration	178
prim., sec., tert. alcohols	logarithmic extrapolation (2nd order)	IR-spectrophotometry	628
organic acids	meth. of proportional equations	titration	629
organic acids	logarithmic extrapolation (2nd order)	titration	630
amines	logarithmic extrapolation (2nd order)	titration	631
amines, amino acids	logarithmic extrapolation (1st, 2nd order)	various	229, 632
amines	single-point method (2nd order)	conductance	633
diazocompounds	logarithmic extrapolation (1st order)	Dumas method	634
nitrogen-containing compounds	logarithmic extrapolation (1st order)	Dumas method	635
organic peroxides	meth. of proportional equations	titration	636
glycols	meth. of proportional equations (2nd order)	various	637
mixt. of amides, amines and other substances	logarithmic extrapolation (1st order)	various	638
mixt. of 1-chloroalkanes	meth. of proportional equations	various	639
hexaalkylditin compds.	logarithmic extrapolation (2nd order)	electrochem.	640
sulphonphthalein dyes	meth. of proportional equations	spectrophotometry	641

TABLE 15 — CONTD.

Mixture	Method	Measuring technique	Ref.
thioacids	computer calculation	stopped-flow method	642
aminopolycarboxylic acids	computer calculation	stopped-flow method	643
methylethane sulphonate ethylmethane sulphonate	logarithmic extrapolation (1st order)	potentiometry	644
mixt. of cations	subst. reaction of DCTA; complexes; computer calculation	stopped-flow method	176, 180
Ca^{2+}, Mg^{2+}	linear extrapolation	square-wave polarog.	645
Co, Ni; Ni, Fe; Zn, Cd; Cu, Pb	logarithmic extrapolation (1st order)	spectrophotometry	646
Ga^{3+}	meth. of proportional equations	spectrophotometry	647
La^{3+}, Nd^{3+}	computer calculation	spectrophotometry	246
Mg^{2+}, Sr^{2+}, Ba^{2+}, Ca^{2+}	subst. reaction of DCTA complexes; computer calculation	stopped-flow method	175
NO, NO_2	logarithmic extrapolation (1st order)	stopped-flow method	648
SiO_3^{2-}, PO_4^{3-}	computer calculation	spectrophotometry	649
rare earths (Sm, Pr, Lu)	meth. of proportional equations	spectrophotometry	650
α- and β-ethynodiodi-acetates	single-point method	titration	823
binary mixtures of penicillin	logarithmic extrapolation	various	824
aliphatic carboxylic acid esters	Robert-Regan	stopped-flow meth.	825
	linear graphical	spectrophotometry	257
Mo + W	meas. at diff. temperature	spectrophotometry	756
Mo + W	on-line computer tech.	stopped-flow meth.	826
Ca + Mg	logarithmic extrapolation	d.c. polarogr.	184
Ca + Mg	logarithmic extrapolation; on-line computer	d.c. polarogr.	827
Zn + Cd	linear extrapolation	extraction-spectro-photometry	828
Cu + Ni + Co	linear extrapolation	stopped-flow meth.	829
La + Nd	graphical extrapolation	spectrophotometry	246

10.3.1.1 Determination of Hydrolases

Choline esterase (CHE). Determination of this enzyme is based on photometric monitoring of the formation of blue indophenol in the reaction [651],

indophenol acetate $\xrightarrow{\text{CHE}}$ indophenol + acetic acid

The rate of hydrolysis of acetyl iodide catalyzed by CHE can be monitored electrochemically [652]. Two cup electrodes are polarized by a constant current of $25\,\mu A$, yielding the dependence given in Fig. 40, where the

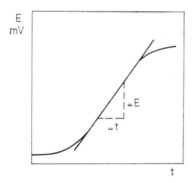

Fig. 40. The ΔE vs Δt dependence obtained for hydrolysis of acetylthiocholine iodide.

$\Delta E/\Delta t$ value is proportional to the CHE concentration. The potential increase is caused by formation of electroactive thiol. The hydrolysis can readily be monitored by measuring the fluorescence. Guilbault and Kramer [653] prepared many fluorogenic substrates, e.g. resorufine butyrate, which do not fluoresce, but on hydrolysis yield intensely fluorescing resorufine. Recently, the use of the choline ester selective electrode was recommended [654].

Cellulase. A determination of cellulase based on the resorufine acetate $\xrightarrow{\text{cellulase}}$ resorufine reaction is advantageous, as the product fluoresces intensely and renders the determination of the enzyme very sensitive; an activity corresponding to 10^{-4} IU can be determined [655].

Chymotrypsine. Determination of chymotrypsine is based on monitoring the rate of the benzoyl-L-tyrosine ethylester hydrolysis [656]; the substrate absorbs light at 256 nm. A more sensitive method uses the hydrolysis of

221

fluorescein dibutyrate producing fluorescein [662]; however, lipase interferes.

Glucosidase. A simple method monitors the rate of hydrolysis of amygdaline, determining the CN^- ions produced potentiometrically with a silver electrode; the dependence obtained [77] corresponds to that in Fig. 40.

β-Glucoronidase. A sensitive determination of this enzyme is based on monitoring the rate of hydrolysis of β-glucoronides of various umbelliferrones [657] with formation of fluorescing umbelliferrone alcohols.

Hyaluronidase. Enzyme hydrolysis of hyaluronic acid is accompanied by variations in a number of physico-chemical parameters, permitting various methods to be used for determining the enzyme. The experimentally simplest and very sensitive method involves monitoring of the hydrolysis of indoxylacetate producing strongly fluorescing indigo white [658]. Fe^{2+} and Cu^{2+} ions may act as inhibitors of this reaction.

Lipase. Many substrates have been proposed for determination of this enzyme, following the hydrolysis photometrically [659]. It is more advantageous to monitor hydrolysis of dibutyryl fluorescein with liberation of fluorescein [75], as the determination is highly selective and rapid. Fluorescence determinations of lipase with various esters of 4-methylumbelliferrone as substrates have also been proposed [660].

Phosphatase. Hydrolysis of p-nitrophenylphosphate catalyzed by phosphatase yields p-nitrophenol that can be monitored spectrophotometrically at 405 nm [661]. More sensitive methods employ substrates whose hydrolysis produces fluorescing compounds, e.g. 4-methylumbelliferrone [662] or flavone-3-diphosphate [663]. From the point of view of selectivity and sensitivity, umbelliferrone phosphate is most suitable, as it is relatively stable and can be used both for acid and for alkaline hydrolysis of phosphoric acid esters; among other enzymes, only β-glucosidase interferes.

Urease. The most convenient determinations of urease use enzyme-selective electrodes, as described in Section 4.4.

10.3.1.2 Determination of Oxidases

Glucodase. Glucose oxidation catalyzed by glucodase yields hydrogen peroxide and thus can readily be monitored polarographically [664] or using an oxygen electrode [665], etc.

The specificity of enzyme catalysis permits indication of the course of the enzyme reaction by following the rate of another reaction, in which the product of the enzyme reaction is a reactant. For example, hydrogen peroxide formed in enzymic oxidation can oxidize a dye, which is colourless in the reduced state, in the presence of another enzyme, peroxidase. An example is the oxidation of 4-hydroxy-3-methoxyphenylacetic acid with an intensely fluorescing reaction product [666].

Peroxidase. Most often, dyes oxidized in the presence of an enzyme are determined; however the potentiometric method can also be used, following the potential of a polarized platinum electrode. The dependence obtained is analogous to that in Fig. 40, the $\Delta E/\Delta t$ value being proportional to the peroxidase concentration [667].

Catalase. This enzyme catalyzes the decomposition of hydrogen peroxide $(2 H_2O_2 \rightarrow 2 H_2O + O_2)$ and thus the oxygen liberated is monitored or concentrations of H_2O_2 are followed spectrophotometrically or electrochemically.

Amino acid oxidases. These enzymes catalyze the decomposition of amino acids with formation of hydrogen peroxide and ammonia. An increase in the H_2O_2 concentration is monitored or NH_3 is measured by an ammonia ion-selective electrode [668].

10.3.1.3 Determination of Dehydrogenases

Dehydrogenases catalyze the oxidation (dehydrogenation) of hydroxy compounds in the presence of coenzymes, such as nicotineamide-adenine dinucleotide (NAD) or nicotineamide-adenine dinucleotide phosphate (NADP). The reduced forms of the coenzymes absorb light at 340 nm and thus can be monitored spectrophotometrically. For example, a determination of clinically very important lactic acid dehydrogenase involves conversion of the acid to pyruvic acid in the presence of the enzyme and NAD. NAD is reduced to fluorescing NADH, the rate of formation of which is proportional to the enzyme concentration [670].

The NAD/NADH and NADP/NADPH systems have been used successfully in the determination of alcohol [671], glycerol [672] and acid dehydrogenases [673].

10.3.1.4 Determination of Other Enzymes

Detailed data concerning these determinations, for which kinetic methods are most important, can be found in Guilbault's book [674] and similar monographs [675].

10.3.2 DETERMINATION OF SUBSTRATES

When using enzymic reactions for determining substrates, constant enzyme concentrations must be employed and the substrate concentration should not exceed the value, $0.2K_M$ (see Section 4.3.1). The determinations are specific and can be used for organic acids, aldehydes, esters, hydroxy compounds, amines, sugars and some inorganic substances (H_2O_2, NH_3, CO_2, NO_3^-, NH_2OH, $P_2O_7^{4-}$ and PO_4^{3-}).

10.3.2.1 Determination of Sugars

Glucose. Great attention has been paid to the determination of glucose, chiefly using glucodase [676] or hexokinase [677] combined with NADP, or peroxidase [678].
Galactose. The most suitable method is the oxidation of the sugar catalyzed by galactose oxidase [679].
Fructose. Three enzymes are used in the determination of fructose. Fructose is converted into fructose-6-phosphate by reaction with adenosine triphosphate in the presence of hexokinase; the product is converted into glucose-6-phosphate by enzyme phosphoglucose isomerase. This substrate is oxidized with formation of NADPH in the presence of dehydrogenase and NADP and the rate of formation of NADPH, which is proportional to the fructose concentration, is monitored fluorimetrically [680].
Saccharose. Saccharose is converted by invertase to glucose and any method for determining glucose can then be used.
Glycogen. Glycogen is a polymer of several glucose units. By hydrolysis, glucose is obtained and any method convenient for the determination of glucose can be used.

A method for analysis of mixtures of sugars has been developed [681]. The mixture is divided into several aliquots, and enzymes specific for one of the sugars present are added to each of them. The glucose formed by hydrolysis is monitored fluorimetrically.

10.3.2.2 Determination of Organic Acids

About fifty organic acids have been determined kinetically, mainly using dehydrogenases. For details see the monographs [674, 675].

10.3.2.3 Determination of Amino Acids

D-amino acid oxidase catalyzes the oxidation of D-amino acids (tyrosine, proline, alanine, valine, tryptophan, serine, leucine, histidine, etc.), with formation of H_2O_2 [682], which is monitored fluorimetrically [683]. The enzyme isolated from pig kidneys is highly selective, so that L-amino acids do not interfere. L-amino acid oxidase, isolated from snake poison, selectively catalyzes the oxidation of L-amino acids and the H_2O_2 formed is monitored [683].

L-amino acid decarboxylases can also be used in the determination of L-amino acids. The corresponding amine is formed and is monitored using another enzyme reaction [684].

10.3.2.4 Determination of Hydroxycompounds

Alcohols. Alcohols are oxidized by dehydrogenases in the presence of NAD. NADH formed can be monitored fluorimetrically [685]. A differential kinetic method has been developed for analyses of mixtures of alcohols [686]. Another method employs oxidation of alcohols catalyzed by oxidases, monitoring the H_2O_2 formed by its reaction with o-dianisidine catalyzed by peroxidase [687].

Phenols. Phenol oxidase catalyzed reactions are used for determination of phenols [688].

Steroid alcohols. These determinations are important in clinical analysis and are based on oxidation catalyzed by hydroxysteroid oxidase [689].

10.3.2.5 Determination of Inorganic Substances as Substrates

Determination of hydrogen peroxide based on peroxidase catalyzed reactions is discussed in Section 10.3.1.2. This determination is important in enzyme analysis, because it is often used as an auxiliary reaction in a more complex enzyme system. Determinations of other inorganic substances $(NH_4^+, NO_3^-, NH_2OH, CO_2)$ are possible, but are of little importance.

10.3.3 DETERMINATION OF ACTIVATORS

The Ba^{2+}, Ca^{2+}, Co^{2+}, K^+, Mg^{2+}, Mn^{2+}, Sr^{2+}, Zn^{2+}, I^-, CN^- and S^{2-} ions are inorganic activators. The principles of their determination are described in Section 4.2.1. The greatest attention has been paid to determination of magnesium ions [690], which activate isocitric acid dehydrogenase. Hence the rate of oxidation of isocitric acid at constant enzyme and NADP concentrations is proportional to the Mg^{2+} concentration from 10^{-6} to 10^{-4} mol.l^{-1}. NADPH is monitored during the process. The

TABLE 16

Substances that can be determined using their inhibitive effect on enzymic reactions

Substance	Enzyme inhibited	Ref.
Ag^+	glucodase	692
	urease	693
	isocitric acid dehydrogenase	578
Al^{3+}	isocitric acid dehydrogenase	578
Be^{2+}	phosphatase	473
Bi^{3+}	phosphatase	473
Ce^{3+}, Cd^{2+}	isocitric acid dehydrogenase	578
Cd^{2+}	urease	693
	peroxidase	678
Co^{2+}	peroxidase	678
	urease	693
Cu^{2+}	isocitric acid dehydrogenase	578
	urease	693
Fe^{2+}, Fe^{3+}	peroxidase	678
Hg^{2+}	glucodase	692
In^{3+}	isocitric acid dehydrogenase	578
Mn^{2+}	urease	693
Ni^{2+}, Pb^{2+}	urease	693
Pb^{2+}	peroxidase	678
Zn^{2+}	urease	693
$Cr_2O_7^{2-}$	peroxidase	678
S^{2-}	peroxidase	678
NH_2OH	peroxidase	678
ascorbic acid	catalase	694

Mn^{2+}, Cu^{2+}, Ca^{2+}, Zn^{2+} and Co^{2+} ions behave similarly. Determination of barium based on reactivation of alkaline phosphatase is more selective and can be carried out in the presence of magnesium and calcium [691]. Although determinations of activators are highly sensitive, they have not yet found wide use in inorganic analysis.

Methods for determination of coenzymes (cofactors) are more important, mainly in biochemistry. Common coenzymes, e.g. NAD and NADP, are necessary for activation of some dehydrogenases. Methods for their determination are comparable to those for inorganic activators.

10.3.4 DETERMINATION OF INHIBITORS

Great attention has been paid to the determination of inhibitors, because these determinations are simple and sensitive down to tenths of µg per ml. Table 16 lists substances inhibiting enzyme activity for which kinetic analytical methods have been developed.

REFERENCES

1. Sandell E. B. and Kolthoff I. M.: *Microchim. Acta* **1**, 9 (1937).
2. Yatsimirskii K. B.: Kincticheskiye Metody Analiza, Gozkhimizdat, Moscow 1963, 2nd Ed.: Khimiya, Moscow 1967.
3. Yatsimirskii K. B.: Kinetic Methods of Analysis, Pergamon Press, Oxford 1966.
4 Mark H. B., Jr. and Rechnitz G. A.: Kinetics in Analytical Chemistry, Interscience, New York 1968.
5. Rechnitz G. A.: *Anal. Chem.* **36**, 453R (1964).
6. Idem ibid.: **38**, 513R (1966).
7. Idem ibid.: **40**, 455R (1968).
8. Guilbault G. G.: *Anal. Chem.* **42**, 334R (1970).
9. Greinke R. A. and Mark H. B., Jr.: *Anal. Chem.* **44**, 295R (1972).
10. Idem ibid.: **46**, 413R (1974).
11. Idem ibid.: **48**, 87R (1976).
12. Idem ibid.: **50**, 70R (1978).
13. Mottola H. A. and Mark, H. B., Jr.: *Anal. Chem.* **52**, 31R (1980).
14. Svehla G.: *Analyst* **94**, 513 (1969).
15. Bontchev R. R.: *Talanta* **17**, 499 (1970).
16. Svehla G.: *Selected Annual Reviews of the Analytical Sciences,* **1**, 235 (1971).
17. Bontchev P. R.: *Talanta* **19**, 675 (1972).
18. Mark H. B., Jr.: *Talanta* **19**, 717 (1972).

19. Malmstadt H. V., Cordos E. A. and Delaney C. J.: *Anal. Chem.* **44**, 26A, 79 A (1972).
20. Gary A. M. and Schwing J. P.: *Bull. Soc. Chim. Fr.* 1972, 3657.
21. Mottola H. A.: *Crit. Rev. Anal. Chem.* **4**, 229 (1975).
22. Weisz H.: *Angew. Chem. Intl. Ed. Engl.* **15**, 150 (1976).
23. Guilbault G. G. in Kolthoff I. M. and Elving P. J. (Editors): Treatise of Analytical Chemistry, Vol. 1. 2nd Ed., Wiley, 1978, p. 663.
24. Fritz J. and Schenk G.: Quantitative Analytical Chemistry, Allyn and Bacon, Boston, 1966.
25. Pecsok R. L. and Schields L. D.: Modern Methods of Chemical Analysis, J. Wiley, New York, 1968.
26. Laitinen H. A. and Harris W. E.: Chemical Analysis, 2nd Ed., McGraw-Hill, New York, 1975, p. 382.
27. Benson S. W.: The Foundations of Chemical Kinetics, McGraw-Hill, New York, 1960.
28. Frost A. A. and Pearson R. G.: Kinetics and Mechanisms, 2nd Ed., J. Wiley, New York, 1961, p. 11.
29. Laidler K. J.: Chemical Kinetics, 2nd. Ed., McGraw-Hill, New York, 1965.
30. Ashmore P. G.: Catalysis and Inhibition of Chemical Reactions, Chap. 1., Butterworths, London 1963.
31. Waters W. A.: Mechanism of Oxidation of Organic Compounds, Methuen, London 1964.
32. Engelsma G.: *Tetrahedron* **2**, 289 (1958).
33. Bontchev P. R., Eliazkova G. B.: *Microchim. Acta* 1967, 116.
34. Dolmanova I. F., Peshkova V. M.: *Zh. Anal. Khim.* **19**, 297 (1964).
35. Kreingold S. U., Bozhevolnov E. A., Sinyaver L. G.: *Zavod. lab.* **31**, 508 (1965).
36. Basolo F., Pearson R. G.: Mechanism of Inorganic Reactions, 2nd Ed., J. Wiley, New York 1967.
37. Rodriguez P. A., Pardue H. L.: *Anal. Chem.* **41**, 1369 (1969).
38. Yatsimirskii K. B., Karatcheva G. A.: *Th. Neorg. Khim.* **3**, 2434 (1958).
39. Bontchev P. R., Yatsimirskii K. B.: *Teor. Eksp. Khim.* **1**, 179 (1965).
40. Bontchev P. R., Nikolov G., Lilova B.: *Ann. Univ. Sofia (Khimiya)* **59**, 87 (1964/65).
41. Yatsimirskii K. B.: *Kinet. Katal.* (SSSR) **4**, 931 (1965).
42. Bontchev P. R., Alexiyev A. A.: *J. Inorg. Nucl. Chem.* **32**, 2237 (1970).
43. Bontchev P. R., Alexiyev A. A., Dimitrova B.: *Mikrochim. Acta* 1970, 1104.
44. Yasinskene E. I., Rasevichiyute N. I.: *Zh. Anal. Khim.* **25**, 458 (1970).
45. Bontchev P. R., Alexiyev A. A., Dimitrova B.: *Talanta* **16**, 597 (1969).
46. Landolt H.: *Ber. Deut. Chem. Ges.* **19**, 1317 (1886).
47. Nikitin E. K.: *Izv. Sektora Fiz.-Khim. Analiza* **13**, 75 (1940).
48. Johnson C. E., Winstein S.: *J. Amer. Chem. Soc.* **74**, 755 (1952).
49. Dahl W. E., Pardue H. L.: *Anal. Chem.* **37**, 1382 (1965).
50. Müller H. and Otto M.: *Z. Chem.* **14**, 159 (1974).
51. Zhabotinskii A. M.: Trudy Vsesoyuz. Simpoziyuma po Kolebatelnym Processam

v Biolog. i Khim. Sistemakh. p. 149, Nauka, Moscow 1967; Nicolis G. and Porthow J.: *Chem. Reviews* **73**, 365 (1973).

52. Zhabotinskii A. M.: *Dokl. Akad. Nauk SSSR* **157**, 392 (1964).
53. Vavilin V. A., Zhabotinskii A. M., Yaguzhinskii L. S.: Trudy Vsesoyuz. Simpoziyuma po Kolebatelnym Processam v Biolog. i Khim. Sistemakh. p. 181, Nauka, Moscow 1967.
54. Vavilin V. A., Gulak P. V., Zhabotinskii A. M., Zayikin A. N.: *Izv. Akad. Nauk SSSR, Ser. Khim.* 1969, 2618.
55. Zhabotinskii A. M.: *Zh. Anal. Khim.* **27**, 437 (1972).
56. Tikhonova L. P., Zakrevskaya L. N., Yatsimirskii K. B.: *Zh. Anal. Khim.* **33**, 191 (1978).
57. Müller H., Werner G.: *Z. für Chemie* **16**, 304 (1976).
58. Otto M., Müller H., Werner G.: *Talanta* **25**, 123 (1978).
59. Yatsimirskii K. B., Raizman L. P.: *Zh. Anal. Khim.* **18**, 29 (1963).
60. Bozhevolnov E. A., Kreingold S. U., Lastovskii R. P., Sidorenko V. V.: *Dokl. Akad. Nauk SSSR* **153**, 97 (1963).
61. Worthington I. B., Pardue H. L.: *Anal. Chem.* **42**, 1157 (1970).
62. Wolff C. M., Schwing J. P.: *Bull. Soc. Chim. France* 1976, 679.
63. Dolmanova I. F., Zolotova G. A., Ushakova N. M., Tchernyavskaya T. N., Peshkova V. M.: Uspekhi Analiticheskoi Khimii, Nauka, Moscow, 1974.
64. Müller H., Tikhonova L. P., Yatsimirskii K. B., Borkovec S.N.: *Zh. Anal. Khim.* **28**, 2012 (1973).
65. Lukovskaya N. M., Kushchevskaya N. F.: *Ukr. Khim. Zh.* **42**, 87 (1976).
66. Otto M., Müller H.: *Talanta* **24**, 15 (1977).
67. Müller H., Schurig H., Werner G.: *Talanta* **21**, 581 (1974).
68. Antonov V. N., Kreingold S. U.: *Zh. Anal. Khim.* **31**, 193 (1976).
69. Otto M., Müller H.: *Anal. Chim. Acta* **90**, 159 (1977).
70. Otto M., Bontchev P. R., Müller H.: *Microchim. Acta* 1977, I 193.
71. Michaelis L., Menten M. L.: *Biochem. Z.* **49**, 333 (1913).
72. Briggs G. E., Haldane J. B. S.: *Biochem. J.* **19**, 338 (1925).
73. Lineweaver H., Burk D. J.: *J. Am. Chem. Soc.* **56**, 658 (1934).
74. Baum P., Czok R.: *Biochem. Z.* **332**, 121 (1959).
75. Guilbault G. G., Kramer D. N.: *Anal. Chem.* **36**, 409 (1964).
76. Kratochvil B., Boyer S. L., Hicks G. P.: *Anal. Chem.* **39**, 45 (1967).
77. Guilbault G. G., Kramer D. N.: *Anal. Biochem.* **18**, 313 (1967).
78. Bowers L. D., Carr P. W.: *Anal. Chem.* **48**, 544A (1976).
79. Weetall H.H.: Immobilized Enzymes, Antibodies and Peptides. M. Dekker, New York 1975.
80. Goldman R., Katchalski E.: *Theor. Biol.* **32**, 243 (1971).
81. Guilbault G. G., Bauman E. K., Kramer D. N., Goodson L. H.: *Anal. Chem.* **37**, 1378 (1965).
82. Guilbault G. G., Kramer D. N.: *Anal. Chem.* **37**, 1675 (1965).
83. Bernfeld P., Wan J.: *Science* **142**, 678 (1963).
84. Hicks G. P., Updike S. J.: *Anal. Chem.* **38**, 726 (1966).

85. Updike S. J., Hicks G. P.: *Science* **158**, 270 (1967).
86. Updike S. J., Hicks G. P.: *Nature* **214**, 986 (1967).
87. Rechnitz G. A.: *Chem. and Eng. News* **53**, (4) 29 (1975).
88. Gray D. N., Keyes M. H., Watson B.: *Anal. Chem.* **49**, 1067A (1977).
89. Weetall H. H.: *Anal. Chem.* **46**, 602A (1974).
90. Lilly M. D., Hornby W. E.: *Biochem. J.* **100**, 718 (1966).
91. Sharp A. K., Kay G., Lilly M. D.: *Biotechnol. Bioeng.* **11**, 363 (1969).
92. Blaedel W. J., Kissel T. K., Boguslavski R. C.: *Anal. Chem.* **44**, 2030 (1972).
93. Treybal R. E.: Mass Transfer Operations, 2nd Ed. McGraw-Hill, New York 1968.
94. Llenado R. A., Rechnitz G. A.: *Anal. Chem.* **43**, 1457 (1971).
95. Siggia S.: Quantitative Organic Analysis via Functional Groups, p. 18. J. Wiley, New York 1949.
96. Blaedel W. J., Petitjean D. L.: *Anal. Chem.* **30**, 1958 (1958).
97. Burgess A. E., Latham J. L.: *Analyst* (London) **91**, 343 (1966).
98. Welle R. P.: *Chem. Rev.* **63**, 171 (1963).
99. Hammett L. P.: Physical Organic Chemistry. McGraw-Hill, New York 1940.
100. Roberts J. D., Regan C.: *Anal. Chem.* **24**, 360 (1952).
101. Wells R. P.: Linear Free Energy Relationship, Interscience, New York 1968.
102. Hammett L. P.: *J. Amer. Chem. Soc.* **59**, 96 (1937).
103. Jaffé H. H.: *Chem. Rev.* **53**, 191 (1953).
104. Taft R. W. in Newman M. S. (Ed): Steric Effects in Organic Chemistry, Chap.13. J. Wiley, New York 1956.
105. Taft R. W.: *J. Amer. Chem. Soc.* **74**, 3120 (1950).
106. Elhafy A., Cram D.: *J. Amer. Chem. Soc.* **75**, 339 (1953).
107. Michaelis L.: *Trans. Electrochem. Soc.* **71**, 107 (1937).
108. Taube H.: *Can. J. Chem.* **37**, 129 (1959).
109. Marcus R. A.: *J. Phys. Chem.* **67**, 853 (1963).
110. Murina A. N., Nesedov Z. D., Svedov V. P.: Radiokhimiya i Khimiya Yadernykh Processov, Goskhimizdat, Leningrad 1960.
111. Marcus R. A.: *Ann. Rev. Phys. Chem.* **15**, 155 (1964).
112. Edicott J. F., Taube H.: *J. Amer. Chem. Soc.* **86**, 1686 (1964).
113. Campion R. J., Purdie N., Sutin N.: *Inorg. Chem.* **3**, 1091 (1964).
114. Candlin J. P., Halpern J.: *Inorg. Chem.* **4**, 766 (1965).
115. Candlin J. P., Halpern J., Trimm D. L.: *J. Amer. Chem. Soc.* **86**, 1019 (1964).
116. Vlček A. A. in Progress in Inorganic Chemistry, vol. V, p. 211. Interscience, New York 1963.
117. Eigen M.: *Angew. Chem., Int. Ed. Engl.* **3**, 1, (1964).
118. Caldin E. F.: Fast Reactions in Solutions. J. Wiley, New York 1964.
119. Hartridge H., Roughton F. J.: *Proc. Roy. Soc.* (London) **A 104**, 376 (1923).
120. Sanderson D., Bittikofer J. A., Pardue H. L.: *Anal. Chem.* **44**, 1934 (1972).
121. Javier A. C., Crouch S. R., Malmstadt H. V.: *Anal. Chem.* **41**, 239 (1969).
122. Chance B.: *Discuss. Faraday Soc.* **17**, 120 (1954).
123. Eigen M.: *Discuss. Faraday Soc.* **17**, 194 (1954).
124. Eigen M.: *Z. Phys. Chem.* (Frankfurt am Main) **1**, 176 (1954).

125. Hart E. J.: *Science* **146**, 19 (1964).
126. Matheson M. S., Rabani J.: *J. Phys. Chem.* **69**, 1324 (1965).
127. Hart E. J., Gordon S., Fielden E. M.: *J. Phys. Chem.* **70**, 150 (1966).
128. Gordon S., Hart E. J., Matheson M. S., Rabani J., Thomas J. K.: *Discuss. Faraday Soc.* **36**, 193 (1963).
129. Thomas J. K., Gordon S., Hart E. J.: *J. Phys. Chem.* **68**, 1524 (1964).
130. Anbar M., Hart E. J.: *J. Phys. Chem.* **69**, 973 (1965).
131. Hart E. J., Gordon S., Thomas J. K.: *J. Phys. Chem.* **68**, 1271 (1964).
132. Asmus K. D., Henglein A.: *Ber. Bunsenges. Phys. Chem.* **68**, 348 (1964).
133. Hart E. J., Fielden E. M.: *Advan. Chem. Ser.* **50**, 253 (1965).
134. Taube H.: Electron Transfer Reactions of Complex Ions in Solutions, Chap. 2, Academic Press, New York 1970.
135. Zwickel A. M., Taube H.: *J. Amer. Chem. Soc.* **83**, 793 (1961).
136. Zwickel A. M., Taube H.: *J. Amer. Chem. Soc.* **81**, 1288 (1959).
137. Eigen M., Wilkins R. G.: *Advan. Chem. Ser.* **49**, 55 (1965).
138. Margerum D. W., Rosen H. M.: *J. Amer. Chem. Soc.* **89**, 1088 (1967).
139. Marks S., Dodgen H. W., Hunt J. P.: *Inorg. Chem.* **7**, 836 (1968).
140. Elving P. J., Zemel B.: *J. Amer. Chem. Soc.* **79**, 5855 (1967).
141. Espenson J. H., Birk J. P.: *Inorg. Chem.* **4**, 527 (1965).
142. Asperger S., Pavlovič D.: *Anal. Chem.* **28**, 1761 (1956).
143. Yatsimirskii K. B., Orlova G. N.: *Zh. Neorg. Khim.* **5**, 2128 (1960).
144. Kraljic I.: *Microchim. Acta* 1960, 586.
145. Kara V., Pinter T.: *Croat. Chem. Acta* **30**, 141 (1958).
146. Margerum D. W.: *Rec. Chem. Progr.* **24**, 237 (1963).
147. Jones J. P.: *Inorg. Chem.* **8**, 4486 (1969).
148. Margerum D. W., Carr J. D.: *J. Amer. Chem. Soc.* **88**, 1639 (1966).
149. Margerum D. W., Bydalek T. J.: *Inorg. Chem.* **1**, 852 (1962).
150. Zabin B. A.: PhD Thesis. Purdue University 1962.
151. Olson D. C., Margerum D. W.: *J. Amer. Chem. Soc.* **85**, 297 (1963).
152. Margerum D. W., Steinhaus K. R.: *Anal. Chem.* **37**, 222 (1965).
153. Carr J. D., Margerum D. W.: *J. Amer. Chem. Soc.* **88**, 1645 (1966).
154. Stará V., Kopanica M.: *Collect. Czech. Chem. Commun.* **38**, 2581 (1973).
155. Toropova V. F., Polyanskaya A. A.: *Issled. Elektrokhim. Magnetokhim. Method Anal.* No. 3, 28 (1970), *Ref. Zh. Khim.* 1970, Abstr. No. 22 V, 170.
156. Stehl R. H., Margerum D. W., Latterell J. J.: *Anal. Chem.* **39**, 1346 (1967).
157. Margerum D. W., Stehl R. H.: *Anal. Chem.* **39**, 1351 (1967).
158. Heim A., Wilmarth H. K.: *J. Amer. Chem. Soc.* **83**, 509 (1961).
159. Harju L., Ringbom A.: *Anal. Chim. Acta* **49**, 205, 221 (1970).
160. Harju L.: *Anal. Chim. Acta* **50**, 475 (1970).
161. Přibil R. et al.: *Talanta* **13**, 1711 (1966); **14**, 266, 313 (1967).
162. Stará V., Kopanica M.: *Collect. Czech. Chem. Commun.* **37**, 2882 (1972).
163. Stará V., Kopanica M.: *Collect. Czech. Chem. Commun.* **37**, 3545 (1972).
164. Haque M. S., Kopanica M.: *Bull. Chem. Soc. Japan* **46**, 3032 (1973).
165. Kopanica M., Stará V.: *Collect. Czech. Chem. Commun.* **37**, 80 (1972).

166. Stará V., Kopanica M.: *Collect. Czech. Chem. Commun.* **38**, 258 (1973).
167. Mottola H. A.: *Anal. Chim Acta* **71**, 443 (1974).
168. Weisz H., Pantel S.: *Anal. Chim. Acta* **62**, 361 (1972).
169. Mottola H. A.: *Talanta* **16**, 1267 (1969).
170. Mottola H. A., Heath G. L.: *Anal. Chem.* **44**, 2322 (1972).
171. Eswara Dutt V. V. S., Mottola H. A.: *Biochem. Med.* **9**, 148 (1974).
172. Margerum D. W., Bydalek J. T.: *Inorg. Chem.* **2**, 683 (1963).
173. Janes D. L., Margerum D. W.: *Inorg. Chem.* **5**, 1135 (1966).
174. Margerum D. W., Medardi P. J., Janes D. L.: *Inorg. Chem.* **6**, 283 (1967).
175. Pausch J. B., Margerum D. W.: *Anal. Chem.* **41**, 226 (1969).
176. Margerum D. W., Pausch J. B., Nyssen G. A., Smith G. F.: *Anal. Chem.* **41**, 233 (1969).
177. Saltzman B. E.: *Anal. Chem.* **31**, 1914 (1959).
178. Siggia S., Hanna J. G.: *Anal. Chem.* **33**, 896 (1961).
179. Kopanica M., Stará V.: *Collect. Czech. Chem. Commun.* **41**, 3275 (1976).
180. Kloosterboer J. G.: *Anal. Chem.* **46**, 1146 (1974).
181. Přibil R.: Analytical Application of EDTA and Related Compounds, Pergamon Press, Oxford 1972.
182. Garmon R. G., Reilley C. N.: *Anal. Chem.* **34**, 600 (1962).
183. Yatsimirskii K. B., Khachatryan A. G., Budarin L. I.: *Dokl. Akad. Nauk SSSR* **211**, 1139 (1973).
184. Gary A. M., Schwing J. P.: *Bull. Soc. Chim. France* 1975, 441.
185. Koutecký J.: *Collect. Czech. Chem. Commun.* **18**, 184 (1953).
186. Brdička R., Hanuš V., Koutecký J.: Progress in Polarography, vol. 3, part 2, p. 145. J. Wiley, New York 1962.
187. Shuman M. S., Shain I.: *Anal. Chem.* **41**, 1818 (1969).
188. Koryta J., Zabranský Z.: *Collect. Czech. Chem. Commun.* **25**, 3153 (1960).
189. Conradi G., Kopanica M., Koryta J.: *Collect.Czech.Chem.Commun.* **30**, 2029 (1965).
190. Koutecký J.: *Collect. Czech. Chem. Commun.* **18**, 311 (1953).
191. Koryta J.: *Z. Elektrochem.* **64**, 23 (1960).
192. Koutecký J.: *Collect. Czech. Chem. Commun.* **20**, 116 (1955).
193. Nürnberg H. W.: Polarography 1964, vol. 1, p. 149. MacMillan, London 1966.
194. Kopanica M., Stará V.: *Chem. Listy* **68**, 525 (1974); *Talanta* **21**, 1073 (1974).
195. Tanaka N., Ebata K.: *J. Electroanal. Chem.* **8**, 120 (1964).
196. Brdička R., Wiesner K.: *Collect. Czech. Chem. Commun.* **12**, 39 (1947).
197. Kapulla H., Berg H.: *J. Electroanal. Chem.* **1**, 108 (1960).
198. Sinyakova S. I., Milyavskii J. S.: *Zavod. Lab.* **37**, 1153 (1971).
199. Kolthoff I. M., Parry E. P.: *J. Amer. Chem. Soc.* **73**, 5315 (1951).
200. Tanaka N., Ito T.: *Bull. Chem. Soc. Jap.* **39**, 1043 (1966).
201. Shapirov R. K., Songina O. A.: *Zh. Anal. Khim.* **20**, 683 (1965).
202. Koryta J., Koutecký J.: *Collect. Czech. Chem. Commun.* **20**, 423 (1955).
203. Kara U., Pinter T.: *Croat. chem. Acta* **30**, 141 (1958).
204. Mark H. B., Jr.: *J. Electroanal. Chem.* **7**, 276 (1964).
205. Mark H. B., Jr.: *Anal. Chem.* **36**, 940 (1964).

206. McCoy L. R., Mark H. B., Jr.: *Anal. Chem.* **37**, 591 (1965).
207. Chaya J., McDonald H. C., Kirova-Eisner E., McCoy L. R., Mark H. B., Jr.: *J. Phys. Chem.* **76**, 1170 (1972).
208. Mark H. B., Jr., Reilley C. N.: *J. Electroanal. Chem.* **4**, 189 (1962).
209. Mark H. B., Jr., Reilley C. N.: *Anal. Chem.* **35**, 195 (1963).
210. Engel A. J., Lawson J., Aikens D. A.: *Anal. Chem.* **37**, 203 (1965).
211. Kůta J.: private communication.
212. Frumkin A. N., Andreeva E. P.: *Dokl. Akad. Nauk SSSR* **40**, 417 (1953).
213. Brdička R.: *Collect. Czech. Chem. Commun.* **1**, 112, 148 (1933).
214. Brdička R., Březina N., Kalous V.: *Talanta* **12**, 1149 (1965).
215. Poluektov N. S.: *Zh. Prikl. Khim.* **14**, 695 (1941).
216. Yatsimirskii K. B.: *Izv. Vyssh. Ucheb. Zaved., Khim. Khim. Tekhnol.* **4**, 315 (1961).
217. Papa L. J., Patterson J. H., Mark H. B., Jr., Reilley C. N.: *Anal. Chem.* **35**, 1889 (1963).
218. Yatsimirskii K. B.: Kineticheskiye Metody Analiza, 2nd. Ed., p. 65, Khimiya, Moscow 1967.
219. Bognár J., Sárosi S.: *Anal. Chim. Acta* **29**, 406 (1963).
220. Bognár J., Jellinek O.: *Microchim Acta* 1964, 317 and 1129.
221. Bognár J., Sárosi S.: *Microchim Acta* 1967, 813.
222. Mottola H. A.: *Talanta* **16**, 1267 (1969).
223. Weisz H., Pantel S.: *Anal. Chim. Acta* **62**, 361 (1972).
224. Yatsimirskii K. B., Fedorova T. I.: *Dokl. Akad. Nauk SSSR* **143**, 143 (1962).
225. Bonting S. C.: *Arch. Biochem. Biophys.* **52**, 272 (1954).
226. Tanaka M., Funahashi S., Shirai K.: *Anal. Chim. Acta* **39**, 437 (1967).
227. Garmon R. G., Reilley C. N.: *Anal. Chem* **34**, 600 (1962).
228. Papa L. J., Mark H. B., Jr., Reilley C. N.: *Anal. Chem.* **34**, 1443 (1962).
229. Siggia S., Hanna J. G.: *Anal. Chem.* **33**, 896 (1961).
230. Papa L. J., Mark H. B., Jr., Reilley C. N.: *Anal. Chem* **34**, 1513 (1962).
231. Siggia S., Hanna J. G., Serencha N. M.: *Anal. Chem.* **35**, 362, 365, 575 (1963); **36**, 638 (1964).
232. Reilley C. N., Papa J. J.: *Anal. Chem.* **34**, 801 (1962).
233. Lee T. S., Kolthoff I. M.: *Ann. N. Y. Acad. Sci.* **53**, 1093 (1951).
234. Schmalz E. O., Geisler G.: *Z. Anal. Chem.* **188**, 241, 253 (1962); **190**, 222, 233 (1962).
235. Roberts J. D., Regan C.: *Anal. Chem.* **24**, 360 (1952).
236. Greinke R. A., Mark H. B., Jr.: *Anal. Chem.* **38**, 340 (1966).
237. Mark H. B., Jr., Backes L. M., Pinkel D.: *Talanta* **12**, 27 (1965).
238. Mark H. B., Jr.: *Anal. Chem.* **36**, 1668 (1964).
239. Mark H. B., Jr., Papa L. J., Reilley C. N.: Advances in Analytical Chemistry and Instrumentation, vol. 2, p. 255, Interscience, New York 1963.
240. Papa L. J., Mark H. B., Jr., Reilley C. N.: *Anal. Chem.* **34**, 1443 (1962).
241. Greinke R. A., Mark H. B., Jr.: *Anal. Chem.* **39**, 1577 (1967).
242. Willeboordse F., Critchfield F. E.: *Anal. Chem.* **36**, 2270 (1964).

243. Willeboordse F., Meeker R. L.: *Anal. Chem.* **38**, 854 (1966).
244. Malmstadt H. V., Cordos E. A., Delaney C. J.: *Anal. Chem.* **44**, 26A (1972).
245. Pinkel D., Mark H. B., Jr.: *Talanta* **12**, 491 (1965).
246. Worthington J. B., Pardue H. L.: *Anal. Chem.* **44**, 767 (1972).
247. Yatsimirskii K. B., Khachatryan A. G., Budarin L. I.: *Dokl. Akad. Nauk SSSR* **211**, 1139 (1973).
248. Ingle J. D., Crouch S. R.: *Anal. Chem.* **43**, 697 (1971).
249. Weisz H., Ludwig G.: *Anal. Chim. Acta* **55**, 303 (1971).
250. Bognar J.: *Microchim. Acta* 1963, 397.
251. Weisz H.: *Angew. Chem.* **88**, 177 (1976).
252. Weisz H., Rothmaier K.: *Anal. Chim. Acta* **82**, 155 (1976).
253. Klockow D., Weisz H., Rothmaier K.: *Z. Anal. Chem.* **264**, 385 (1973).
254. Weisz H., Ludwig H.: *Anal. Chim. Acta* **60**, 385 (1972).
255. Wilson A. M.: *Anal. Chem.* **38**, 1784 (1966).
256. Weisz H., Ludwig H.: *Anal. Chim. Acta* **62**, 125 (1972).
257. Connors K. A.: *Anal. Chem.* **48**, 87 (1976).
258. Connors K. A.: *Anal. Chem.* **49**, 1650 (1977).
259. Carr P. W.: *Anal. Chem.* **50**, 1602 (1978).
260. Mieling G. E., Pardue H. L.: *Anal. Chem.* **50**, 1611 (1978).
261. Loach P. A., Lloyd R. J.: *Anal. Chem.* **38**, 1709 (1966).
262. Vaughan A., Guilbault G. G., Hackney D.: *Anal. Chem.* **43**, 721 (1971).
263. Weisz H., Rothmaier K.: *Anal. Chim. Acta* **68**, 93 (1974).
264. Ludwig H., Weisz H., Lenz T.: *Anal. Chim. Acta* **70**, 359. (1974).
265. Rechnitz G. A., Lin Zvi-Feng: *Anal. Chem.* **39**, 1406 (1967).
266. Malmstadt H. V., Piepmeier E. M.: *Anal. Chem.* **37**, 34 (1965).
267. Charlton G., Dread D., Reed J.: *J. Appl. Physiol.* **18**, 1247 (1963).
268. Thompson H., Rechnitz G. A.: *Anal. Chem.* **46**, 246 (1974).
269. Guilbault G. G., Montalvo J.: *J. Amer. Chem. Soc.* **91**, 2164 (1969).
270. Sand J. B., Huber C. O.: *Anal. Chem.* **42**, 238 (1970).
271. Weisz H., Klockow D., Ludwig H.: *Talanta* **16**, 921 (1969).
272. Simon R. K., Christian G. D., Purdy W. C.: *Clin. Chem.* **14**, 463 (1968).
273. Norris J. F., Ware V. W.: *J. Amer. Chem. Soc.* **61**, 1418 (1939).
274. Johnson D. E., Enke C. G.: *Anal. Chem.* **42**, 329 (1970).
275. Pantel S., Weisz H.: *Anal. Chim. Acta* **68**, 311 (1974).
276. Blaedel W. J., Hicks G. P. in Reilley C. N. (Editor): Advances in Analytical Chemistry and Instrumentation, vol. 3, p. 105. Interscience, New York 1969.
277. Crouch S. R. in Mattson J. S., Mark H. B., Jr. and MacDonald H. C. (Editors): Computers in Chemistry and Instrumentation, vol. 3, p. 107. Dekker, New York 1973.
278. Kalvoda R.: The Use of Operational Amplifiers in Chemical Instrumentation (in Czech), SNTL, Prague 1974.
279. Cordons E. M., Crouch S. R., Malmstadt H. V.: *Anal. Chem.* **40**, 1812 (1968).
280. Ingle J. D., Jr., Crouch S. R.: *Anal. Chem.* **42**, 1055 (1970).
281. Pardue H. L., Frings C. S., Delangey C. J.: *Anal. Chem.* **37**, 1426 (1965).

282. Pardue H. L., Parker R. A., Willis G. B.: *Anal. Chem.* **42**, 56 (1970).
283. Pinkel D., Mark H. B., Jr.: *Talanta* **12**, 491 (1965).
284. James G. E., Pardue H. L.: *Anal. Chem.* **41**, 1618 (1969).
285. Hicks G. P., Eggert A. A., Toren E. C., Jr.: *Anal. Chem.* **42**, 729 (1970).
286. Willis B. G., Bittikofer J. A., Pardue H. L., Margerum D. W.: *Anal. Chem.* **42**, 1340 (1970).
287. Willis B. G., Woodruff W. H., Frysinger J. F., Margerum D. W., Pardue H. L.: *Anal. Chem.* **42**, 1350 (1970).
288. Janata J.: ref. 277, p. 209.
289. Johnson D. W., Callis J. G., Christian G. D.: *Anal. Chem.* **49**, 747A (1977).
290. Thomas L. C., Christian G. D.: *Anal. Chim. Acta* **89**, 83 (1977).
291. Wightman R. M., Scott R. L., Reilley C. N., Murray R. W., Burnet J. N.: *Anal. Chem.* **46**, 1492 (1974).
292. Nieman T. A., Enke C. G.: *Anal. Chem.* **48**, 619 (1976).
293. Ridder G. M., Margerum D. W.: *Anal. Chem.* **49**, 2098 (1977).
294. Clark L., Lyons A.: *Ann. N.Y. Acad. Sci.* **102**, 29 (1962).
295. Guilbault G. G., Hrabánková E.: *Anal. Chim. Acta* **56**, 285 (1971).
296. Guilbault G. G., Nagy G., Kvan S. K.: *Anal. Chim. Acta* **67**, 195 (1973).
297. Anfält T., Granéli A., Jagner D.: *Anal. Lett.* **6**, 969 (1973).
298. Mell L. D., Maloy J. T.: *Anal. Chem.* **47**, 299 (1975).
299. Hansen E. H., Růžička J.: *Anal. Chim. Acta* **72**, 353 (1974).
300. Larsen N. R., Hansen E. H., Guilbault G. G.: *Anal. Chim. Acta* **79**, 9 (1975).
301. Johansson G., Ögren L.: *Anal. Chim. Acta* **84**, 23 (1976).
302. Johansson G., Edström K., Ögren L.: *Anal. Chim. Acta* **85**, 55 (1976).
303. Jordan J., Grime J. K., Waugh D. M., Miller C. D., Cullis H. M., Lohr D.: *Anal. Chem.* **48**, 427A (1976).
304. Adams R. E., Betso S. R., Carr P. W.: *Anal. Chem.* **48**, 1989 (1976).
305. Weisz H., Rothmaier K.: *Anal. Chim. Acta* **75**, 119 (1975).
306. Pantel S., Weisz H.: *Anal. Chim. Acta* **70**, 391 (1974); **89**, 47 (1977)
307. Pantel S., Weisz H.: *Anal. Chim. Acta* **74**, 275 (1975).
308. Lazarou L. A., Hadjiioannou T. P.: *Anal. Chem.* **51**, 790 (1979).
309. Efstathiou G. H., Hadjiioannou T. P.: *Anal. Chem.* **47**, 864 (1975).
310. Murray R. W., Heineman W. R., O'Dom G. W.: *Anal. Chem.* **39**, 1666 (1967).
311. Heineman W. R., Norris B. J., Goetz J. F.: *Anal. Chem.* **47**, 79 (1975).
312. Weisz H., Ludwig H.: *Anal. Chim. Acta* **75**, 181 (1975).
313. Eswara Dutt V. V. S., Mottola H. A.: *Anal. Chem.* **47**, 357 (1975).
314. Růžička J., Hansen E. H.: *Anal. Chim. Acta* **78**, 145 (1975).
315. Eswara Dutt V. V. S., Mottola H. A.: *Anal. Chem.* **49**, 776 (1977).
316. Eswara Dutt V. V. S., Scheeler D., Mottola H. A.: *Anal. Chim. Acta* **94**, 289 (1977).
317. Wolff C. M., Mottola H. A.: *Anal. Chem.* **50**, 94 (1978).
318. Eswara Dutt V. V. S., Eskander-Hanna A., Mottola H. A.: *Anal. Chem.* **48**, 1207 (1976).
319. Papadakis N., Coolen R. B., Dye J. L.: *Anal. Chem.* **47**, 1644 (1975).

320. Porterfield R. I., Olson C. L.: *Anal. Chem.* **48**, 556 (1976).
321. Irachi E. S., Malmstadt M. V.: *Anal. Chem.* **45**, 1766 (1973).
322. Wilson R. L., Ingle J. D.: *Anal. Chim. Acta* **83**, 203 (1976).
323. Chlapowski E. W., Mottola H. A.: *Anal. Chim. Acta* **76**, 319 (1975).
324. Coombs L. C., Vasiliades J., Margerum D. W.: *Anal. Chem.* **44**, 2325 (1972).
325. Mieling G. E., Taylor R. W., Hargis L. G., English J., Pardue H. L.: *Anal. Chem.* **48**, 1686 (1976).
326. Likovskaya N. M., Markova L. V., Yevtushenko N. F.: *Zh. Anal. Khim.* **29**, 767 (1974).
327. Jost P., Lagrangé P., Wolf M., Schwing J. P.: *J. Less-Common Metals,* **36**, 169 (1974).
328. Alekseyeva I. I., Gromova A. D., Rysev A. P., Khvorostukhina N. A., Yakshinskii A. I.: *Zh. Anal. Khim.* **29**, 1017 (1974).
329. Ludwig H., Weisz H., Lenz T.: *Anal. Chim. Acta* **70**, 359 (1974).
330. Hadjiioannou T. P., Piperaki E. A., Papastathopoulos D. S.: *Anal. Chim. Acta* **68**, 447 (1974).
331. Rozovskii G. I., Poshkute Z. A., Prokopchik A. Yu.: *Zh. Anal. Khim.* **24**, 512 (1974).
332. Romanov V. F., Konishevskaya G. A., Yatsimirskii K. B.: *Tr. Khim. Khim. Tekhnol.* **13**, 65 (1973).
333. Ramanauskas E., Bunikiene L., Grigoriene K., Neverdauskiene Z.: *Tr. Khim. Khim. Tekhnol.* **13**, 119 (1973).
334. Witekova S., Kaminski W.: *Zesz. Nauk. Politech., Lodž. Chem. Spożyw* (1973) 163.
335. Shcherbov D. P., Inyutina O. D., Ivankova A. I.: *Zh. Anal. Khim.* **28**, 1372 (1973).
336. Tiginyanu Ya. D., Opryia V. I.: *Zh. Anal. Khim.* **28**, 2206 (1973).
337. Kreingold S. U., Panteleimonova A. A., Poponova R. V.: *Zh. Anal. Khim.* **28**, 2179 (1973).
338. Morozova R. P., Ilenko L. V.: *Zh. Anal. Khim.* **28**, 1835 (1973).
339. Pilipenko A. T., Terletskaya A. V.: *Zh. Anal. Khim.* **28**, 1135 (1973).
340. Alekseyeva I. I., Rysev A. P., Yakshinskii A. I., Ignatova N. K.: *Zh. Anal. Khim.* **28**, 1368 (1973).
341. Ke P. J., Thibert R. J., Walton R. J., Soules D. K.: *Microchim. Acta* 1973, 569.
342. Vzorova I. F., Kreingold S. U., Kharchenko L. A.: *Tr. Vses. Nauch.-Issled. Inst. Khim. Reaktiv. Osobo Chist. Khim. Veshchestv* 1972, no. 34, 147.
343. Kharlamov I. P., Mancevich A. D.: *Tr. Centr. Nauch.-Issled. Tekhnol. Mashinostr.* 1972, No. 110, 40.
344. Kataoka M., Kambara T.: *Bunseki Kiki* **10**, 773 (1972).
345. Grosse Yu. I., Miller A. D.: *Metody Anal. Redkometal. Miner. Rud. Gorn. Porod.* 1971, No. 2, 52.
346. Welch R. M., Allaway W. A.: *Anal. Chem.* **44**, 1644 (1972).
347. Yatsimirskii K. B., Fedorova T. I.: *Zh. Anal. Khim.* **18**, 1300 (1963).
348. Bognár J., Sárosi S.: *Microchim. Acta* 1963, 1072.

349. Kurzawa Z., Solecki R.: *Chem. Anal.* (Warsaw) **5**, 893 (1960).
350. Kozlovskii M. T., Grushina N. V.: *Zh. Anal. Khim.* **18**, 585 (1963).
351. Dolmanova I. F., Peshkova V. M.: *Zh. Anal. Khim.* **19**, 297 (1964).
352. Czarnecki K.: *Chem. Anal.* (Warsaw) **5**, 377 (1960).
353. Krause A., Zelinski S., Fedorov R.: *Z. Anal. Chem.* **187**, 350 (1962).
354. Peshkova V. M., Dolmanova I. F., Semenova I. M.: *Zh. Anal. Khim.* **18**, 1228 (1963).
355. Bognár J., Sipos L.: *Microchim. Acta* 1963, 442.
356. Jedrzejewski W., Gelo H.: *Chem. Anal.* (Warsaw) **7**, 753 (1962).
357. Yasinskiene E. I., Birmantas I. I.: *Izv. Vyssh. Ucheb. Zaved. Khim. Khim. Tekhnol.* **6**, 918 (1963).
358. Budanov V. V.: *Izv. Vyssh. Ucheb. Zaved., Khim. Khim. Tekhnol.* **5**, 47 (1962).
359. Michalski E., Czarnecki K., Pietrucha K.: *Chem. Anal.* (Warsaw) **8**, 713 (1963).
360. Michalski E., Gelowa H.: *Chem. Anal.* (Warsaw) **8**, 643 (1963).
361. Czarnecki K.: *Chem. Anal.* (Warsaw) **5**, 875 (1960).
362. Bognár J., Sárosi S.: *Chem. Abstr.* **60**, 6205g (1964).
363. Miller A. D., Shneyder L. A.: *Zh. Anal. Khim.* **18**, 371 (1963).
364. Oradovskii S. G.: *Zh. Anal. Khim.* **19**, 864 (1964).
365. Korenman I. M., Lebedeva A. N.: *Tr. Khim. Khim. Tekhnol.* **2**, 257 (1962).
366. Jedrzejewski W.: *Chem. Anal.* (Warsaw), **5**, 207 (1960).
367. Anolkhina L. G., Sopin A. Yu.: *Chem. Abstr.* **61**, 27f (1964).
368. Yatsimirskii K. B., Filipov A. P.: *Zh. Neorg. Khim.* **9**, 2096 (1964).
369. Bognár J., Sipos L.: *Microchim. Acta* 1963, 1066.
370. Yatsimirskii K. B., Parkhomenko N. V.: *Zh. Anal. Khim.* **18**, 229 (1963).
371. Michalski E., Walewski L.: *Chem. Anal.* (Warsaw) **6**, 273 (1961).
372. Lazarev A. I., Lazarevova V. I., Rozhayevskii V. V.: *Zh. Anal. Khim.* **18**, 202 (1963).
373. Michalski E., Wtorkowska A.: *Chem. Anal.* (Warsaw) **7**, 691 (1962).
374. Kurzawa Z.: *Chem. Anal.* (Warsaw) **5**, 551 (1960).
375. Tanaka M., Kawashima T.: *Bull. Soc. Japan* **37**, 1085 (1964).
376. Michalski E., Wtorkowska A.: *Chem. Anal.* (Warsaw) **7**, 783 (1962).
377. Kurzawa Z.: *Chem. Anal.* (Warsaw) **5**, 567 (1960).
378. Kurzawa Z.: *Chem. Anal.* (Warsaw) **5**, 731 (1960).
379. Michalski E., Wtorkowska A.: *Chem. Anal.* (Warsaw) **6**, 635 (1961).
380. Budarin L. I., Rumyantseva T. I., Shérina G. G.: *Zh. Anal. Khim.* **19**, 470 (1964).
381. Yatsimirskii K. B., Morozova R. P., Voronova T. A., Gershkovich R. M.: *Zh. Anal. Khim.* **19**, 705 (1964).
382. Li-Ching-Tien, Wen-Ping Chang: *Chem. Abstr.* **61**, 6392g (1964).
383. Bulgakova A. M., Zalyubovskaya N. P.: *Zh. Anal. Khim.* **18**, 1475 (1963).
384. Pantaler R. P.: *Zh. Anal. Khim.* **18**, 603 (1963).
385. Yatsimirskii K. B., Rayzman L. P.: *Zh. Anal. Khim.* **18**, 829 (1963).
386. Habig R. L., Pardue H. L., Worthington J. B.: *Anal. Chem.* **39**, 600 (1967).
387. Pardue H. L.: *Rec. Chem. Progr.* **27**, 151 (1966).
388. Shiokawa T.: *Chem. Abstr.* **45**, 2815 (1951).

389. Alekseyeva I. I., Smirnova I. B.: Analiz i tekhnologiya blagorodnych metallov. Metallurgiya, Moscow 1971.
390. Alekseyeva I. I., Smirnova I. B., Yatsimirskii K. B.: *Zh. Anal. Khim.* **25**, 539 (1970).
391. Kuznecov V. I.: *Dokl. Akad. Nauk SSSR* **70**, 629 (1950).
392. Filipov A. P., Zyatkovskii V. M., Yatsimirskii K. B.: *Zh. Anal. Khim.* **25**, 1769 (1970).
393. Zyatkovskii V. M., Filipov A. P., Yatsimirskii K. B.: Analiz i tekhnologiya blagorodnych metallov. Metallurgiya, Moscow 1971.
394. Gregorowicz Z., Suwinska Z.: *Microchim. Acta* 1967, 547.
395. Kulberg A. M.: *Zh. Obshch. Khim.* **8**, 1139 (1938).
396. Bognár J., Sárosi S.: *Magy. Kem. Foly.* **67**, 193 (1961).
397. Kalinina V. E., Morozova R. P., Yatsimirskii K. B., Ignateva O. N.: *Zh. Neorg. Khim.* **16**, 1097 (1971).
398. Reznik B. E., Bednyak N. A.: *Zh. Anal. Khim.* **23**, 1502 (1968).
399. Alekseyeva I. I., Rysev A. P., Ignatova I. K., Yakshinskii A. I.: *Zh. Neorg. Khim.* **16**, 1654 (1971).
400. Morozova R. P., Yatsimirskii K. B.: *Zh. Anal. Khim.* **24**, 1183 (1969).
401. Kalinina V. E., Yatsimirskii K. B., Zimina T. S.: *Zh. Anal. Khim.* **24**, 1178 (1969).
402. Ottaway J. M., Fuller C. W., Allan J. J.: *Analyst* (London) **94**, 522 (1969).
403. Babko A. K., Terletskaya A. V., Dubovenko L. I.: *Zh. Anal. Khim.* **23**, 932 (1968).
404. Kralyich J.: *Microchim. Acta* 1960, 586.
405. Litvinenko V. A.: *Ukr. Khim. Zh.* **35**, 763 (1969).
406. Yamane T., Suzuki T., Mukoyama T.: *Anal. Chim. Acta* **70**, 77 (1974).
407. Alekseyeva I. I., Ignatova N. K., Rysev A. P., Yakshinskii A. I.: *Zh. Anal. Khim.* **29**, 335 (1974).
408. Grosse Yu. I., Miller A. D.: *Zavod. Lab.* **40**, 262 (1974).
409. Dubovenko L. I., Belshickii N. V.: *Zh. Anal. Khim.* **29**, 111 (1974).
410. Kreingold S. U., Vasnev A. N., Serebryakova G. V.: *Zavod. Lab.* **40**, 6 (1974).
411. Yasinskiene E., Umbraziunaite O.: *Zh. Anal. Khim.* **28**, 2025 (1973).
412. Rychkova V. I., Rychkov A. A.: *Zavod. Lab.* **39**, 1053 (1973).
413. Alekseyeva I. I., Reiznov L. P., Chernysheva L. M., Khachatryan E. G.: *Izv. Vyssh. Ucheb. Zaved., Khim. Khim. Tekhnol.* **16**, 1445 (1973).
414. Fukasava T., Yaman T.: *Jap. Analyst* **22**, 168 (1973).
415. Markova L. V., Kaplan M. Z.: *Zh. Anal. Khim.* **28**, 1018 (1973).
416. Wood G. P.: *Analyst London* **98**, 525 (1973).
417. Dolmanova I. F., Ushakova N. M., Peshkova V. M.: *Zh. Anal. Khim.* **28**, 1131 (1973).
418. Kalinina V. E., Boldyreva O. I.: *Zh. Anal. Khim.* **28**, 1159 (1973).
419. Costache D., Sasu S.: *Rev. Roum. Chim.* **18**, 913 (1973).
420. Pilipenko A. T., Markova L. V., Maksimenko T. S.: *Zh. Anal. Khim.* **28**, 1544 (1973).
421. Dolmanova I. F., Zolotova G. A., Voronina R. D., Peshkova V. M.: *Zavod. Lab.* **39**, 281 (1973).

422. Tikhonova L. P., Yatsimirskii K. B., Svarkovskaya I. P., Zakrevskaya L. N.: *Zh. Anal. Khim.* **28**, 561 (1973).

423. Utsumi S., Okutani T.: *Nippon Kagaku Kaishi* 1973, 75.

424. Ke P. J., Thibert R. J.: *Microchim. Acta* 1973, 417.

425. Ke P. J., Thibert R. J.: *Microchim. Acta* 1973, 15.

426. Ke P. J., Thibert R. J.: *Microchim. Acta* 1973, 768.

427. Toropova V. F.: Rybkina A. A.: *Izv. Vyssh. Ucheb. Zaved., Khim. Khim. Tekhnol.* **16**, 148 (1973).

428. Khvostova V. P., Shlenskaya V. I., Kadyrova G. I.: *Zh. Anal. Khim.* **28**, 328 (1973).

429. Dolmanova I. F., Yatsimirskaya N. T., Peshkova V. M.: *Zh. Anal. Khim.* **28**, 112 (1973).

430. Pilipenko A. T., Pavlova V. K.: *Zh. Anal. Khim.* **27**, 1253 (1972).

431. Pilipenko A. T., Markova L. V., Maksimenko T. S.: *Zh. Anal. Khim.* **27**, 2009 (1972).

432. Feldman D. P., Matseyevskii B. P.: *Zh. Anal. Khim.* **27**, 1906 (1972).

433. Janjic T. J., Milovanovic G. A.: *Glas. Hem. Drus.* (Beograd) **37**, 173 (1972).

434. Costache D.: *An. Univ. Bucuresti, Khim.* **21**, 147 (1972).

435. Zhelyazkova B. G., Tsventanova A. L., Yatsimirskii K. B.: *Zh. Anal. Khim.* **27**, 795 (1972).

436. Orav M. T., Kokk Kh. Yu., Suit L. R.: *Zh. Anal. Khim.* **27**, 54 (1972).

437. Dolmanova I. F., Bolshova T. A., Shekhovcova T. N., Peshkova V. M.: *Zh. Anal. Khim.* **27**, 1848 (1972).

438. Kalinina V. E., Kosyakova A. S.: *Izv. Vyssh. Ucheb. Zaved., Khim Khim Tekhnol.* **15**, 131 (1972).

439. Tikhonova L. P., Yatsimirskii K. B., Kudinova G. V.: *Zh. Anal. Khim.* **27**, 1331 (1972).

440. Costache D.: *Farmacia* (Bucharest) **20**, 545 (1972).

441. Yamane T., Kitamura T., Fukasawa T., Suzuki T.: *Jap. Analyst* **21**, 799 (1972).

442. Pets L. I.: *Tr. Tallinsk. Politekh. Inst.* Ser. A 1972, 79.

443. Yasinskiene E., Rauckiene N.: *Liet. TSR Aukst. Mokyklu Mokslo Darb., Chem. Technol.* **13**, 105 (1971).

444. Lazerev A. I., Lazereva V. I., Shkotnikova G. A., Poddubienko G. A.: *Novye Metody Khim. Anal. Mater.* 1971, 74.

445. Costache D., Popa G.: *An. Univ. Bucuresti, Chim.* **20**, 83 (1971).

446. Blazys I., Paeda R., Yurevicius R.: *Nauch. Tr. Vyssh. Ucheb. Zaved. Litev. SSR, Khim. Khim. Tekhnol.* **13**, 35 (1971).

447. Zhukov Yu. A., Milyukova A. P.: *Izv. Vyssh. Ucheb. Zaved., Khim. Khim. Tekhnol.* **15**, 449 (1972).

448. Kreingold S. U., Antonov V. N.: *Metody Anal. Galogenidov Shchelochn. Shchelochnozemeln. Metal. Vys. Chist.* 1971, No. 2, 72.

449. Kreingold S. U., Marcokha V. I.: *Sb. Nauch. Tr. Vses. Nauch.-Issled. Inst. Lyuminoforov Osobo Chist. Veshchestv* 1971, No. 6, 129.

450. Datta K., Das J.: *Indian J. Chem.* **10**, 116 (1972).

451. Morgen E. A., Vlasov N. A., Kozhemyakina L. A.: *Zh. Anal. Khim.* **27**, 2064 (1972).
452. Tabata M., Funahashi S., Tanaka M.: *Anal. Chim. Acta* **62**, 289 (1972).
453. Vajgand V. J., Gaal F. F., Zrnic-Zeremski L. P., Sörös V. I.: Therm. Anal., Proc. 3rd ICDTA, vol. 2, p. 437, Davos 1971.
454. Costache D.: *An. Univ. Bucuresti, Chim.* **21**, 145 (1972).
455. Pantaler R. P., Chernomord L. D.: *Metody Anal. Galogenidov Shchelochn. Shchelochnozemeln. Metal. Vys. Chist.* 1971, No. 2, 92.
456. Fukasava T., Iwatsuki M., Uesugi K.: *Jap. Analyst* **21**, 1169 (1972).
457. Goncharik V. F., Yatsimirskii K. B., Tikhonova L. P.: *Zh. Anal. Khim.* **27**, 1348 (1972).
458. Alekseyeva I. I., Davydova Z. P.: *Zh. Anal. Khim.* **26**, 1786 (1971).
459. Costache D., Sasu S.: *Rev. Roum. Chim.* 1971, 1211.
460. Funahashi S., Yamada S., Tanaka M.: *Anal. Chim. Acta* **56**, 371 (1971).
461. Funahashi S., Tabata M., Tanaka M.: *Anal. Chim. Acta* **57**, 311 (1971).
462. Kreingold S. U., Antonov V. A., Pantaleymonova A. A., Demin A. A., Bozhevolnov E. A.: *Metody Anal. Khim. Reaktiv. Prep.* 1971, No. 18, 23.
463. Rozenblyum V. P., Shafran I. G., Shteinberg G. A.: *Metody Anal. Khim. Reaktiv. Prep.* 1971, No. 18, 175.
464. Klochkovskii S. P., Chistota V. D.: *Sb. Nauch. Tr., Magnitogorsk. Gorno-Met. Inst.* 1971, No. 87, 70.
465. Dubovenko L. I., Guta A. M.: *Visn. Lvin. Univ., Ser. Khim.* 1971, No. 12, 39, 83.
466. Kreingold S. U., Bozhevolnov E. A., Antonov V. N., Sosenkova L. I.: *Metody Anal. Khim. Reaktiv. Prep.* 1971, No. 18, 32.
467. Vershinin V. I., Chuyko V. T., Reznik B. E.: *Zh. Anal. Khim.* **26**, 1710, 1971.
468. Crisan I. A., Popa E.: Lucr. Conf. Nat. Chim. Anal., 3rd, vol. 2, p. 131, Brasov 1971.
469. Crisan I. A., Albu C., Popa E.: Lucr. Conf. Nat. Chim. Anal., 3rd, vol. 3, p. 79, Brasov 1971.
470. Drotsenidze N. E., Makharashvili L. Sh.: *Soobshch. Akad. Nauk. Gruz. SSR* **67**, 331 (1972).
471. Dolmanova I. F., Shekhovcova T. N., Peshkova V. M.: *Zh. Anal. Khim.* **27**, 1981 (1972).
472. Yankauskiene E. K., Yasinskiene E. J.: *Zh. Anal. Khim.* **24**, 527 (1969).
473. Guilbault G. G., Sadar M., Zimmer M.: *Anal. Chim. Acta* **44**, 361 (1969).
474. Dubovenko L. I., Chan Ty Chen: *Ukr. Khim. Zh.* **35**, 637 (1969).
475. Hadjiioannou T.: *Talanta* **15**, 535 (1968).
476. Bilidiene R., Yasinskiene E.: Elem. Mikrokiekin Nustatymas Fiz. Khem. Metod., Liet. TSR Khem. Anal. Mokslines Konf. Darb., 2, p. 76, Vilnius 1969.
477. Jankiewicz B., Soloniewicz R.: *Chem. Anal.* (Warsaw) **17**, 1341 (1972).
478. Kriss E. E., Savichenko Ya. S., Yatsimirskii K. B.: *Zh. Anal. Khim.* **24**, 875 (1969).
479. Kreingold S. U., Sosenkova L. I.: *Zh. Anal. Khim.* **26**, 332 (1971).
480. Alekseyeva I. I., Nemzer I. I.: *Zh. Anal. Khim.* **24**, 1393 (1969).

481. Dubovenko I. I., Bogoslavskaya T. A.: *Ukr. Khim. Zh.* **37**, 1057 (1971).
482. Yatsimirskii K. B., Tikhonova L. P., Goncharik V. P., Kudinova G. V.: *Chem. Anal.* (Warsaw) **17**, 789 (1972).
483. Oiwa K., Kimura T., Makino H.: *Jap. Analyst* **17**, 805 (1968).
484. Rodriguez P. A., Pardue H. L.: *Anal. Chem.* **41**, 1376 (1969).
485. Tamarchenko L. M.: *Zh. Anal. Khim.* **25**, 567 (1970).
486. Sychev A., Tiginyanu V.: *Zh. Anal. Khim* **24**, 1842 (1969).
487. Bartkus P., Yasinskiene E.: *Liet. TSR Aukst. Mokyklu Mokslo Darb., Khim. Khim. Tekhnol.* **11**, 83 (1970).
488. Pavlova V. K., Yatsimirskii K. B.: *Zh. Anal. Khim.* **24**, 1347 (1969).
489. Fedorova T. I., Shvedova L. V., Yatsimirskii K. B.: *Zh. Anal. Khim.* **25**, 307 (1970).
490. Yasinskiene E., Kalesnikaite S.: Elem. Mikrokiekin Nustatymas Fiz. Khem. Metod., Liet. TSR Khem. Anal. Mokslines Konf. Darb., 2, p. 204, Vilnius 1969.
491. Yasinskiene E., Kalesnikaite S.: *Liet. TSR Aukst. Mokyklu Mokslo Darb., Khim. Khim. Tekhnol.* **12**, 5 (1970).
492. Yasinskiene E., Kalesnikaite S.: *Zh. Anal. Khim.* **25**, 87 (1970).
493. Morozova R. P., Yatsimirskii K. B., Yegorova I. T.: Anal. Tekhnol. Blagorod. Metal. Tr. Soveshch. 8, p. 119, Ivanovo 1969.
494. Pilipenko A. T., Markova L. V., Kaplan M. I.: *Zh. Anal. Khim.* **25**, 2414 (1970).
495. Litvinenko V. A.: *Zh. Anal. Khim.* **23**, 1807 (1968).
496. Lazarev A. I.: *Opred. Mikroprimesey* **1**, 95 (1968).
497. Yatsimirskii K. B., Kalinina V. E.: *Zh. Anal. Khim.* **24**, 390 (1969).
498. Lazarev A. I., Lazareva V. I.: *Zh. Anal. Khim.* **24**, 395 (1969).
499. Filippov A. P., Zyatkovskii V. M., Yatsimirskii K. B.: *Ukr. Khim. Zh.* **35**, 451 (1969).
500. Costache D.: *Rev. Roum. Chim.* **16**, 849 (1971).
501. Litvinenko V. A.: *Ukr. Khim. Zh.* **35**, 311 (1969).
502. Toropova V. F., Tamarchenko L. M.: *Zh. Anal. Khim.* **22**, 576 (1967).
503. Nedvěd J.: *Chem. listy* **62**, 598 (1968).
504. Kreingold S. V., Bozhevolnov E. A., Drapkina D. A.: *Zh. Anal. Khim.* **22**, 218 (1967).
505. Michalski E., Gelowa H.: *Chem. Anal.* (Warsaw) **12**, 147 (1967).
506. Toropova V. F., Tamarchenko L. M.: *Zh. Anal. Khim.* **22**, 234 (1967).
507. Babko A. K., Lisetskaya G. S., Tsarenko G. A.: *Zh. Anal. Khim.* **23**, 1342 (1968).
508. Bartkus P. J., Yasinskiene E. J.: *Zh. Anal. Khim.* **23**, 1622 (1968).
509. Yasinskiene E. J., Kalesnikaite S. Z.: *Zh. Anal. Khim.* **23**, 1169 (1968).
510. Reznik B. E., Bednyak N. A.: *Zh. Anal. Khim.* **23**, 1502 (1968).
511. Lazarev A. I.: *Zh. Anal. Khim.* **22**, 1836 (1967).
512. Malor D., Townshend A.: *Anal. Chim. Acta* **39**, 235 (1967).
513. Proskuryakova G. F.: *Zh. Anal. Khim.* **22**, 802 (1967).
514. Babko A. K., Maksimenko T. S.: *Zh. Anal. Khim.* **22**, 570 (1967).
515. Yasinskiene E., Bilidiene E.: *Zh. Anal. Khim.* **22**, 741 (1967).
516. Babko A. K., Litvinenko V. A.: *Zh. Anal. Khim.* **21**, 302 (1966).

517. Jedrzeyewski W., Yatsimirskii K. B.: *Zh. Anal. Khim.* **21**, 314 (1966).
518. Yasinskiene E., Yankauskiene E.: *Zh. Anal. Khim.* **21**, 940 (1966).
519. Babko A. K., Markova L. V., Prikhodko M. U.: *Zh. Anal. Khim.* **21**, 935 (1966).
520. Yachiyo K., Namiki M., Goto H.: *Talanta* **13**, 1561 (1966).
521. Hadjiioannou T. P.: *Anal. Chim. Acta* **35**, 351 (1966).
522. Ginzburg S. I., Juzko M. I.: *Zh. Anal. Khim.* **21**, 79 (1966).
523. Bognár J., Toth K.: *Microchim. Acta* 1966, 562.
524. Fishman M. J., Skougstad M. W.: *Anal. Chem.* **36**, 1643 (1964).
525. Malmstadt H. V., Hadjiioannou T. P.: *Anal. Chem.* **35**, 2157 (1963).
526. Guilbault G. G., Kramer D. N., Haeckley E.: *Anal. Biochem.* **18**, 241 (1967).
527. Alekseyeva I. I., Ignatova N. K.: *Uch. Zap. Mosk. Inst. Tonkoj Khim. Tekhnol.* 1970, No. 2, 59.
528. Gleu K., Katthan W.: *Ber. Deutsch. Chem. Ges.* **86**, 1077 (1953).
529. Kawashima T., Tanaka M.: *Anal. Chim. Acta* **40**, 137 (1968).
530. Peshkova V. M., Astakhova E. K., Dolmanova I. F., Savotskina V. M.: *Acta Chim.* (Budapest) **53**, 121 (1967).
531. Hargis L. G.: *Anal. Chem.* **42**, 1497 (1970).
532. Kawashima R., Nakano S., Tanaka M.: *Anal. Chim. Acta* **49**, 443 (1970).
533. Páll A., Svehla G., Erdey L.: *Talanta* **17**, 211 (1970).
534. Jankiewicz B., Soloniewicz R.: *Chem. Anal.* (Warsaw) **17**, 101 (1972).
535. Aleksiyev A. A., Bonchev P. R.: *Microchim. Acta* 1970, 13.
536. Janjic T., Milovanovic G., Celap M.: *Anal. Chem.* **42**, 27 (1970).
537. Knapp G.: *Microchim. Acta* 1970, 467.
538. Yatsimirskii K. B., Rosolowski S., Kriss E. E.: *Zh. Anal. Khim.* **25**, 324 (1970).
539. Klockow D., Ludwig H., Girando M. A.: *Anal. Chem.* **42**, 1682 (1970).
540. Mikhailova V., Yevtimova B., Bonchev P.: *Microchim. Acta* 1968, 922.
541. Hadjiioannou T. P., Valkana C.: *Hem. Hron.* **32**, 89 (1967).
542. Hadjiioannou T. P.: *Anal. Chim. Acta* **35**, 360 (1966).
543. Yeliazkova B. G., Bonchev P. R., Aleksiyev A. A.: *Microchim. Acta* 1972, 896.
544. Shigematsu T., Munakata M.: *Bull. Inst. Chem. Res.*, Kyoto Univ. **48**, 198 (1970).
545. Tikhonova L. P., Yatsimirskii K. B., Svarkovskaya I. P.: *Zh. Anal. Khim.* **25**, 1766 (1970).
546. Fuller C. W., Ottaway J. M.: *Analyst* (London) **95**, 41 (1970).
547. Mottola H. A., Harrison C. R.: *Talanta* **18**, 683 (1971).
548. Svehla G., Erdey L.: *Microchem. J.* **7**, 221 (1963).
549. Thompson H., Svehla G.: *Z. Anal. Chem.* **247**, 244 (1969).
550. Sha-May Lin: *Bull. Inst. Chem.*, Acad. Sinica 1971, 133.
551. Bognár J., Jellinek O.: *Microchim. Acta* 1966, 453.
552. Bognár J., Jellinek O.: *Microchim. Acta* 1968, 1013.
553. Bognár J., Sárosi S.: *Microchim. Acta* 1969, 463.
554. Christian G. D.: *Anal. Lett.* **4**, 187 (1971).
555. Yatsimirskii K. B., Fedorova T. I.: *Dokl. Akad. Nauk SSSR* **143**, 143 (1962).
556. Weisz H., Klockow D.: *Z. Anal. Chem.* **232**, 321 (1967).
557. Weisz H., Pantel S., Ludwig H.: *Z. Anal. Chem.* **262**, 269 (1972).

558. Burton K. C., Irwing H. M. N. H.: *Anal. Chim. Acta* **52**, 441 (1970).
559. Mottola H. A., Freiser H.: *Anal. Chem.* **40**, 1266 (1968).
560. Weisz H., Muschelknautz U.: *Z. Anal. Chem.* **215**, 17 (1966).
561. Bognár J., Jellinek O.: *Microchim. Acta* 1962, 746.
562. Mottola H. A.: *Anal. Chem.* **42**, 630 (1970).
563. Fedorova T. I., Yatsimirskii K. B.: *Zh. Anal. Khim.* **22**, 283 (1967).
564. Bognár J., Sárosi S.: *Microchim. Acta* 1965, 1004; 1966, 543.
565. Weisz H., Janjic T.: *Z. Anal. Chem.* **215**, 17 (1966).
566. Abe S., Takanashi R., Matsuo T.: *Nippon Kagaku Kaishi* 1973, 963.
567. Reznik B. E., Chuyko V. T., Vershinin V. I.: *Zh. Anal. Khim.* **27**, 395 (1972).
568. Weisz H., Kiss T., Klockow D.: *Z. Anal. Chem.* **247**, 248 (1969).
569. Weisz H., Kiss T.: *Z. Anal. Chem.* **249**, 302 (1970).
570. Klockow D., Garcia Beltrán L.: *Z. Anal. Chem.* **249**, 304 (1970).
571. Vajgand V. J., Gaál F. Γ., Zrnic-Zeremski L. P., Sörös V. I.: *Thermal Analysis*, Vol. 2, 1971, 437.
572. Janjic T. J., Milovanovic G. A.: *Anal. Chem.* **45**, 390 (1973).
573. Mottola H. A., Haro M. S., Freiser H.: *Anal. Chem.* **40**, 1263 (1968).
574. Eswara Dutt V. V. S., Mottola H. A.: *Biochem. Med.* **9**, 148 (1974).
575. Mottola H. A., Freiser H.: *Anal. Chem.* **39**, 1294 (1967).
576. Eswara Dutt V. V. S., Mottola H. A.: *Anal. Chem.* **46**, 1090 (1974).
577. Eswara Dutt V. V. S., Mottola H. A.: *Anal. Chem.* **46**, 1777 (1974).
578. Kratochvil B., Boyer S. L., Hicks G. P.: *Anal. Chem.* **39**, 45 (1967).
579. Linde H. W.: *Anal. Chem.* **31**, 2092 (1951).
580. Guilbault G. G., McCurdy W. II., Jr.: *Anal. Chem.* **33**, 580 (1961).
581. Vangham G. A., Swithenbank J. J.: *Analyst* (London) **90**, 594 (1965).
582. Vajgand V. J., Gaál F. F.: *Talanta* **14**, 345 (1967).
583. Engel A. J., Lawson J., Aikens D. A.: *Anal. Chem.* **37**, 203 (1965).
584. Sharipov R. K., Songina O.: *Zh. Anal. Khim.* **21**, 800 (1966).
585. Stará V., Kopanica M.: *J. Electroanal. Chem.* **52**, 251 (1974).
586. Violanda A. T., Cooke W. D.: *Anal. Chem.* **36**, 2287 (1964).
587. Toropova V. F., Averko-Antonovich A. A.: *Zh. Anal. Khim.* **27**, 116 (1972).
588. Yezerskaya N. A., Kiseleva I. N.: *Zh. Anal. Khim.* **28**, 316 (1973).
589. Buckley J. P.: *Anal. Chim. Acta* **52**, 379 (1970).
590. Hojman J., Stefanovic A., Stankovic B., Zuman P.: *J. Electroanal. Chem.* **30**, 469 (1971).
591. Khikryzova E. G., Mashinskaya S. Ya.: *Zh. Anal. Khim.* **26**, 1105 (1971).
592. Toropova V. F., Zabbarova R. S.: *Zh. Anal. Khim.* **25**, 1059 (1970).
593. Toropova V. F., Vekslina V. A., Chovnyk N. G.: *Zh. Anal. Khim.* **27**, 346 (1972).
594. Habashi F., Thurston G.: *Anal. Chem.* **39**, 243 (1967).
595. O'Shea T. A., Parker G. A.: *Anal. Chem.* **44**, 184 (1972).
596. Turyan J., Saksin E.: *Zh. Anal. Khim.* **25**, 988 (1970).
597. McCoy L. R., Mark H. B., Jr.: *Anal. Chem.* **37**, 591 (1965).
598. Milyavskii Yu. S., Sinyakova S. S.: *Zh. Anal. Khim.* **23**, 1183 (1968).
599. Sinyakova S. S., Stepanova I. K.: *Zh. Anal. Khim.* **23**, 1405 (1968).

600. Yezerskaya N. A., Kiseleva I. N.: *Zh. Anal. Khim.* **24**, 1684 (1969).
601. Toropova V. E., Anisimova L. A., Pavlichenko L. A., Bankovskii J. A.: *Zh. Anal. Khim.* **24**, 1031 (1969).
602. Katz M., Kaufman N., Garcia de Bullande M. E.: *Arch. Bioquim. Farm.* **17**, 19 (1971).
603. Karayannis M. I., Tzouwara-Karayannis S. M., Hadjiioannou T. R.: *Anal. Chim. Acta* **70**, 351 (1974).
604. Bognár J., Czekkel J.: *Microchim. Acta* 1970, 572.
605. Funahashi S., Yamada S., Tanaka M.: *Anal. Chim. Acta* **56**, 37 (1971).
606. Crisan J. A., Popa E., Vatulescu R.: *Lucr. Conf. Nat. Chim. Anal.* 3rd, vol. 2, p. 141, Brasov 1971.
607. Crisan J. A., Oprea A.: *Lucr. Conf. Nat. Chim. Anal.*, 3rd, vol. 2, p. 137, Brasov 1971.
608. Schwerdffeger E.: *Mitteilungsbl. Ges. Deut. Chem., Fachgruppe Lebensmittel Chem., Gericht. Chem.* **28**, 100 (1974).
609. Crouch S. R., Malmstadt M. V.: *Anal. Chem.* **39**, 1084 (1967).
610. Crouch S. R., Malmstadt M. V.: *Anal. Chem.* **40**, 1901 (1968).
611. Kreuger R. H., Vas S., Jaselskis B.: *Talanta* **18**, 116 (1971).
612. Lamothe P. J., Cormick P. G.: *Anal. Chem.* **45**, 1906 (1973).
613. Rao G. N., Gupta R. K.: *Curr. Sci.* **41**, 880 (1972).
614. Render R., Jacob P.: *Anal. Lett.* **5**, 143 (1972).
615. Babko A. K., Markova L. V., Maksimenko T. S.: *Zh. Anal. Khim.* **23**, 1268 (1968).
616. Pardue H. L., Sheperd S.: *Anal. Chem.* **35**, 21 (1963).
617. Rao G. N.: *Z. Anal. Chem.* **264**, 414 (1973).
618. Burgess A. A., Latham J. L.: *Analyst* (London) **91**, 343 (1966).
619. Tikhonova V. I.: *Zh. Anal. Khim.* **23**, 1720 (1968).
620. Hewitt T. E., Pardue H. L.: *Clin. Chem.* **19**, 1128 (1973).
621. Rao G. G., Rao B. M.: *Anal. Chim. Acta* **59**, 461 (1972).
622. Yaselskis B., Kreuger R. H.: *Talanta* **13**, 945 (1966).
623. Belskii V. E., Vinnik M. I.: *Zh. Anal. Khim.* **19**, 375 (1964).
624. Kolthoff I. M., Lee T. S., Mairs M. A.: *J. Polym. Sci.* **2**, 199, 206, 220 (1947).
625. Siggia S., Hanna J. G., Serencha N. M.: *Anal. Chem.* **35**, 362 (1963).
626. Papa L. J., Mark H. B., Jr., Reilley C. N.: *Anal. Chem.* **34**, 1443 (1962).
627. Reilley C. N., Papa L. J.: *Anal. Chem.* **34**, 801 (1962).
628. Willeboordse F., Critchfield F. E.: *Anal. Chem.* **36**, 2270 (1964).
629. Garmon R. G., Reilley C. N.: *Anal. Chem.* **34**, 600 (1962).
630. Berka A., Korečková J.: *Anal. Lett.* **6**, 1113 (1973).
631. Hanna J. G., Siggia S.: *Anal. Chem.* **34**, 547 (1962).
632. Shresta I. L., Das M. N.: *Anal. Chim. Acta* **50**, 135 (1970).
633. Greinke R. A., Mark H. B., Jr.: *Anal. Chem.* **38**, 1001 (1966).
634. Siggia S., Hanna J. G., Serencha N. M.: *Anal. Chem.* **35**, 575 (1963).
635. Block J., Morgan E., Siggia S.: *Anal. Chem.* **35**, 573 (1963).
636. Hawk J. P., McDaniel E. L., Parish T. D., Simmons K. E.: *Anal. Chem.* **44**, 1315 (1972).

637. Benson D., Fletcher N.: *Talanta* **13**, 1207 (1966).
638. Lohman F. H., Mulligan T. F.: *Anal. Chem.* **41**, 243 (1969).
639. Jordan D. E.: *Anal. Chem.* **40**, 1717 (1768).
640. Zaia P., Peruzzo V., Lazzogua G.: *Anal. Chim. Acta* **51**, 317 (1970).
641. Ellis G. E., Mottola H. A.: *Anal. Chem.* **44**, 2037 (1972).
642. Sanderson D., Bittikofer J. A., Pardue H. L.: *Anal. Chem.* **44**, 1934 (1972).
643. Coombs L. C., Vasiliades J., Margerum D. W.: *Anal. Chem.* **44**, 2325 (1972).
644. Brook A. J. W., Munday K. C.: *Analyst* (London) **94**, 909 (1969).
645. Kopanica M., Stará V.: *Collect. Czech. Chem. Commun.* **41**, 3275 (1976).
646. Tanaka M., Funahashi S., Shirai K.: *Anal. Chim. Acta* **39**, 437 (1967).
647. Mark H. B., Jr., Rechnitz G. A.: Kinetics in Analytical Chemistry, p. 91, Interscience, New York 1968.
648. Coetzee J. F., Balya D. R., Chattopadhyay P. K.: *Anal. Chem.* **45**, 2266 (1973).
649. Ingle J. D., Crouch R. S.: *Anal. Chem.* **43**, 7 (1971).
650. Budarin L. I., Yatsimirskii K. B., Khachatryan A. G.: *Zh. Anal. Khim.* **26**, 1499 (1971).
651. Kramer D. N., Gamson R. M.: *Anal. Chem.* **30**, 251 (1958).
652. Kramer D. N., Cannon P. L., Guilbault G. G.: *Anal. Chem.* **34**, 842 (1962).
653. Guilbault G. G., Kramer D. N.: *Anal. Chem.* **36**, 1662 (1964).
654. Baum G., Ward F. B., Yaverbaum S.: *Clin. Chim. Acta* **36**, 405 (1972).
655. Guilbault G. G., Heyn A.: *Anal. Lett* **1**, 163 (1967).
656. Humme B. C. W.: *Can. J. Biochem. Physiol.* **37**, 1393 (1959).
657. Mead J. A., Smith J. N., Williams R. T.: *Biochem. J.* **61**, 569 (1955).
658. Guilbault G. G., Kramer D. N., Hackley F.: *Anal. Biochem.* **18**, 241 (1967).
659. Kramer S. P., Bartalos M., Karpa J. N., Midel J. M., Chang A., Seligman A. M.: *J. Surgical Res.* **4**, 23 (1964).
660. Jacks T. J., Kirchner H. W.: *Anal. Biochem.* **21**, 270 (1967).
661. Neuman H., Van Vreedendaal M.: *Clin. Chim. Acta* **17**, 183 (1967).
662. Guilbault G. G., Sadar S. H., Glazer R., Haynes J.: *Anal. Lett.* **1**, 333 (1968).
663. Land D. B., Jackim E.: *Anal. Biochem.* **16**, 481 (1966).
664. Makino Y., Koono K.: *Rinsko Byori* **15**, 391 (1967).
665. Updicke S. J., Hicks G. P.: *Nature* **214**, 986 (1967).
666. Guilbault G. G., Brignac P., Zimmer M.: *Anal. Chem.* **40**, 190 (1968).
667. Guilbault G. G.: *Anal. Biochem.* **14**, 61 (1966).
668. Guilbault G. G., Montalvo J.: *J. Amer. Chem. Soc.* **91**, 2164 (1969).
669. Orlova M. N.: *Vop. Med. Chim.* **18**, 16 (1972).
670. Bergerman J.: *Clin. Chem.* **12**, 797 (1966).
671. Vallee B. L., Hick F. L.: *Proc. Mat. Acad. Sci. USA.* **41**, 327 (1955).
672. Burton R. M.: Methods in Enzymology, vol. 1, Academic Press, New York 1955.
673. Hemler A. M., Kornberg A., Grisola S., Ochoa S.: *J. Biol. Chem.* **174**, 961 (1948).
674. Guilbault G. G.: Enzymatic Methods of Analysis, Pergamon Press, Oxford 1970.
675. Bergmeyer H. V.: Methods of Enzymatic Analysis, Academic Press, New York 1965.

676. Guilbault G. G., Tyson B., Cannon P., Kramer D. N.: *Anal. Chem.* **35**, 582 (1963).
677. Kaufman S.: *J. Biol. Chem.* **216**, 153 (1955).
678. Guilbault G. G., Brignac P., Zimmer M.: *Anal. Chem.* **40**, 190 (1968).
679. Guilbault G. G., Brignac P., Juneau M.: *Anal. Chem.* **40**, 1256 (1968).
680. Guilbault G. G.: Enzymatic Methods of Analysis. Pergamon Press, Oxford 1970, p. 123.
681. Guilbault G. G., Sadar M. H., Peres K.: *Anal. Biochem.* **31**, 91 (1969).
682. Krebs H. A.: The Enzymes, vol. 2, p. 508. Academic Press, New York 1951.
683. Guilbault G. G., Hieserman J.: *Anal. Biochem.* **26**, 1 (1968).
684. Guilbault G. G.: Enzymatic Methods of Analysis. Pergamon Press, Oxford 1970, p. 129.
685. Guilbault G. G., Kramer D. N.: *Anal. Chem.* **37**, 1219 (1965).
686. Mark H. B., Jr.: *Anal. Chem.* **36**, 1668 (1964).
687. Guilbault G. G.: Enzymatic Methods of Analysis. Pergamon Press, Oxford 1970, p. 156.
688. Drawert F., Gebling H., Ziegler A.: *J. Chromatogr.* **30**, 259 (1964).
689. Hurlock B., Talalay P.: *Arch. Biochem.* **80**, 468 (1959).
690. Baum P., Czok R.: *Biochem. Z.* **332**, 121 (1959).
691. Townshend A., Vaugham A.: *Anal. Lett.* **1**, 913 (1968).
692. Toren F. C., Burger F.: *Microchim. Acta* 1968, 538.
693. Toren F. C., Burger F.: *Microchim. Acta* 1968, 1049.
694. Orr C.: *Biochim. Biophys. Res. Commun.* **23**, 854 (1966).
695. Wilson R. L., Ingle J. D., Jr.: *Anal. Chem.* **49**, 1066 (1977).
696. Nikolelis D. P., Hadjiioannou T. P.: *Analyst* **102**, 591 (1977).
697. Efstathiou C. E., Hadjiioannou T. P.: *Anal. Chem.* **49**, 414 (1977).
698. Nomura T., Nakagawa G.: *Talanta* **24**, 467 (1977).
699. Fukasawa T., Yamane T.: *Anal. Chim. Acta* **88**, 147 (1977).
700. Bonchev P. R., Raikova D., Aleksiyev A. A., Yankova D.: *Clin. Chim. Acta* **57**, 37 (1974).
701. Aleksiyev A. A., Bonchev P. R., Raikova D.: *Microchim. Acta* 1974, 751.
702. Dubovenko L. I., Tananaiko M. M., Drokov V. G.: *Ukr. Khim. Zh.* **40**, 758 (1974).
703. Pilipenko A. T., Pavlova V. K.: *Zh. Anal. Khim.* **29**, 1165 (1974).
704. McGlothlin C. D., Purdy W. C.: *Anal. Chim. Acta* **73**, 216 (1974).
705. Sanchez-Pedreno G., Arias J. J.: *Quim. Analit.* **28**, 184 (1974).
706. Rayzman L. P., Senyagina T. V., Uvarova N. P.: *Zh. Anal. Khim.* **29**, 1157 (1974).
707. Kataoka M., Kambara T.: *Japan Analyst* **23**, 1157 (1974).
708. Vershinin V. I., Reznik B. E., Statsenko V. P.: *Zh. Anal. Khim.* **29**, 380 (1974).
709. Lopéz-Cueto G., Casado-Riobó J. A., Lucena-Conde F.: *Talanta* **21**, 669 (1974).
710. Lukavskaya N. M., Terletskaya A. V., Bogoslavskaya T. A.: *Zh. Anal. Khim.* **29**, 2268 (1974).
711. Trofimov N. V., Kanaev N. A., Busev A. I.: *Zh. Anal. Khim.* **29**, 2001 (1974).
712. Trofimov N. V., Akashina V. F., Kanaev N. A., Busev A. I., Zolotova S. V.: *Zavod. Lab.* **41**, 1177 (1975).

713. Feys R., Devynck J., Tremillon B.: *Talanta* **22**, 17 (1975).

714. Abe S., Takahashi K., Matsuo T.: *Anal. Chim. Acta* **80**, 135 (1975).

715. Yasinskiene E. I., Umbraziunaite O. P.: *Zh. Anal. Khim.* **30**, 962 (1975).

716. Yasinskiene E. I., Umbraziunaite O. P.: *Zh. Anal. Khim.* **30**, 1590 (1975).

717. Pantaler R. P., Afimova L. D., Bulgakova A. M., Pulyaeva I. V.: *Zh. Anal. Khim.* **30**, 946 (1975).

718. Lauber K.: *Anal. Chem.* **47**, 769 (1975).

719. Müller H., Otto M.: *Microchim. Acta* 1975, 519.

720. Pantaler R. P., Afimova L. D., Bulgakova A. M.: *Zh. Anal. Khim.* **30**, 1584 (1975).

721. Rigin V. I., Aleksieva G. I.: *Zh. Anal. Khim.* **30**, 2372 (1975).

722. Tikhonova L. P., Borkovets S. N., Revenko L. N.: *Ukr. Chim. Zh.* **42**, 869 (1976).

723. de Oliviera Meditsch J.: *Revta Quim. Ind.* (Rio de J.) **45**, 7 (1976).

724. Hirayama K., Sawaya T.: *Nippon Kagaku Kaishi* 1976, 1401.

725. Ionescu G., Duca A., Aelenei N.: *Bul. Inst. Politeh. Iasi*, Sect. 2, **22**, 23 (1976).

726. Khain V. S., Volkov A. A., Fomina E. V.: *Zh. Anal. Khim.* **31**, 1500 (1976).

727. Rysev A. P., Alekseeva I. I., Koryakova S. F., Zhitenko L. P., Yakshinskii A. I.: *Zh. Anal. Khim.* **31**, 508 (1976).

728. Tashkhodzhaev A. T., Zel'cer L. E., Talipov S. T., Khikmatov Kh.: *Zh. Anal. Khim.* **31**, 485 (1976).

729. Yatsimirskii K. B., Tikhonova L. P., Borkovets S. N.: *Zh. Anal. Khim.* **31**, 339 (1976).

730. Rigin V. I., Bachmurov A. S.: *Zh. Anal. Khim.* **31**, 93 (1976).

731. Fukasawa T., Yamane T., Yamazaki T.: *Bunseki Kagaku* **26**, 200 (1977).

732. Perez-Bendito D., Valcarel M., Ternero M., Pino F.: *Anal. Chim. Acta* **94**, 405 (1977).

733. Fukasawa T., Yamane T.: *Bunseki Kagaku* **26**, 692 (1977).

734. Pikhugina G. V., Khvostova V. P., Alimarin I. P.: *Izv. Akad. Nauk SSSR, Ser. Khim.* 1977, 2190.

735. Rigin V. I., Blokhin A. I.: *Zh. Anal. Khim.* **32**, 312 (1977).

736. Rigin V. I., Blokhin A. I.: *Zh. Anal. Khim.* **32**, 2340 (1977).

737. Fedorova T. I., Yatsimirskii K. B., Vasileva E. V., Markichev V. G.: *Zh. Anal. Khim.* **32**, 1951 (1977).

738. Vasilikiotis G. S., Papadopoulos C., Sofoniou M., Themelis D.: *Microchim. J.* **22**, 541 (1977).

739. Kreingold S. U., Vasnev A. H.: *Zavod. Lab.* **44**, 265 (1978).

740. Kreingold S. U., Sosenkova L. I., Panteleimonova A. A., Lavrelashvili L. V.: *Zh. Anal. Khim.* **33**, 2168 (1978).

741. Dolmanova I. F., Mel'nikova O. I., Shekhovtsova T. N.: *Zh. Anal. Khim.* **33**, 2096 (1978).

742. Skorobogatyi Ya. P., Zinchuk B. K.: *Zh. Anal. Khim.* **33**, 1587 (1978).

743. Dolmanova I. F., Zolotova G. A., Ratina M. A.: *Zh. Anal. Khim.* **33**, 1356 (1978).

744. Kalinina V. E., Lyakushina V. M., Petrova N. V.: *Zh. Anal. Khim.* **33**, 959 (1978).

745. Kreingold S. U., Antonov V. N., Vasnev A. N., Kamentseva L. G., Loginova E. V.: *Zh. Anal. Khim.* **33**, 928 (1978).
746. Tamarchenko L. M.: *Zh. Anal. Khim.* **33**, 824 (1978).
747. Lukovskaya N. M., Terletskaya A. V., Kushchevskaya N. F.: *Zh. Anal. Khim.* **33**, 750 (1978).
748. Dolmanova I. F., Zolotova G. A., Shekhovtsova T. N., Bubelo V. D., Kurdyukova N. A.: *Zh. Anal. Khim.* **33**, 274 (1978).
749. Rudenko V. K., Zhukova L. I.: *Zh. Anal. Khim.* **34**, 605 (1979).
750. Koupparis M. A., Hadjiioannou T. P.: *Anal. Chim. Acta* **96**, 31 (1978).
751. Kalinina V. E., Lyakushina V. M., Rybina A. E.: *Zh. Anal. Khim.* **33**, 125 (1978).
752. Pilipenko A. T., Maksimenko T. S., Lukovskaya N. M.: *Zh. Anal. Khim.* **34**, 523 (1979).
753. Eswara Dutt V. V. S., Mottola H. A.: *Anal. Chem.* **48**, 80 (1976).
754. Kurzawa Z., Matusiewicz H., Matusiewicz K.: *Chem. Anal.* (Warsaw) **21**, 797 (1976).
755. Kurzawa J., Kurzawa Z., Swit Z.: *Chem. Anal.* (Warsaw) **21**, 791 (1976).
756. Wolff C. M., Schwing J. P.: *Bull. Soc. Chim. France* 1976, 679.
757. Wolff C. M., Schwing J. P.: *Bull. Soc. Chim. France* 1976, 675.
758. Kurzawa Z., Kubaszewski E.: *Chem. Anal.* (Warsaw) **21**, 565 (1976).
759. Truesdale V. W., Chapman P.: *Mar. Chem.* **4**, 29 (1976).
760. Weisz H., Pantel S., Meiners W.: *Anal. Chim. Acta* **82**, 145 (1976).
761. Ohashi K., Sagawa T., Goto E., Yamamoto K.: *Anal. Chim. Acta* **92**, 209 (1977).
762. Efstathiou C. E., Hadjiioannou T. P.: *Anal. Chem.* **49**, 414 (1977).
763. Klockow D., Auffarth J., Kopp C.: *Anal. Chim. Acta* **89**, 37 (1977).
764. Kawashima T., Kai S., Takashima S.: *Anal. Chim. Acta* **89**, 65 (1977).
765. Sekketa M. A., Milovanovic G. A., Janjic T. J.: *Microchim. Acta* 1978, I, 297.
766. Weisz H., Pantel S., Vereno I.: *Microchim. Acta* 1975, II, 287.
767. Kurzawa Z., Szukalska A.: *Chem. Anal.* (Warsaw) **21**, 297 (1976).
768. Toropova V. F., Budnikov G. K., Medyantseva E. P.: *Zh. Anal. Khim.* **30**, 1435 (1975).
769. Ruvinskii O. E., Neverova A. K., Kabatskii A. V.: *Zh. Anal. Khim.* **30**, 2409 (1975).
770. Budnikov G. K., Toropova V. P., Ulakhovich N. A., Viter I. P.: *Zh. Anal. Khim.* **29**, 752 (1974).
771. Korolezuk M., Matysik J., Franczak A.: *J. Electroanal. Chem.* **71**, 229 (1976); **88**, 421 (1978).
772. Astafeva V. V., Prokhorova G. V., Salikhdzhanova R. M.: *Zh. Anal. Khim.* **31**, 260 (1976).
773. Nomura T., Nakagawa G.: *Anal. Lett.* **8**, 161 (1975).
774. Turyan Ya. I., Makarova L. M.: *Elektrokhimiya* **15**, 203 (1979).
775. Rigin V. I., Bachmurov A. S., Blokhin A. I.: *Zh. Anal. Khim.* **30**, 2413 (1975).
776. Pavlova V. H., Pilipenko A. T., Voevutskaya R. N.: *Zh. Anal. Khim.* **30**, 2190 (1975).
777. Sychev A. Ya., Isak V. G., Pfannmeller U.: *Zh. Anal. Khim.* **33**, 1351 (1978).

778. Toropova V. F., Vekslina V. A.: *Zh. Anal. Khim.* **30**, 326 (1975).
779. Khikryzova E. G., Kiriyak L. G.: *Zh. Anal. Khim.* **27**, 1747 (1972); **29**, 899 (1974).
780. Shapirova N. S., Songina O. A.: *Zh. Anal. Khim.* **28**, 2348 (1973).
781. Toropova V. F., Vekslina V. A., Khovnyk N. G.: *Zh. Anal. Khim.* **28**, 967 (1973).
782. Khikryzova G. G., Bardina S. M.: *Zh. Anal. Khim.* **29**, 2414 (1974).
783. Stach B., Schoene K.: *Microchim. Acta* 1977, II, 567.
784. Budnikov G. K., Toropova V. F., Ulakhovich N. A., Medyantseva E. P., Frolova V. P.: *Zh. Anal. Khim.* **32**, 212 (1977).
785. Stepanova I. K., Sinyakova S. I.: *Zh. Anal. Khim.* **27**, 2219 (1972).
786. Itabashi E., Yusa I.: *Bull. Chem. Soc. Japan* **49**, 122 (1976).
787. Alexander P. W., Hoh R., Smythe L. E.: *Talanta* **24**, 543, 549 (1977).
788. Ignatova N. K., Zaitsev P. M., Gornostaeva M. Yu.: *Zh. Anal. Khim.* **33**, 2140 (1978).
789. Donoso N. G., Chadwick W. I., Santa Ana V. M. A.: *Anal. Chim. Acta* **77**, 1, (1975).
790. Miftakhova A., Zabbarova R. S., Sabitova Z. A., Toropova V. F.: *Izv. Vyssh. Ucheb. Zaved. Khim. Khim. Tekhnol.* **19**, 1364 (1976).
791. Manok F., Varhelyi C., Kerecsendi G.: *Monatsh. Chem.* **104**, 1307 (1973).
792. Nikolayeva T. D., Zhdanov S. I.: *Zh. Anal. Khim.* **34**, 326 (1979).
793. Nomura T., Nakagawa G.: *Talanta* **24**, 467 (1977).
794. Rachinskaya G. G., Vakhobova R. U., Milyavskii Yu. S.: *Zh. Anal. Khim.* **31**, 1712 (1976).
795. Rachinskaya G. G., Vakhobova R. U., Milyavskii Yu. S.: *Zh. Anal. Khim.* **31**, 1482 (1976).
796. Demkin A. M.: *Zh. Anal. Khim.* **30**, 939 (1975).
797. Zaitsev P. M., Sukhomlinov A. B., Turyan Y. I.: *Zh. Anal. Khim.* **28**, 2243 (1973).
798. Khikryzova E. G., Kiriyak L. G.: *Zh. Anal. Khim.* **29**, 2420 (1974).
799. Turyan Y. I., Ilina L. F., Kudinov P. I., Kochubei N. D.: *Zh. Anal. Khim.* **30**, 1842 (1975).
800. Turyan Y. I., Strizhov N. K.: *Zh. Anal. Khim.* **27**, 1423 (1972).
801. Strizhov N. K., Turyan Y. I.: *Zh. Anal. Khim.* **28**, 1615 (1973).
802. Lopez Fonseca J. M., Sanz Pedrero P., Tutor J. C.: *An. Quim.* **69**, 455 (1973).
803. Utsumi S., Okutani T., Yamada T.: *Bunseki Kagaku* **24**, 789 (1975).
804. Senn D. R., Carr P. W., Klatt L. N.: *Anal. Chem.* **48**, 954 (1976).
805. Okutani T., Kotani H., Utsumi S.: *Bunseki Kagaku* **26**, 116 (1977).
806. Pantaler R. P., Pulyayeva I. V.: *Zh. Anal. Khim.* **32**, 394 (1977).
807. Mentasti E., Pellizetti E.: *Anal. Chim. Acta* **78**, 227 (1975).
808. Kriss E. E., Kurbatova G. T., Yatsimirskii K. B.: *Zh. Anal. Khim.* **31**, 598 (1976).
809. Richmond J., Rainey C., Meloan C. C.: *Anal. Lett.* **9**, 105 (1976).
810. Antonov V. N., Kreingold S. U.: *Zh. Anal. Khim.* **31**, 193 (1976).
811. Janjic T. J., Milovanovic G. A.: *Anal. Chim. Acta* **85**, 169 (1976).
812. Sheehan T. L., Hercules D. M.: *Anal. Chem.* **49**, 446 (1977).
813. Klochkovskii S. P., Neimysheva L. P.: *Zh. Anal. Khim.* **29**, 929 (1974).

814. Rohm T. J., Purdy W. C.: *Anal. Chim. Acta* **72**, 177 (1974).
815. Fedorova T. I., Yatsimirskii K. B., Shvedova L. V., Ermolayeva T. G.: *Zh. Anal. Khim.* **30**, 59 (1975).
816. Klochkovskii S. P., Klochkovskaya G. D.: *Zh. Anal. Khim.* **32**, 736 (1977).
817. Kuroda K., Saito T., Kiuchi T., Oguma K.: *Z. Anal. Chem.* **277**, 29 (1975).
818. Hadjiioannou T. P., Koupparis M. A., Efstathiou C. E.: *Microchim. Acta* 1977, I, 61.
819. Hadjiioannou T. P., Piperaki E. A.: *Anal. Chim. Acta* **90**, 329 (1977).
820. Hadjiioannou T. P., Koupparis M. A., Efstathiou C. E.: *Anal. Chim. Acta* **88**, 281 (1977).
821. Greenhow E. J.: *Analyst* **102**, 584 (1977).
822. Gaal F. F., Abramovic B. F., Canic V. D.: *Talanta* **25**, 113 (1978).
823. Csizér E., Görög S.: *Anal. Chim Acta* **86**, 217 (1976).
824. Koprivic L., Polla E., Hranilovic J.: *Acta Pharm. Suec.* **13**, 421 (1976).
825. Mentasti E., Pelizzetti E., Saini G.: *Anal. Chim. Acta* **86**, 303 (1976).
826. Collin J. P., Lagrange P.: *Bull. Soc. Chim. France* 1976, 1309.
827. Gary A. M., Jost P., Schwing J. P.: *Bull. Soc. Chim. France* 1976, 1609.
828. Haraguchi K., Ito S.: *Japan Analyst* **24**, 405 (1975).
829. Ito S., Haraguchi K., Nakagawa K., Yamada K.: *Bunseki Kagaku* **26**, 554 (1977).
830. Kurzawa Z., Zietkiewicz M.: *Chem. Anal.* (Warsaw) **21**, 13 (1976).
831. Gregorowicz Z., Gorka P., Suwinska T.: *Z. Anal. Chem.* **271**, 354 (1974).
832. Nikolelis D. P., Karayannis M. I., Hadjiioannou T. P.: *Anal. Chim. Acta* **94**, 415 (1977).
833. Landis J. B., Robec M., Pardue H. L.: *Anal. Chem.* **49**, 785 (1977).
834. Nikolelis D. P., Hadjiioannou T. P.: *Anal. Chem.* **50**, 205 (1978).
835. Gijsbers J. C., Kloosterboer J. G.: *Anal. Chem.* **50**, 455 (1978).
836. Wilson R. L., Ingle J. D., Jr.: *Anal. Chim. Acta* **92**, 417 (1977).
837. Lazarou L. A., Siskos P. A., Koupparis M. A., Hadjiioannou T. P.: *Anal. Chim. Acta* **94**, 475 (1977).
838. Dolmanova I. F., Zolotova G. A., Marko T. N.: *Zh. Anal. Khim.* **32**, 1025 (1977).
839. Nikolelis D. P., Karayannis M. I., Kordi E. V., Hadjiioannou T. P.: *Anal. Chim. Acta* **90**, 209 (1977).
840. Tamarchenko L. M.: *Zh. Anal. Khim.* **30**, 127 (1975).
841. Veres S., Czányi L.: *Anal. Chem.* **49**, 980 (1977).
842. Oteiza R. M., Krottinger D. L., McCrachen M. S., Malmstadt H. V.: *Anal. Chem.* **49**, 1586 (1977).
843. Nieman T. A., Hoeler F. J., Enke C. G.: *Anal. Chem.* **48**, 899 (1976).
844. Hasty R. A., Lima E. J., Ottaway J. M.: *Analyst* **102**, 313 (1977).
845. Yap S. K., Rhodes C. T., Fung H. L.: *Anal. Chem.* **47**, 1183 (1975).

Application of Computers in Analytical Chemistry

by

KAREL ECKSCHLAGER ET AL.

Contents

1 The Analytical Result

1.1 Analysis as a Process of Obtaining Information

Chemical or physico-chemical instrumental analyses are carried out to obtain information about the chemical composition of a test sample or the material from which the sample was taken. The information thus obtained answers the question, "what?", through qualitative analysis and the question, "how much?", through quantitative analysis. The results of some methods, e.g. spectral or chromatographic, answer both questions; however, sometimes several methods must be combined to obtain the required information.

Analysis, as a process of obtaining information about the chemical composition of a test sample, comprises two successive steps. In the first, information is obtained experimentally about the unknown but fixed content of analyte X_i, $i = 1, 2, ..., k$, where k is the number of components determined simultaneously. The information is coded in an analytical signal or, for $k > 1$, in a sequence of signals or in a continuous spectrum, i.e. in an "analytical report". The analytical report can be recorded in analog form, i.e. can be given by the continuously varying value of a certain quantity. If it is plotted graphically, it is called the recording of an analytical signal or spectrum. An analytical report can also be recorded digitally, in the form of pairs of discontinuous numerical values of the two variables. Any analytical signal has a certain intensity, η_i, $i = 1, 2, ..., k$; if several components are to be determined simultaneously, i.e. for $k > 1$, the analytical report can consist of a sequence of isolated signals or of a continuous spectrum. The signal intensity at point z_j, $j = 1, 2, ..., m$, is then given by the sum of the intensity contributions from the individual components, i.e. $\eta(z_j) = \sum_{i=1}^{k} \eta_i$; $j = 1, 2, ..., m$, $i = 1, 2, ..., k$. The value of m is the number of positions in the signal at which the measurement is carried out and k is the number of components for which $\eta_i > 0$

at the given position. The analytical signal intensity, η_i, is a continuous random quantity which has a certain probability distribution $p(y)$ and on repeated measurement attains values y_i, $i = 1, 2, ..., n$, where n is the number of measurements. The signal intensity contains information on the amount (quantity) of the analyte. The position of the signal in a spectrum or chromatogram is either a fixed non-random quantity and is then denoted as z, or is a continuous random quantity denoted as ζ_i, which attains values z_i, $i = 1, 2, ..., n$, on repeated measurements. Its probability distribution is then $p(z)$. The signal position contains information about the kind of test component, i.e. on the qualitative composition.

In the second step, the information contained in the analytical signal is decoded and appears in the form of a numerical result ξ_i, which has the character of a continuous random quantity. On repeated measurements the result attains values $x_{i,j}$, $j = 1, 2, ..., n$, where n is the number of measurements; ξ_i has probability distribution $p(x)$. Finding the most probable value of random quantity ξ, i.e. of value $E[\xi]$, is the goal of quantitative analysis and determination of i, i.e. assigning the signal to a particular component, is the goal of qualitative analysis.

The two steps of obtaining information on the chemical composition of a sample take place in independent systems. The first step, the experimental obtaining of information, generally takes place in an analytical instrument. Decoding of analytical signals to obtain the result takes place as a series of logic and computing operations in a conceptual (thought) system or in a real system, e.g. in a computer, using a program. The two systems, namely, that in which the information is obtained and that in which it is decoded, can sometimes be connected through a feedback, by means of which the second system (computer) controls the operation of the first system (instrument) to ensure that the whole information gathering process is opti-

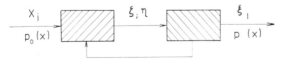

Fig. 41. Information chain.

mized. The process of obtaining analytical information is sometimes depicted by the information chain (Fig. 41).

The problems of experimental gathering of information on the chemical composition of a test sample is the subject of analytical chemistry. This chapter will discuss the problems connected with decoding information contained in the analytical signal, principally using computing techniques.

1.2 Decoding of an Analytical Signal

The analytical signal is processed, using a computer or otherwise, to decode the information it contains. This decoding is a substantial part of the whole process of obtaining information; sometimes it is simply the method of processing the results that differentiates analysis from physico-chemical measurements of constants of substances, their structure, etc. Actually, the question of whether some physico-chemical method can be used analytically is decided, in addition to other factors, by the possibility of decoding the signal to obtain analytical information on the qualitative and/or quantitative composition. Of course, in analytical practice not all the information contained in the signal is decoded, but only the part relevant to further use. Irrelevant information is not decoded and, indeed, it is often not known which irrelevant information is contained in the signal. It is evident that the concept of information relevance is relative; it often happens that information originally considered irrelevant becomes important later. Therefore, it is advantageous to store the analytical signal for later decoding of further information if necessary.

An analytical "report" chiefly contains the following information:

(1) Which components are present in the test sample, i.e. information on the qualitative composition (answer to the question "what?").

(2) The amount of the components present, i.e. information on the quantitative composition (answer to the question "how much?").
Information concerning the presence or the content of components that are not to be determined is considered irrelevant.

Especially after repeated measurements, calibration, subtraction of the blank, etc., the analytical signal also provides information on the properties of analytical signals and results, mainly:

(1) the precision, i.e. the agreement of the results or the signal intensities after repeated measurements; its characteristic was given by E. C. Wood in Vol. IA, p. 78 and by P. Móritz (Vol. XI, p. 1 and 3); the latter approach will be treated in Section 1.3.3;

(2) the accuracy, i.e. the agreement of the results with the actual value; this value is given by the mean error, i.e. the difference $\delta = |X_i - E[\xi_i]|$ (see also Section 1.3.3) and can be determined only by analysis of a reference material with a known X_i value;

(3) the agreement of the results obtained by various analytical methods;

(4) the shape of the signals, i.e. the dependence of the signal intensity on the "position", i.e. time, wavelength, etc., and the overlap;

(5) the limit of determination, given chiefly by the baseline noise and the sensitivity.

Other properties of analytical signals and results are also sometimes determined.

A relatively frequent and practically important part of the analytical "report" is determination of a relationship between two quantities, the dependence of the signal intensity on the analyte concentration (calibration), the sensitivity of the determination of a component to the concentration of another component (matrix effect), the dependence of the results on another quantity, etc. A common part of processing of results using a regression dependence is also the determination of statistical estimates of parameter characteristics, the correlation coefficient, the best fit characteristics, etc. Use of computers is now especially important in these processing methods.

If information on the properties of the analytical signal can be obtained and the analytical instrument is connected by a feed-back with a computer (using e.g. an on-line computer), some properties of the results can be improved by judicious signal processing, namely,

(1) the precision and accuracy, which can be improved by calibration, exclusion of outlying results, subtraction of the blank, smoothing of the signal, etc.;

(2) the selectivity, which can be improved by mathematical separation of overlapping signals;

(3) a decrease in the determination limits, e.g. by improving the signal-to-noise ratio.

Sometimes it is possible to automatically control the whole process of gathering relevant analytical information so that it proceeds under optimal conditions.

An analytical signal is decoded, i.e. analytical information and that on the properties of the results and of the signal itself are obtained, according

260

to a certain algorithm (a program), representing a sequence of successive computing and logic operations. Such a program is either single-purpose or has several alternatives. Programs for processing analytical data are discussed in greater detail below.

1.2.1 SIGNAL DECODING FOR QUALITATIVE ANALYSIS

The results of qualitative analysis have the character of alternative information; e.g., component i $(i = 1, 2, ..., k)$ is or is not present in the test sample, or its content in the sample is or is not greater than the detection limit of the method. Thus even in qualitative analysis the signal intensity, y_i, at position z_i is evaluated and if $y_i \geqq y_{min}$, component i is considered present and if $y_i < y_{min}$, where y_{min} is the lowest value of the signal intensity distinguishable from the background noise at position z_i, component i is considered not present. If a component is to be identified according to the signal position, the operator relationship

$$i = A\{z_i\}$$

can be formally considered, operator A converting the set of signal positions to the set of components to which the signal corresponds in the method used. Operator A can have the form of tables, provided that z_i is a fixed value. When the signal position is a random value, ζ_i, operator A is, for a given procedure, obtained experimentally by comparing the position of the signal with the positions of the signals of known "standard" substances.

1.2.2 SIGNAL DECODING FOR QUANTITATIVE ANALYSIS

Results of quantitative analysis have the character of numerical information. The random continuous quantity ξ_i, which, on repeated measurement, attains values x_j, $j = 1, 2, ..., n$, is determined from signal intensity η_i, which, on repeated measurement, attains values y_i, according to the relationship

$$x_i = f_i(y_i) \tag{1.1}$$

Analytical function f_i is known, e.g. from stoichiometry, or is found by

261

calibration. It rather frequently occurs that

$$x_i = \frac{1}{S_i} y_i$$

where S_i is the sensitivity defined as $S_i = dy_i/dx_i$. The $1/S_i$ value can be given as a stoichiometric constant or by the slope of the calibration curve. If the method yields a single signal specific for the analyte, or it enables quite selective determination of the component in the presence of other components, i.e. the signals corresponding to component i and to the other components do not overlap, and the value of S_i is known or determined by calibration, then the signal is readily decoded using Eq. (1.1). However, if the signals of the components partially overlap, mathematical separation can be carried out or the individual component contents decoded by solving the set of equations in the form, $\sum_{ij} S_{i,j} x_i = y(z_j)$, where $i = = 1, 2, \ldots, k$, k is the number of components and z_j is the position at which the measurement is performed. If the number of measurements is greater than the number of test components, the condition of the smallest sum of the squares of the deviations must be introduced, i.e.

$$[\sum (y(z_j) - S_{i,j} x_i)^2 / s_j^2] = \min$$

where s_j^2 is the estimate of the measurement variance at position z_j necessary for the set of equations to have an unambiguous solution. Computers are generally used for this purpose. If this decoding of the analytical signal yields insufficiently accurate results, prior physical or chemical separation must be carried out and the individual components must be determined separately.

1.3 Analytical Signals and Analytical Results

The analytical signal is always represented by a change in the physical state. For a specific determination of a single component $k = 1$, all the analytical information is contained in the signal intensity; for qualitative detection, the presence of a signal with an intensity distinguishable from zero is decisive; for quantitative determination the signal intensity is proportional to the analyte concentration. The shape of the signal is important only when it affects the precision of the determination of its intensity. For selective simultaneous determinations of several components ($k > 1$), the analytical information is contained in both the position and the intensity of the signal. The shape of the signal, which is part of the analytical "report", e.g. in graphical representation of a recording, has a relatively strong effect on the properties of the analytical information, i.e. the result. The shape of the signal will be dealt with in Section 1.3.1.

The signal, or a sequence of signals, as the carrier of the analytical information, has certain semiotic properties or fulfils some semiotic functions. The use of semiotics in analytical chemistry has been studied chiefly by Prof. H. Malissa [1].

Semiotics originally developed as a science dealing with signs (symbols) and their systems, but some results of semiotics can also be used in the theory of analytical signals. Modern semiotics can be divided into three main fields: syntax, semantics and pragmatics. *Syntax* treats the internal structure of systems of symbols and their interrelationships. For expression of the information properties of a sequence of analytical signals, their interrelations are important and it should also be borne in mind that e.g. reaction schemes, programming languages, etc. have a certain, frequently very rigorous syntactic structure. *Semantics* is a science dealing with the relationships of symbols (or signals) to the subjects they describe; for example, the semantic function of the analytical signal lies in its unambiguous relationship to the chemical composition of the test sample. The *pragmatic function* of the analytical information (signal, result) is given by its relationship to its recipient, especially when deciding on further procedure to be employed.

Processing of an analytical signal yields more or less of the inherent information, the properties of the analytical result being affected considerably by the properties of the signal processed. Therefore, a general assumption can be introduced that the intensity and sometimes also the position of the signal and the results of analysis are continuous random quantities and some of their properties can be studied together (e.g. the probability distribution, precision, information properties, etc., see Sections 1.3.3 and 1.3.4); other properties of signals and results are only loosely related or are not related at all and must be studied independently.

1.3.1 PROPERTIES OF ANALYTICAL SIGNALS

An analog two-dimensional recording of a sequence of signals obtained as the output in selective simultaneous determination of several components has the shape either of a stepped curve or of a series of more or less well separated peaks. The two shapes can be interconverted: differentiation of a stepped curve yields peaks and integration of peaks yields a stepped curve (Fig. 42). The signal shape is decisive for computer treatment. Some signal shapes are considered in Section 3.1.2.

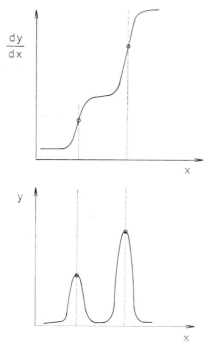

Fig. 42. Stepped curve and peaks.

1.3.2 ANALYTICAL SIGNAL AND RESULT AS PROBABILITY QUANTITIES

Phenomena that can, but need not occur, depending on chance, are termed *random phenomena* and quantities that can randomly attain various values are called *random quantities*. Adjustment of conditions under which a random phenomenon can occur or a random quantity can attain various values is called a *random experiment*. Chance plays a role here because, in addition to basic controlled conditions, other, often unknown circumstances also affect the result. Chemical analysis is also a random experiment.

A random phenomenon occurs or a random quantity attains a certain value with a certain probability. The probability that phenomenon A occurs is expressed classically as

$$P(A) = \frac{n(A)}{N} \tag{1.2}$$

where $n(A)$ is the number of occurrences of A and N is the number of all possible cases. The values of $n(A)$ and N can sometimes, e.g. in statistical thermodynamics, be determined by combinatorial analysis. However, this is impossible in common practice and thus a statistical definition of probability has been introduced,

$$P(A) = \lim_{n \to \infty} \frac{n(A)}{n} \tag{1.3}$$

where n is the number of random experiments carried out and the ratio, $n(A)/n$, is the *relative frequency of phenomenon A*. This definition is based on the statistical regularity, i.e. on the fact that the relative frequency of a certain result of a random experiment approaches a constant value on multiple repetition of the experiment.

The modern *axiomatic probability theory* (for more details see [2]) defines the probability, as number $P(A)$ assigned to random process A by a probability function, fulfilling the following three axioms:
1) $0 \leq P(A) \leq 1$,
2) the probability of a certain phenomenon, $P(S) = 1$,
3) if A_1, A_2, ... is a set of disjunct phenomena, then $P(U_i A_i) = \sum_i P(A_i)$.

For a uniform probability space, i.e. for the same probability of all random phenomena, Eq. (1.2), i.e. the classical definition of probability, can be used as the probability function.

As a random phenomenon is completely characterized by the probability of its occurrence, a random quantity attaining various values with different probabilities is characterized by a describing, i.e. either frequency or cumulative, distribution function. A random quantity can be either discrete or continuous. *Cumulative distribution function $F(x)$* expresses the probability that random quantity $\xi \leq x$, i.e. $F(x) = P\{\xi \leq x\}$. With a discrete random quantity that can attain only certain, e.g. integral values, the function can be represented by a stepped curve, and a continuous random quantity by a continuous curve. Another descriptive function is the probability density, $p(x)$, for which $p(x) = dF(x)/dx$. The probability that discrete random quantity ξ attains value $p\{a < \xi \leq b\} = \sum\limits_{a < x_i \leq b} p(x_i)$ and that continuous random quantity ξ attains this value is $P\{a < \xi \leq b\} = \int_a^b p(x)\,dx$. The probability density is sometimes also termed the *frequency function*, in view of the validity of Eq. (1.3).

Describing functions are suitable for calculation of the probability that $a < \xi \leq b$; however, in practice it is more often required to determine (a) the most probable expected value that can be attained by random quantity ξ and (b) the scatter of the other values around the most probable values. The expected value of discrete random quantity ξ is $E[\xi] = \sum\limits_{i=1}^{N} x_i\, P(x_i)$, where N is the number of all values attainable by the quantity, so that $\sum\limits_{i=1}^{N} P(x_i) = 1$. The expected value of a continuous random quantity is $E[\xi] = \int_{-\infty}^{\infty} x\, p(x)\,dx$, while $\int_{-\infty}^{\infty} p(x)\,dx = 1$. The variance, i.e. the characteristic of the scatter of values attained by the random quantity around the expected value is $V[\xi] = \sum\limits_{i=1}^{N} (x_i - E[\xi])^2\, P(x_i)$ for a discrete random quantity and $V[\xi] = \int_{-\infty}^{\infty} (x - E[\xi])^2\, p(x)\,dx$ for a continuous random quantity. These values are often denoted as $E[\xi] = \mu$ and $V[\xi] = \sigma^2$; the μ and σ^2 values have been mentioned in Vol. IA, p. 80 and the following pages.

An important distribution of a discrete random quantity is the *binomial distribution*, given by

$$f(k) = \binom{n}{k} P(A)^k \left(1 - P(A)\right)^{n-k} \tag{1.4}$$

where the binomial coefficient is $\binom{n}{k} = n!/k!\,(n-k)!$. The binomial distribution holds when phenomenon A with probability $P(A)$ occurs in k cases of n possible cases. It is a principal discrete probability distribution and is important e.g. in the determination of a maximal error. It can be denoted by symbol $Bi(P(A))$. The *Poisson distribution* is also derived from the binomial distribution,

$$f(k) = \frac{\lambda^k \cdot e^{-\lambda}}{k!} \tag{1.5}$$

and has a single parameter λ. It expresses the probability distribution of e.g. the number of particles in a certain volume, the number of occurrences per time interval, etc., if the mean number is λ. This distribution describes

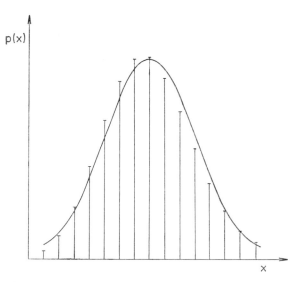

Fig. 43 Normal and Poisson distribution.

e.g. low-level β-counting; however, for larger λ values it is often approximated by a continuous normal distribution (Fig. 43). It can be briefly denoted as $Po(\lambda)$.

Continuous distributions find greater use in analytical chemistry. The results of quantitative analysis and the analytical signal intensity mostly have *normal (Gaussian) distribution* with the probability density,

$$p(x) = \frac{1}{\sigma \sqrt{(2\pi)}} e^{-\frac{1}{2}(x-\mu/\sigma)^2} \tag{1.6}$$

where μ is the mean value and σ^2 is the variance. It is briefly denoted as $N(\mu, \sigma^2)$. The plot of the Gauss curve and the significance of parameters μ

and σ^2 for the position and shape of the curve is given by Wood (vol. IA, p. 79 and following) and by Móritz (vol. XI, p. 22). If $\mu = X$, the results are accurate. The results of trace analysis or the signal intensity at values close to y_{min} are more often described by the *logarithmic-normal distribution*, i.e. not the values themselves but their logarithms are normally distributed. If the determination limit is x_0, then the distribution of the results is characterized by the probability density,

$$
p(x) = \begin{cases} = \dfrac{e^{-\frac{1}{2\vartheta^2}[\log(x-x_0)-\log w.x_0]^2}}{(x-x_0)\,\vartheta\,\sqrt{(2\pi)}} & x > x_0 \\ = 0 & x \leqq x_0 \end{cases} \tag{1.7}
$$

This distribution has three parameters: x_0 — the coefficient of shift, w — the coefficient of asymmetry and ϑ — the coefficient of scatter. The expected value is then $\mu = x_0(1 + we^{\frac{1}{2}\vartheta^2})$ and the variance, $\sigma^2 = x_0^2 w^2 (e^{\vartheta^2} - 1)\, e^{\vartheta^2}$;

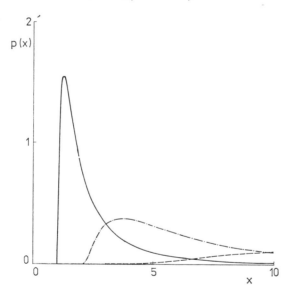

Fig. 44. Logarithmic-normal distribution.

the modus, i.e. the most probable value of this distribution is $M = = x_0(1 + we^{-\vartheta^2})$. The curve of the log-normal distribution is asymmetrical and its asymmetry increases with decreasing coefficient of asymmetry w.

268

The curve becomes almost symmetrical for large values of w. Several curves of log-normal distribution, briefly denoted as $LN(x_0, w, \vartheta^2)$ are given in Fig. 44. Another continuous probability distribution which is, however, mostly used as a characteristic of preliminary assumptions about the value of X_i, is the *uniform (rectangular) distribution* with the probability density

$$p(x) = \begin{cases} = \dfrac{1}{x_2 - x_1} & x \in (x_1, x_2) \\ = 0 & \text{for other } x \end{cases} \tag{1.8}$$

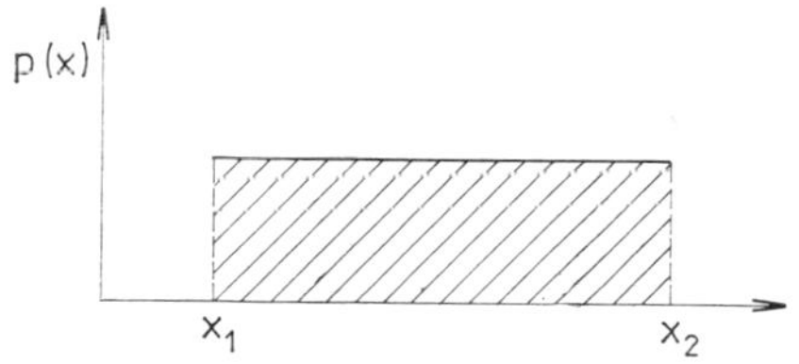

Fig. 45. Uniform distribution.

which is depicted in Fig. 45. The expected value of the distribution is then $\mu = \frac{1}{2}(x_1 + x_2)$ and the variance, $\sigma^2 = \frac{1}{12}(x_2 - x_1)^2$. This distribution is denoted as $U(x_1, x_2)$.

1.3.3 STATISTICAL CHARACTERISTICS OF THE PROPERTIES OF ANALYTICAL SIGNALS AND RESULTS

As shown in Section 1.2 and according to E. C. Wood, Vol. IA, p. 85 and following, the expected value $E[\xi]$ and variance $V[\xi]$ are important characteristics. Difference $\delta = |X - E[\xi]|$, termed the mean error, characterizes the accuracy of the results and $\sigma_x^2 = V[\xi]$ characterizes the precision of the results or $\sigma_y^2 = V[\eta]$ characterizes the precision of the signal intensity. If the signal position is also a random quantity, then the precision of the position is $\sigma_z^2 = V[\zeta]$. The expected value and the variance are themselves random quantities, cannot be precisely determined for random selections and must be estimated. Various relationships can be used to estimate random quantities, which lead to estimates of somewhat different values and properties. *Point estimates*, characterizing the value estimated by a single numerical value, or interval estimates, characterizing it by a value interval, are used. The relationships for estimation are mostly determined by the maximum probability method of **R. A. Fischer**, a special case of which is the least squares method treated in Section 3.3.2.1. In computer processing of analytical data, estimates other than the most probable ones are sometimes useful.

The most probable point estimate of the expected value of the normal distribution is the *arithmetic mean*,

$$\bar{x} = \frac{1}{n}(x_1 + x_2 + \ldots + x_n) = \frac{1}{n}\sum_{j=1}^{n} x_j \tag{1.9}$$

Calculation of this estimate is very easy and many table-top and pocket calculators contain a microprogram for it. Another, less common estimate of the expected value of the normal distribution is median $\tilde{x}$, i.e. the middle result in a series of results ordered according to their magnitude, or the mean of the two middle results if the number of results is even. The estimate of the expected value of the log-normal distribution is the geometric mean of the set.

The most common estimate of the variance of normally distributed values is the *standard deviation* which, for n parallel determinations on a single sample, is given by

$$s = \sqrt{\left(\frac{1}{n-1}\sum_{j=1}^{n}(\bar{x} - x_j)^2\right)} \tag{1.10a}$$

When m samples $l = 1, 2, \ldots, m$ are analyzed n_l times each, then

$$s = \sqrt{\left(\frac{1}{N-m}\sum_{j=1}^{n_l}(\bar{x}_l - x_{j,l})^2\right)} \tag{1.10b}$$

where the number of all determinations $N = \sum_{l=1}^{m} n_l$ and the mean of the analyses of the l-th sample is $\bar{x}_l$. The sum of the squares of the deviations in Eqs. 1.10a, b can be calculated as

$$\sum_{j=1}^{n}(\bar{x} - x_j)^2 = \sum_{j=1}^{n} x_j^2 - \frac{1}{n}\left(\sum_{j=1}^{n}(x_j)\right)^2 ,$$

especially when a computer or a programmable calculator is available. The value of s for the logarithmic-normal distribution is calculated from Eq.

270

(1.10a) or (1.10b), but x_j is replaced by $\log x_j$ and $\bar{x}$ or $\bar{x}_l$ by the *geometric mean*. Another very simple, though seldom used estimate of the variance of normally distributed values is *range* $R = x_n - x_1$, where $x_1 < x_2 < \ldots$ $\ldots < x_n$, i.e. the difference between the largest and smallest value of a series of parallel determinations.

Point estimates, as random quantities, have a certain probability distribution and thus also a variance characterizing the precision of the estimation. For example, the precision of the arithmetic mean is characterized by the standard deviation of the mean,

$$s_{\bar{x}} = \frac{1}{\sqrt{n}} \, s = \sqrt{\left(\frac{1}{n(n-1)} \sum_{j=1}^{n} (\bar{x} - x_j)^2 \right)} \tag{1.10c}$$

Hence a point estimate and its precision must be characterized by two values, e.g. the estimate of the expected value of normally distributed results by $\bar{x}$, and the precision of the estimation by $s_{\bar{x}}$.

For this reason, *interval estimates* are often used. For example, the *confidence interval* is used for estimation of the expected value of normally distributed results, for which it holds that accurate result X lies in the interval, $\bar{x} \pm t(v, \alpha) \, s_{\bar{x}}$, with probability $(1 - \alpha)$, which can be written as

$$P\{\bar{x} - t(v, \alpha) \, s_{\bar{x}} \leqq X \leqq \bar{x} + t(v, \alpha) \, s_{\bar{x}}\} = (1 - \alpha) \tag{1.11a}$$

The critical values of the Student t-distribution for a chosen significance level α and the number of degrees of freedom $v = n - 1$, i.e. the $t(v, \alpha)$ values, are given in Table II, vol. IA, p. 88, for $\alpha = 0.10$, 0.05 and 0.01. Approximate calculations will be treated below (Appendix D). Eq. (1.11a) is valid only if the result is not subject to a systematic error (unbiased), i.e. when $\mu = X$. The confidence interval is then symmetrical with respect to μ. If the result is subject to a systematic error, $\delta \neq 0$, the confidence interval for $\delta > 0$ is given by

$$P\{\bar{x} - 2\delta - t(v, \alpha, \delta) \, s_{\bar{x}} \leqq X \leqq \bar{x} + t(v, \alpha, \delta) \, s_{\bar{x}}\} = (1 - \alpha) \tag{1.11b}$$

and for $\delta < 0$ by

$$P\{\bar{x} - t(v, \alpha, \delta) \, s_{\bar{x}} \leqq X \leqq \bar{x} - 2\delta + t(v, \alpha, \delta) \, s_{\bar{x}}\} = (1 - \alpha) \tag{1.11c}$$

where $t(v, \alpha, \delta)$ is the critical value of the non-central t-distribution for $v = n - 1$, significance level α and the δ value. The critical values of the non-central t-distribution are tabulated or can be found by approximation (see Appendix E). The confidence interval according to Eqs. (1.11b, c) is symmetrical with respect to the accurate X value, but not with respect to μ, i.e. the determined expected value.

The mean standard deviation, mean error and the confidence interval are statistical characteristics of analytical results that quite suffice for common practice.

1.4 Information Characteristics of Analytical Signals, Results and Methods

The purpose of chemical analyses is obtaining information on the chemical composition of test samples; therefore, analytical results and methods of analysis have also recently been studied from the point of view of their information properties. Any information, including that obtained by analysis, has both qualitative and quantitative aspects.

The *qualitative aspect of information* is given e.g. by its relevance; it is often more important than the quantitative aspect, but cannot be expressed numerically. It is usually sufficient if we can decide that information $i \in I$ obtained from an analysis is relevant or irrelevant in a given problem, i.e. if we can assign it to the subset of relevant or irrelevant information. If the subset of relevant information is sharp, i.e. $I_r \subset I$, the competence of i to I_r is described by a function $\varphi(i)$ such that $\varphi(i) = 0$ for $i \notin I_r$, and $\varphi(i) = 1$ for $i \in I_r$. Then we have only two alternatives: information is either completely relevant or completely irrelevant. However, if we hold the subset of relevant information as fuzzy, i.e. $I_r \subsetneq I$, the competence of i to the fuzzy subset I_r can be characterized by a function $\tilde{\varphi}(i)$ which continuously assumes values from 0 through 1 and is the closer to one, the more i belongs to the fuzzy subset I_r. If we can attribute a particular value c of the function $\tilde{\varphi}(i)$ to an individual result, e.g. from experience, these items of information can be ordered, or at least grouped, into several classes according to their relevance. Relevance is often also dependent on the order of obtaining information. The analysis, as a process of obtaining information, can sometimes be controlled or optimized, so that the most relevant information is obtained with greatest possible accuracy and precision.

The *quantitative aspect of information* is expressed numerically, in terms of exactly defined dimensionless units. Originally [3], the information content of the result of an observation or measurement was expressed as

272

ld N, where N is the number of possible differentiations and ld is $\log_2$. This method, applicable, e.g., for expressing the information obtained by carrying out a simple qualitative proof, is not suitable for the results of instrumental qualitative, identifitication and quantitative analyses. Therefore, newer methods of determining the quantitative aspect of analytical information, based on mathematical information theory, will be discussed.

1.4.1 INFORMATION GAIN, INFORMATION CONTENT AND THE AMOUNT OF INFORMATION

Information is defined as a decrease in the uncertainty of knowledge about a certain real process or object. The uncertainty of a random variable is given by the *Shannon entropy*, which, for ξ as a discrete random variable attaining values x_i, $i = 1, 2, \ldots$ with probability $P(x_i)$, is given by

$$H(P) = - \sum_i P(x_i) \log P(x_i) \tag{1.12a}$$

For ξ as a continuous random variable with probability density $p(x)$, the Shannon entropy is given by

$$H(p) = - \int_{x_1}^{x_2} p(x) \log p(x) \, dx \tag{1.12b}$$

It must hold in Eq. (1.12a) that $\sum_i P(x_i) = 1$ and in Eq. (1.12b) integration limits x_1 and x_2 must be chosen so that $\int_{x_1}^{x_2} p(x) \, dx = 1$. Entropy is a quantity with a broad significance and always characterizes uncertainty or disorder in a system.

For two dependent random variables ξ and η, i.e. for ξ attaining values x_i, $i = 1, 2, \ldots, n$ with probability $P(x_i)$ and for η, attaining values y_j, $j = 1, 2, \ldots, m$ with probability $P(y_j)$, the conditional probability $P(x_i/y_j)$, that ξ attains value x_i, if η attains value y_j, is given. Shannon entropy [2, 4]

$$H(P(x/y_j)) = - \sum_{i=1}^{n} P(x_i/y_j) \log P(x_i/y_j) \tag{1.12c}$$

characterizes the uncertainty of random variable ξ, given value y_j of η. The expected value of the uncertainty of a random variable ξ for given η is characterized by the equivocation

$$H(P(x/y)) = \sum_{j=1}^{m} P(y_j)\, H(P(x/y_j)) =$$

$$= -\sum_{j=1}^{m} \sum_{i=1}^{n} P(y_j)\, P(x_i/y_j) \log P(x_i/y_j) \tag{1.12d}$$

D. F. Kerridge introduced the quantity

$$H(q, p) = -\int_{x_1}^{x_2} q(x) \log p(x)\, \mathrm{d}x \tag{1.13}$$

and termed it the *inaccuracy*. This quantity characterizes the uncertainty of a continuous random variable when it is assumed that its probability density is $p(x)$, whereas its actual distribution is $q(x)$. The inaccuracy thus involves the degree of uncertainty of random variable ξ and the error of the determination of its distribution as $p(x)$, whereas the correct distribution is $q(x)$.

The *information gain* obtained by carrying out the analysis is given by the difference in the a priori uncertainty $H(p_0)$, where the a priori distribution $p_0(x)$ is a probability distribution characterizing the a priori knowledge of the identity i or of the value X and distribution $p(x)$ is a posteriori distribution. The entropy of the a posteriori distribution characterizes the residual uncertainty after the analysis was made in order to obtain some specific analytical report. The difference in the Shannon entropy for the a priori and a posteriori distribution, i.e. $H(p_0) - H(p)$, can be used for evaluation of the results of qualitative analyses [5]. This difference is, however, not a very suitable criterion of the information gain of quantitative analyses and measurements, because it cannot differentiate whether analytical results confirm the a priori assumption. The difference $I(p, p_0) = H(p, p_0) - H(p)$, which, for a continuous random variable, can be written as

$$I(p, p_0) = \int_{x_1}^{x_2} p(x) \log p(x)\, \mathrm{d}x - \int_{x_1}^{x_2} p(x) \log p_0(x)\, \mathrm{d}x =$$

$$= \int_{x_1}^{x_2} p(x) \log \frac{p(x)}{p_0(x)}\, \mathrm{d}x \tag{1.14a}$$

is better suited for the practice of quantitative analysis and is termed the

274

divergence measure of the information gain, because it also expresses the divergence, i.e. the dissimilarity between the a priori $[p_0(x)]$ and the a posteriori $[p(x)]$ distribution. The divergence measure of the information gain was derived in a different way to that of the difference between the Kerridge inaccuracy of the a priori distribution and the Shannon entropy of the a posteriori distribution, but this expression is very lucid and is always logically justified: before the analysis, the analyst assumes that quantity ξ has distribution $p_0(x)$ and the analysis shows that the real distribution is $p(x)$. Such an ambiguity of a probability distribution is characterized by

Information Gain of Analytical Results

$p_0(x)$	$p(x)$	$I(p, p_0)$	Note
$\dfrac{1}{k_0}$	$\dfrac{1}{k}$	$\ln \dfrac{k_0}{k}$	Information gain of results of qualitative analyses; k_0 and k are the numbers of possible, but still unidentified components before and after detection, resp.
$U(x_1, x_2)$	$N(\mu, \sigma^2)$	$\ln \dfrac{x_2 - x_1}{\sigma \sqrt{(2\pi e)}}$	Information gain of results of quantitative analyses, when it is assumed that $X \in \langle x_1, x_2 \rangle$ and the result confirms this assumption.
$N(\mu_0, \sigma_0^2)$	$N(\mu, \sigma^2)$	$\ln \dfrac{\sigma_0}{\sigma} + \dfrac{1}{2}\left[\left(\dfrac{\mu - \mu_0}{\sigma_0}\right)^2 + \dfrac{\sigma^2 - \sigma_0^2}{\sigma_0^2}\right]$	Information gain of the results of a refined analysis, when the results of preliminary analysis are normally distributed with parameters μ_0 and μ_0^2 and the results of refined analysis are also normally distributed with parameters μ and σ^2 and it holds that $\sigma^2 \leqq \sigma_0^2$.
$U(0, x_1)$	$LN(x_0, w, \vartheta^2)$	$\ln \dfrac{x_1}{x_0}$	Information gain of the results of trace analysis, when $X < x_0$, where x_0 is the determination limit.
		$\ln \dfrac{x_1}{x_0} \dfrac{1}{w\vartheta \sqrt{(2\pi e)}}$	Information gain of the results of trace analysis when $X > x_0$.

quantity $H(p, p_0)$. As the divergence is $I(p, p_0) \geq 0$, the equality sign holds if $p_0(x) = p(x)$. The divergence measure makes it possible to introduce relationships expressing the information gain specifically for various analyses. These relationships are given and discussed in more detail in the monograph by Eckschlager and Štěpánek [2]; only a few of them, for the most common types of analytical result, are surveyed below.

An analytical procedure can be characterized better by its expected value, i.e. by information content I, than by the information gain. The information gain, e.g. in an instrumental qualitative analysis, is given by the difference between the entropy of the a priori distribution,

$$H\left(P_0\left(x_i\right)\right) = -\sum_{i=1}^{n} P_0(x_i) \cdot \log P_0(x_i),$$

and the entropy of the conditional probability, i being the a priori expected identity, when the signal in position z_j is measured, i.e.
$H(P(x_i/z_j)) = -\sum_{i=1}^{n}\sum_{j=1}^{m} P(x_1/z_j) \log P(x_i/z_j)$. The expected value of the
information gain, the information content, si given by the difference between the entropy of the a priori distribution, $H(P_0(x_i))$, and the equivocation, $H(P(x/z))$, i.e. by the difference

$$I = -\sum_{i=1}^{n} P_0(x_i) \log P_0(x_i) + \sum_{j=1}^{m}\sum_{i=1}^{n} P(z_j) P(x_i/z_j) \log P(x_i/z_j) =$$

$$= \sum_{j=1}^{m}\sum_{i=1}^{n} P(z_j) P(x_i/z_j) \log \frac{P(x_i/z_j)}{P_0(x_i)} = \sum_{j=1}^{m} P(z_j) I(P(x_i/z_j), P_0/(x_i)) \qquad (1.14b)$$

This expression shows that the information content can also be regarded as the expected value of the information gain expressed by the divergence measure

$$I(P(x_i/z_j), P_0(x_i)) = \sum_{i=1}^{n} P(x_i/z_j) \log \frac{P(x_i/z_j)}{P_0(x_i)} \qquad (1.14c)$$

More details about the information gain and content of instrumental qualitative analyses are given by Cleij and Dijkstra in the paper [5].

If several components $(k > 1)$ are determined e.g. by an instrumental method, the amount of information obtained by the analysis is determined

276

from

$$M(p, p_0) = \sum_{i=1}^{k} I(p, p_0)_i \tag{1.15a}$$

For example, in spectral analysis simultaneously determining major, minor and trace components, some of which are not found at all in the test sample, the amount of information obtained is

$$M(p, p_0) = \sum_{i=1}^{k_1} \ln \frac{(x_2 - x_1)_i}{\sigma_i \sqrt{(2\pi e)}} + \sum_{i=1}^{k_2} \ln \frac{x_{1,i}}{x_{0,i}} + \sum_{i=1}^{k_3} \ln \frac{x_{1,i}}{x_{0,i} w_i \vartheta_i \sqrt{(2\pi e)}} \tag{1.15b}$$

where k_1 is the number of major and minor components determined, k_2 is the number of components assumed but not found and k_3 is the number of trace components determined; $k = k_1 + k_2 + k_3$.

The information gain, content or the amount of information used need not always be related to the distribution of the analytical results and the a priori assumptions; it is, e.g., possible to determine the information gain of an analytical signal or the amount of information contained in an analytical "report". For example, if it is known that the signal intensity is $y \in \langle y_{\min}, y_{\max} \rangle$ and the σ_y value is also known, then

$$I(p, p_0) = \ln (y_{\max} - y_{\min})/(\bar{\sigma}_y \sqrt{(2\pi e)})$$

and the maximum amount of information obtainable from the analysis is

$$M(p, p_0) = \frac{z_{\max} - z_{\min}}{\Delta z} \ln \frac{y_{\max} - y_{\min}}{\bar{\sigma}_y \sqrt{(2\pi e)}}$$

where $y_{\min}, z_{\min}, y_{\max}$ and $z_{\max}$ are the lowest and highest values of the signal intensity and position, respectively, that can be recorded by the instrument and $\bar{\sigma}_y$ is the mean value of the standard deviation of the signal intensity. The Δz value, i.e. the smallest distance between the signals necessary for resolution, depends on the signal profile and on whether the peak area or the $y_{\max}$ value is used as the signal intensity (see Section 1.3.1).

The information gain, content and the amount of information are in dimensionless units, dependent on the base of the logarithm in Eq. (1.14). In more complex cases the information units are characterized by pairs of variables that correspond to a curve connecting points for which $I(p, p_0) = 1$ [2].

1.4.2 OTHER INFORMATION QUANTITIES

In practice analytical information is required that not only has a sufficient content, but can be obtained in a short time and at the lowest possible cost. Therefore, various quantities are introduced, including the *information performance*,

$$L(p, p_0) = \frac{1}{t_A} M(p, p_0) \tag{1.16}$$

where t_A is the analysis time, and the *specific information price*,

$$C(p, p_0) = \frac{\tau_A}{M(p, p_0)} \tag{1.17}$$

where τ_A is the cost of the analysis. These two quantities make it possible to evaluate analytical methods absolutely, i.e. without respect to their use for solution of a concrete analytical problem.

If an analytical method is to be selected or evaluated relatively, i.e. with regard to its concrete practical use, a characteristic quantity termed the *information profitability* is introduced,

$$P(p, p_0) = \frac{M(p, p_0)}{\tau_A} E \sum_{i=1}^{k} c_i = \frac{E}{\tau_A} \sum_{i=1}^{k} I_i c_i \tag{1.18}$$

where c_i is the relevance coefficient (see Section 1.4); for the i-th component, the efficiency $E = \varepsilon_1 \cdot \varepsilon_2 \ldots \varepsilon_n = \prod_{i=1}^{n} \varepsilon_i$, $0 \leqq E \leqq 1$. Efficiency coefficients ε_i are given by the ratio of the values of the characteristics of the required properties to those actually yielded by the method [7].

Information evaluation of analytical results and methods is advantageous in that it can also be used for methods whose results cannot be characterized statistically (e.g. qualitative analysis, identification methods, structural analysis methods, etc.), permitting comparison of these methods. Information quantities such as response functions have a certain importance for optimization of analytical procedures. Applications of information theory in analytical chemistry are surveyed in the monograph [2] and in [6].

1.4.3 THE EFFECT OF PROCESSING OF RESULTS ON THEIR INFORMATION PROPERTIES

Processing, i.e. decoding, of analytical signals or their sequences, whether carried out by modern computing techniques or by simpler means, does not create new information.

Note: The question of whether data processing creates new information was discussed when computing techniques were first introduced; see e.g. Leon Brillouin [3].

The method of processing the analytical signal and the results has a substantial effect on the information properties for the following reasons:

(1) Suitable data processing, e.g. using a computer, enables decoding of information which is contained in this data, but which would be lost through improper handling, most frequently information on the properties of the analytical results, but also on the presence of a trace component, etc. Hence the $M(p, p_0)$ value is affected directly.

(2) A suitable and sufficiently rapid method of handling the analytical signal affects the parameters that influence the information content or the amount of information: the density of data reading, calibration even with a non-linear dependence and smoothing of the analytical signal affect the value of σ, an improvement in the signal-to-noise ratio decreases the determination limit and subtraction of the blank decreases mean error δ, thus increasing the information content of the signal and of the result calculated from the signal. Separation of overlapping signals improves the selectivity of the determination, thus increasing the amount of information contained in a sequence of analytical signals. Automated data handling mostly substantially shortens the analysis time, t_A, which includes the time required for data processing. In this way, the information flow and performance $L(p, p_0)$ of the analytical method are affected. On the other hand, the use of a computer generally increases cost τ_A of obtaining the information required.

(3) On-line coupling of an analytical instrument with a computer and computer control of the instrument operation to optimize the information obtaining process affect the information profitability of the whole instrument-computer system.

Therefore, the suitability of introducing automated data processing is sometimes evaluated by comparing particular information quantitites, most often $P(p, p_0)$, for processing with and without the use of computing techniques. Of course, it is necessary that all the properties of the analytical results and methods that are affected by various techniques of result processing be evaluated and expressed in terms of coefficients ε_i (Section 1.4.2).

1.5 Testing of Analytical Results

One of the most important uses of the methods of mathematical statistics in analytical practice is result testing; its principles are explained and some tests given by E. C. Wood in Vol. IA, pp. 94–97. Here only the basis of tests for checking whether a result from a series of parallel determinations is subject to a gross error will be given. First of all, a test for finding whether a result is outlying will be given, because this test is a common part of most programs for processing experimental data, and a check based on comparison of the actual and maximum possible difference between parallel determinations, which is very advantageous in cumulative treatment of data pairs, will be described.

It holds for result x_j, $j = 1, 2, \ldots, n$, where n is the number of repeated determinations and x_j belongs in a normal selective set with distribution

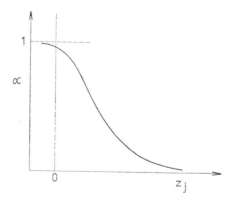

Fig. 46. Operating characteristic curve for statistical testing.

$N(\mu, \sigma^2)$, that variable $t_j = |(\mu - x_j)/\sigma|$ has greater probability that it will not be exceeded, the smaller its value (see Fig. 46). Occurrence of a large z_j value is so improbable that it is generally explained by a gross error in the determination of x_j. This value is then termed *outlying*, is excluded from the

280

set of results and is not used in the calculation of the mean, the coefficients of a regression equation, etc. In practice, t_j is estimated, termed the *Grubbs characteristic of outlying results*,

$$T_n = \left| \frac{\bar{x} - x_j}{s_n} \right| \qquad (1.19)$$

where the estimate of parameter μ is found as $\bar{x} = (1/n) \sum_{j=1}^{n} x_j$ and that of

σ as $s_n = \sqrt{\left(\frac{1}{n} \sum_{j=1}^{n} (\bar{x} - x_j)^2 \right)}$

In testing, the T_n value is calculated and compared with the critical $T(\alpha, n)$ value, which is tabulated or determined by approximation (see Appendix D) for a chosen significance level α and a given number of results n. If $T_n <$ $< T(\alpha, n)$, the result x_j is not outlying; if $T_n \geq T(\alpha, n)$, result x_j is outlying at significance level α and should be excluded. For a different test of outlying results, see Vol. XI, p. 31 − 35. It is evident that, in a set of results ordered according to their magnitude, $x_1 < x_2 < \ldots < x_n$, only the limiting values, i.e. x_1 and x_n are tested to find if they are outlying. When using a computer, it is normally faster to test all the results. On excluding one or several results from a set of data, the $\bar{x}$ and especially the s_n values change somewhat; it is therefore necessary to determine them again and to test the remaining results once more.

Note: The s_n value in the denominator of Eq. (1.19) is not identical with estimate s in. Eq. (1.10a) which is commonly determined as a characteristic of the precision of the results, for determination of the confidence inerval, etc. The Grubbs characteristic is therefore often calculated in practice as

$$T_n = \left| \frac{\bar{x} - x_j}{s} \right| \sqrt{\frac{n - 1}{n}} \, ,$$

where s is substituted from Eq. (1.10a).

In analytical practice, especially in quality control, two parallel determinations x_j, $j = 1, 2$, are often carried out. Two results cannot be treated by the Grubbs test and thus the maximum permitted difference between two parallel determinations, R_{max}, is often determined as

$$R_{max} = a_R \cdot \bar{R} = b_s \cdot s \, ,$$

281

where $\bar{R}$ is the mean range of two parallel determinations, s is the standard deviation estimated from Eq. (1.10b) or, in calculation for a pair of parallel determinations, from the equation

$$s = \frac{1}{2m} \sum_{i=1}^{m} R_i^2 .$$

The following coefficients are used for calculation of the permitted difference of two parallel determinations at various significance levels α:

α	a_R	b_s
0.01	2.06	2.33
0.05	2.46	2.77
0.10	3.23	3.64

The test is then carried out by comparing the real range $R = |x_1 - x_2|$, i.e. the difference in two parallel determinations, with the R_{max} value. If $R \geq R_{max}$, a third determination must be carried out and, if necessary, a check by the Grubbs test, to discover if any of these three results is outlying.

2 Application of Computers in Analytical Chemistry

2.1 Fields of Use of Computing Techniques

Analysis is the process of obtaining information on the composition of substances and mixtures. The quality of the resultant information depends on the experimental methods used, control of the analytical experiments, the quality of recording of the experimental information and, lastly, on the quality of the data processing.

Analytical chemistry is closely related to other physical sciences in which progress is based on measurement of specific properties. The most important form of experimentally obtained information is qualitative information expressed numerically. Numerical data can correspond to isolated facts, e.g. physical properties of substances, or can be a measure of a relationship between two or more quantities, such as the refractive index of a liquid and its temperature.

The increasing use of instrumental methods in analytical chemistry and development of new analytical methods result in extensive sets of experimental data. Improving parameters of analytical instruments, such as increasing the sensitivity, expanding the measuring range or increasing the recording speed, lead to an increase in the volume of information obtained. The more productive and effective the instruments used, the greater the amount of information that must be handled. The use of semi-automatic and automatic instruments facilitates the collection of experimental data, enables simple repetition of experiments and results in extensive sets of data, from which more precise information can be obtained, suppressing the effect of random errors on the results. Evaluation of the information from sets of experimental data necessitates introduction of new procedures and methods, because the time requirements on their processing rapidly increase.

The use of computers for recording, storage and processing of the information obtained is then the only possible solution. The application of computers for direct control of analytical processes, not only for maintaining the parameters affecting the course of the analytical experiment at a required level, but also for their programmed variation, is equally important. The development of computers has extended the potentialities of analytical chemistry, has strongly affected the style of work in the analytical laboratory and has vastly increased its productivity [8]. Introduction of many modern analytical methods depends on the existence of computing techniques, although the principles of these methods were known much earlier.

These facts unambiguously indicate the advantages of computing techniques in the analytical laboratory, but these techniques must not be applied without meeting all the necessary conditions. To be able to correctly decide when the application of a computer for information processing is useful, its potentialities must be considered:

(A) The rate of mathematical operations. This rate for large computing systems is of the order of 10^6 instructions per second, but even minicomputers used in laboratories are capable of carrying out hundreds of thousands of operations per second. The rate of computation is not only important from the point of view of handling experimental data in real time, but also for the extent of the theoretical and simulation computations that can be performed.

(B) Errorless performance of mathematical operations for virtually unlimited time.

(C) Ability to perform logic operations, which is used not only in processing data according to a selected algorithm, but also in decisions and control of the experiment by an on-line computer.

(D) The computer input can be directly interfaced to the output of an analytical instrument and the information can be processed immediately, provided that the information can be very rapidly transferred not only inside the computer, but also from the laboratory instruments, i.e. from the peripheries.

(E) Computers can control operation of analytical instruments.

(F) Computers are capable of storing in a memory and rapid recalling of large sets of numerical data. The data can be stored not only in the computing memory of the computer, but also in various external media. The large memory capacity of computers has enabled computerized collections of spectra and structural data to be drawn up, which are used in identification of unknown samples.

(G) The input data and the results obtained can be transmitted to any distance and presented in various forms. Computer terminals located in laboratories can be used for interaction with computers, thus assuring a substantial increase in the effectiveness of the experimental data handling.

This survey is, of course, not exhaustive and lists only the principal advantages of application of computers in analytical practice. However, computerized information processing also entails some difficulties:

(A) The input data must be prepared in the form in which the computer can handle it.

(B) A suitable procedure (algorithm) must be found for handling the information and a general program must be worked out and stored in the computer memory.

It is sometimes rather difficult to find a suitable algorithm and work out an effective program for handling the information that is to be treated, i.e. on the basis of visual perception using previous experience.

2.2 Advantages of Information Collection and Processing by Computers

Computers at present find application in analytical chemistry not only in combination with complex and expensive instruments, but also in many other tasks, such as:

1) Automatic collection of experimental information combined with fast handling of extensive sets of experimental data, which facilitates repetition of analytical experiments and results in improved reliability and precision of final data.

2) The capacity of the memory enables use of large collections of data, again leading to improved reliability of the resultant information.

3) Control of analytical experiments by a computer shortens the analysis time, thus improving the exploitation of the instrument, and optimization of the control improves the reliability of the results.

4) Errorless and rapid performance of mathematical operations enables selection of more complicated mathematical procedures, thus increasing the information efficiency of the experimental results.

Some analytical experiments could not be carried out in practice without computer control, because they require simultaneous control of many parameters. The mathematical treatment of information from some experimental methods is also so complicated that it cannot be carried out without a computer.

From the point of view of information transfer between the analytical instrument and the computer, two principal applications of computers are differentiated. One involves general computations, during which the formation and the processing of information are separated in time. This is

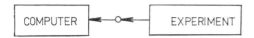

Fig. 47. Off-line use of a computer.

off-line use of a computer (Fig. 47). Any universal computer with a suitable interface and an output device for presentation of the results can handle data in this way. The other method of application is on-line connection, which enables handling of information immediately after its collection and also control of the measuring apparatus on the basis of the processed experimental information. This procedure is sometimes called computation in real time, i.e. in the time during which the experiment takes place. These two application variations are characterized below in more detail.

2.3 Off-line Problem Solution

This group includes all analytical chemistry problems, in which sets of data stored on various media are treated during or after the experiment. Examples are the treatment of X-ray diffractograms, simulation of spectra or statistical mathematical treatment of analytical results.

Sets of input data can be obtained in various ways. A common example is a set of data taken from an industrial laboratory diary, which e.g. consists of titrant consumption − sample weight data pairs for determination of an active component in samples taken at preset time intervals from the production line. If this set of data is complemented by the titrant concentration data, all the information required for computation is available.

Another example is a data set recorded directly on a punched tape by a puncher connected with the digital output of an analyzer for oxygen in water.

It can be seen from these examples that the sets of data are obtained in very different ways, but they always have the character of numerical information stored in a medium in a form suitable for computer treatment. The data can then be stored for any length of time and treated in batches after obtaining a greater amount of data or according to the capacity of the computing time available. The computation can be arbitrarily interrupted and restarted without loss of information.

The versatility of computations is increased by terminal networks of many computing centres which simplify access to the computer and enable problem solution to be carried out on a computer of adequate power.

2.4 On-line Coupling of an Instrument with a Computer

This connection has two variants, namely, an open-loop system for data collection and a closed-loop system with feedback. Examples of these variants are e.g. selective data collection from a gas chromatograph, programmed temperature change during an experiment and regulation of the homogeneity of the magnetic field in NMR spectrometers, on the basis of continuous data treatment.

In open-loop systems, the analytical instrument is actually part of the computer periphery, but operates independently of the computer. The computer secures transfer, recording and treatment of the experimental information at preset time intervals determined by the program or at the instant when the data is formed in the experiment. The final mathematical evaluation of the data is usually carried out after completion of the experiment. The demands on the computation rate are usually not high.

In closed-loop systems the data must be transferred from the instrument to the computer at selected time intervals; it is also necessary to transfer regulating impulses from the computer to the instrument. The computer must operate in real time, i.e. must handle the information obtained in a time short enough for control of the experiment taking place. The computation cannot be interrupted without loss of information. Hence the requirements on the computing rate are higher than in the previous case and the computing algorithm must be chosen considering the duration of the experiment, even at the cost of some simplifications or approximations made in the rigorous computing procedure.

On-line coupling can involve a single-purpose computer (controlling computer), constructed especially for the given analytical instrument (this is often a microprocessor), or the analytical instrument can be coupled with a time-shared computing system. The prices of minicomputers as evaluating devices are at present acceptable compared with the prices of the instruments themselves and thus data can be handled in real time or several instruments can be controlled simultaneously.

An important problem associated with closed-loop systems is the availability of digitally operated regulating elements for control of the analytical experiment.

This type of computer use has revolutionized the style of work in analytical laboratories and has paved the way for new principles and methods.

In places where the conditions ensure effective application of on-line computers, networks of interconnected analytical instruments and computers are often formed. Some connecting variants will be depicted schematically:

1) A single-purpose computer connected to a laboratory instrument;

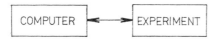

Fig. 48.

this instrument is designed for a single application (Fig. 48). The software is very simple and the whole procedure of obtaining and decoding the in-

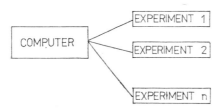

Fig. 49.

formation is optimized. Various commercial automatic analyzers are typical examples. If the systems are not used routinely (steel-mill analyses, clinical diagnosis, environmental pollution monitoring, etc.), their use is not very effective, because the configuration must meet maximal demands,

288

whereas most common requirements are substantially lower. The preparation of programs is limited by the given configuration and thus it is usually carried out on a larger computer. This originally used and later abandoned variant is recommended now in view of the availability of microprocessors at constantly decreasing prices.

2) Time-shared computers: the cost decreases with an increasing number of instruments (Fig. 49) connected to a single computer. A computer with wider possibilities and peripheries can be selected. The reliability of the whole system may become a problem. The software is more complex and thus errors can occur. The response time to the instrument signal is longer.

3) The star network is a combination of the two previous variants (Fig. 50).

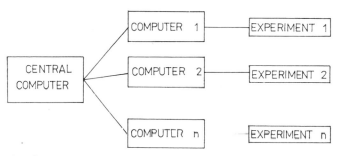

Fig. 50.

Here, small, single-purpose computers carry out the operation for which they were constructed, i.e. control the measuring process, the collection and consistency of the data, etc. The more complex evaluating mathematical procedures that cannot be carried out on a small computer are then performed on the central computer, which also serves for preparation of programs for single-purpose computers and can be used for other external computations. The communication between the central computer and the instruments is minimized.

4) Hierarchical and general networks and special laboratory computing systems make the use of computers and instruments more economical; the interconnection of instruments and computers has many variants (Fig. 51). Of course, formation of such a network is rather costly. Recent development is leading to introduction of some international standards ensuring communication among instruments and computers from various manu-

289

facturers, programming languages for work in real time are being developed and software facilitating formation of various open laboratory computing systems that can be supplemented and extended is being introduced.

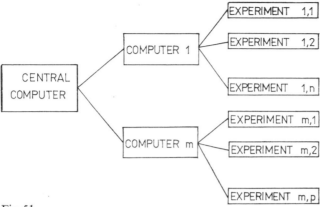

Fig. 51.

2.5 Computing Devices

Computing devices and their technical level, volume and price are constantly developing and only continuous knowledge of scientific and manufacturer's literature can keep workers in touch with the present situation. The elements of computing systems are being replaced by new, more perfect types, which are often based on new principles. As a result of this development, microprocessors, microcomputers and minicomputers have emerged, table-top calculators operate faster than did medium computers twenty years ago and pocket calculators are often capable of carrying out all routine calculations in analytical laboratories. The progress in production technology of the elements of these systems and introduction of large-scale integration (LSI) in electronics has brought about a drastic decrease in the cost of hardware, input-output equipments and information storage devices. At present, the price of peripheral devices is usually a substantial part of the overall price of a computing system. Production of small, cheap and reliable minicomputers marked the beginning of wide application of computers in chemical laboratories and strongly affected the direction of further development of analytical chemistry. Microcomputers based on LSI circuits are becoming an inseparable part of analytical instruments and we are experiencing a revolution in analytical chemistry similar to that thirty years ago when electronics penetrated into analytical chemistry in connection with NMR and mass spectrometry.

2.5.1 PRINCIPAL CONCEPTS IN THE FIELD OF COMPUTING SCIENCE

A computer is a device capable of accepting information, treating it in a certain way and recording the results of this process. According to the method of representing information, computers can be classified into analog, digital and hybrid types.

In the analog treatment, the information is assigned to a continuously variable physical quantity, e.g. the resistance of a platinum thermometer is related to the temperature of the medium in which the thermometer is immersed. In an analog treatment, a physical model of the problem is formed in the computer, mostly using electric circuits. This system behaves analogously to the system studied and the computer is actually a system whose mathematical description is the problem to be solved.

In the digital treatment, numerical information is treated according to a certain algorithm, observing the rules of numerical mathematics.

Digital computers have three main advantages over analog computers, namely:
— better precision
— the ability to solve logic problems
— greater universality.

The main advantage of analog computers is their high computing rate, which finds use in on-line application for preliminary treatment of instrumental output information. Most analog computers are constructed for solving a single type of problem.

Hybrid computers exploit the advantages of the two approaches; part of the information treatment is performed by the analog method and another part digitally. They are used chiefly for solving certain types of scientific and technological problems, and especially for solving differential equations when rapid calculation is required. In analytical chemistry, on-line connection of a digital computer with an analytical instrument, during which analog filtration or other signal transformation occurs, can be considered as a simple hybrid system.

We shall further discuss digital computers that are universal and most frequently used for solving analytical problems. Workers must be able to formulate their requirements, consider the capabilities of a given computing system and discuss these questions with computer specialists.

From the point of view of the hardware, i.e. the construction and technical aspects, the complex of elements collectively termed a computer can be divided into the following principal parts: an operation unit, a control logic, a memory and peripheral equipment. Peripheral equipment serves for feeding the data into the computer and for obtaining output information in a form as understandable and lucid for the user as possible. Many types of equipment have been developed, differing in the output form, rate and price. Common parts are a card or tape puncher, line or mosaic printer, magnetic tape data memories, discs and floppy discs and a plotter. Among less common peripheral devices, multiline displays or screens, magnetic tape casettes and digitizers for entering the coordinates of points of planar representations can be mentioned.

The memory is the part of the computer used for information storage, i.e. program instructions, input and output data and intermediate results during the computation. Memories are evaluated chiefly on the basis of the amount of information that can be stored in them and of the length of the recollection (access) time. The operation unit performs operations with numbers, logic operations and other elementary instructions. Combined with the control logic that controls the operation of the peripheries and of the operation unit, it forms the central processing unit (CPU), which is now manufactured as several variants of an integrated compact unit and forms one part of the whole called a microprocessor.

Software is the program equipment of the computer, without which the computer is incapable of performing the program instructions, and consists chiefly of the operational system, the programming languages and system program library. An analytical chemist, as a user, is usually not interested in classification of computers according to hardware, but differentiates them according to their applicability to processing of analytical information. From this point of view, three main groups of computers can be distinguished:

— *Midicomputers* are systems operating under control of a complex operating system, have a powerful processing unit and make use of both operational and disc memories with large capacities. Fast printers and plotters serve for communication with users.

— *Minicomputers* operate under the control of a much less sophisticated operating system. They can rapidly respond to signals from the peripheries, they are flexible in application, no specialized staff is required for their operation and they can be operated directly by the user.

— *Microcomputers* are the cheapest means of computing, they can control analytical instruments, carry out simple data acquisition operations and give reports to the user.

2.5.2 MICROPROCESSORS

These are sets of LSI elements, performing the functions equivalent to those of a computer processor and consist of one or several chips. The manufacture of microprocessors was started in 1972 and since that time about 50 types have been brought on the market. Their prices have dropped considerably because of production in large series and thus microprocessors have become powerful elements in automation of analytical instruments.

There are many types of microprocessors, but all of them contain:

— a central processing unit — CPU
— input/output control — I/O
— a memory.

Two memory types are used, ROM (read only memory) and RAM (random access memory). The set of microprocessor instructions is formed by a microprogram stored in the ROM. This ROM is usually part of the CPU chip and is inaccessible for the programmer. The information in this memory can only be read, but cannot be erased. The access time is very short (around 200 ns). Memories of this type that can be reprogrammed using special procedures have also been developed and are denoted as PROM, REPROM, etc. The memory is available in the 1K, 4K and 16K bit form or at the board level that contains a complete memory system.

Microprocessors are used as building parts of microcomputers, but even a microprocessor itself can often be used to solve simple problems of controlling apparatus and instruments, on the basis of a fixed program stored in a ROM memory.

A microcomputer contains a microprocessor, another memory, I/O interfaces, peripheral devices, power and a clock. Its configuration can be chosen by the user considering the character of its function and the micro-computer thus assembled then becomes an inseparable part of the measuring apparatus controlling their operation and processing the experimental results.

2.5.3 COMPUTER NETWORKING

Computers find use in analytical chemistry in two main directions which differ greatly in their requirements on the computing techniques. On one hand, computers are used to automate the operation of an analytical laboratory. Automation does not merely mean automated data collecting and modification for printing out, but also control of the analytical experiment, i.e. starting the instrument operation, calibration, sample preparation, etc. These requirements determine the type and configuration of the computer used.

On the other hand, computers can process large sets of data, carry out complex correlation computations and manipulate extensive data bases to obtain additional information from the analysis. These procedures are often controlled by very complicated algorithms and sophisticated graphics software and hardware are used to present the results.

The equipment of computing centres and analytical laboratories in research institutions and universities has developed according to the two above requirements. On one hand, there are maxicomputers provided with a wide scale of peripheral devices and exploiting very complex and extensive data base management programs, such as IMAGE, TOTAL, etc., and, on the other hand, cheap microcomputers built into classical analytical instruments control the operation, collect the data and transmit them to the user.

The present increasing demands on processing of primary information resulting from automation of analytical laboratories and introduction of new analytical methods can only be met by amalgamating the two extreme approaches, such as data base management and I/O control, so that they can interact. Laboratory computing systems are now designed so that the computing facilities can share sources, capabilities and data. Such an arrangement makes it possible to place smaller, single-purpose computing systems inside analytical instruments, which can also utilize the powerful components of large computing systems. This combination of various means of computing is termed networking or distributed processing and enables formation of open-ended computing systems that can be extended further. In these systems the advantages of various types of computer are utilized and their interaction is made possible.

"Intelligent instruments" controlled by microprocessors and micro-computers must be provided with programs for carrying out the experiments; these programs are prepared and tested on larger computers. Minicomputers are capable of carrying out common analytical calculations, but cannot cooperate with large data bases and perform cross-correlation. This job is usually performed by a midicomputer that can be simultaneously used for commercial calculations. Various manufacturers have developed software support packages utilizing the network protocols.

Programs stored in microcomputer ROM memories determine the sequence of operations and communicate with the operator; they are mostly entered into the memory during the production of the instrument. A number of these instruments can be controlled by a minicomputer that collects the data on the samples analyzed by various procedures, but for further treatment the data must often be transferred to a large host computer, which compares it with a data base and can carry out extensive correlation computations, store the data on disc memories and present the results in a suitable form. It can also be used to prepare and store programs for micro-computers and to test and initiate them. Programs can be formulated in these host computer systems in an ASSEMBLER, BASIC or FORTRAN, then down-line loaded in the machine code to a satellite and executed. There are many purpose built variants of computer connection based on modern laboratory equipment. Some examples were given in Section 2.4.

2.6 Programming of Computations

It follows from the computer properties that, for a problem to be solvable, it must be divisible into a finite number of elementary operations, solvable by the particular type of computer construction. The sequence of these operations, starting with the initial data and leading to the final solution, is called an *algorithm*. An algorithm must be
- finite
- unambiguous
- general.

Algorithms are generally represented by block schemes to elucidate interrelations of the individual instructions or blocks of the program. The block scheme is quite general and does not consider the computer type or the programming language. The program is the algorithm of a computation, written considering the capabilities of a particular computer in a code that is acceptable for the computer compiler. The program consists of individual instructions, i.e. descriptions of individual operations in some programming language. At present, higher programming languages, such as FORTRAN, ALGOL, COBOL, PL/1 or BASIC, are most frequently used. The program formulated in one of these languages is usually first translated by a special compile routine into the machine language of the computer and only then can it be used as an instruction network for the computation. Some minicomputers store instructions directly in their memories in the programming language and interpret them during the computation. An advantage of this procedure is the possibility of ready modification of the program and of interactive solution of the problem. A drawback of this procedure is a substantially lower rate of computation and thus this procedure is mostly used with laboratory minicomputers.

2.7 Computer Access

The simplest way of solving a problem by a computer is the stand-alone method, used for non-standard unrepeated computations. According to the operator's instructions, the program is stored in the computer memory and used immediately for the actual computation. This way of using a computer is evidently rather ineffective, because the CPU is virtually not used during read-out of the data and output of the computation results. It is used mostly with minicomputers.

In the batch-mode operation, the storage of the program in the memory and the computation itself are controlled by a program monitor without interference from operators. The organization of the computation is determined by the user instructions in job control language. A higher degree of this mode of operation is the time-sharing system, in which a particular program is exploited by a number of users through a network of terminals at various laboratories. This method is most commonly used for data-banks of physico-chemical data or spectra. It is also possible to store several user programs in the operation memory simultaneously and to use them in a priority order to ensure maximal utilization of the CPU.

296

Computers can be interconnected (computer networking) and the computing strategy can be optimized to effectively use the CPU and peripheries.

2.8 Examples of Computer Application in Analytical Chemistry

This section gives only a brief survey of the uses of computers in analytical instrumentation; for details, see Chapter 4.

2.8.1 X-RAY DIFFRACTOMETRY

Manual measurement of crystals is very tedious and subsequent structure determination took several months before the introduction of computers. Computers can control the measuring process, make intensity corrections and record data. Further treatment of the experimental data for finding the structure of the substance, i.e. the bond angles, interatomic distances and electron density, can be carried out off-line on a larger computer using standard procedures. The result can be plotted graphically or displayed. Treatment of experimental data by iteration procedures used only to be possible on medium computers, because of the high demands on the memory capacity. Introduction of improved algorithms and progress in computing techniques now enable structure solution for relatively uncomplicated molecules on minicomputers.

2.8.2 GAS CHROMATOGRAPHY

This method was computerized early, as it is widely used in industrial control laboratories. Computers are used at many levels. In addition to single-purpose microcomputers coupled with chromatographs, more complicated systems involving a microcomputer connected to several chromatographs can be encountered. The microcomputer can simply collect the data and pretreat it or can also control the instrument operation. Mathematical treatment of the results is then carried out off-line on a larger computer, or the chromatographs are connected on-line with a time-shared computer. In closed-loop systems in real time, a computer can regulate the column temperature, control the sample collection, etc. Various uses of computers in gas chromatography are depicted in Fig. 52.

2.8.3 MASS SPECTROMETRY

Development of the method and its computerization has been hastened by coupling with gas chromatography for analyses of mixtures of substances. During chromatographic separation, mass spectra are taken at

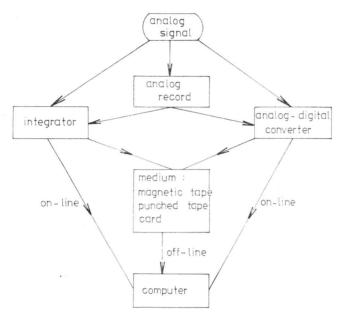

Fig. 52. Various uses of computers in gas chromatography.

regular time intervals, stored in the computer memory, the total ionic current is recorded and a mass spectrogram is constructed from all the data on completion of the experiment. Many computations can be carried out after the experiment. The computer then is used not only for treatment of large sets of data, bur also for their recording, either by photometrically evaluating a photographic recording or by directly monitoring the detector electric signal. The analog signal is transformed to digital form by an A/D converter and stored in the computer memory. The catalogues of spectra are also stored in the computer.

2.8.4 NUCLEAR MAGNETIC RESONANCE SPECTROMETRY

The use of computers in recording NMR spectra has led to an improvement in the instrument resolution. Repeated signal sampling in digital form, the time averaging technique, improves the signal-to-noise ratio and thus the spectra of samples containing NMR-active isotopes of some elements can be recorded at concentrations corresponding to their natural abundance.

Digital recording of impulse NMR spectra and Fourier analysis transferring them from the time domain into the frequency domain have made it possible to determine relaxation times and thus obtain important information on the structure of some solutions.

On-line use of a computer in a closed-loop system enables control of the intensity and homogeneity of the magnetic field. Similar to other spectral methods, computers are used to store spectra and to compare experimental and catalogued spectra. Without computers, it would be impossible to calculate theoretical spectra that are compared with experimental ones to confirm or reject working hypotheses.

Interactive computations are very effective but place great demands on the capacity of the computer memory and on the computation rate.

2.8.5 OTHER FIELDS OF COMPUTER APPLICATION

The questions discussed above are also solved when using computers in IR and Raman spectroscopy. Computers make various corrections to the monitored signal, enable data recording in Raman flash spectroscopy, compare spectra with those stored in the memory and are used off-line to compute structures and force constants. The use of computers in UV and visible spectroscopy is analogous.

Extensive programs have been worked out for off-line data treatment in DTA, calorimetry, polarography and photometry. Numerical treatment always improves the sensitivity and resolution of the instruments and their effective exploitation and considerably shortens the time required for data processing. Concrete examples and problems of these computations are discussed in the following chapter.

2.9 Preparation of Data for Computer Treatment

Although the increasing specialization of workers in analytical laboratories often necessitates cooperation with mathematicians or specialists in computing techniques, knowledge of the rules of data preparation is very advantageous. The initial data for computer treatment is analytical information obtained by various procedures and coded in a set of numerical values. The sets of data are obtained in two ways.

Non-extensive sets of data are taken from laboratory diaries and are prepared for treatment in the form of tables or are filled in on forms for the computing centre. The sequence of the initial data and its form must be decided by the author of the program for its treatment or by the workers at the computing centre when a standard program is used, or the description of the initial data preparation, which should be part of any general program, must be carefully followed.

Preparation of large sets of data in this way would be tedious and thus devices for automatic conversion of the information into digital form are used whenever possible and the digital data obtained is stored directly on a medium suitable for immediate computer treatment. Measuring instruments yield information in analog or in digital form. Analog representation

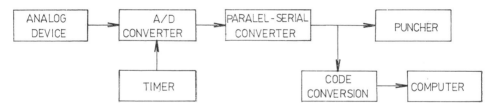

Fig. 53. Automated data collection.

must be transformed into digital form in a suitable converter. Information is then recorded on a punched tape or on a magnetic tape data memory combined with a buffer memory because of the simplicity of this procedure. The data are recorded in series, i.e. number after number, on a punched tape. Because the output of the instrument or converter is parallel, a parallel-series converter must be connected before the tape puncher input.

In the on-line treatment, the data are stored directly in the computer operation memory using a suitable interface, i.e. a hardware configuration facilitating data transfer between the measuring instrument and the computer. The main interface components are an analog-digital converter, a timer and a multiplexer when a greater number of measuring instruments is involved. In treating digitized information, a component for converting the sign code into a series form must often be employed. Fig. 53 shows one of the systems for automated data collection.

Some problems of automated recording of numerical data and their form for off-line or on-line treatment are discussed below.

2.9.1 SIGNAL QUALITY

In most analytical instruments, the measured quantity is represented by a voltage and thus transmission of a voltage signal and its digitization are required. The measured signal always has a certain noise, caused by the statistical character of the measured quantity itself, by instability of the experimental arrangement and by electric impulses in the measuring electronic circuits. Part of the noise can be eliminated by damping RC-filters that are contained in most instruments, but the noise cannot be completely suppressed.

The quality of analog information is affected by the extent of the non-linearity of the measuring system response to the measured quantity, by the response reproducibility and especially by the frequency of spikes in the analog signal. When the analog signal is recorded by a recorder, most spikes do not appear in the graphical recording because of the large time constant of the recording system, but they can seriously affect the quality of information after analog-to-digital conversion. Some of the spikes can be removed by good electric insulation of the measuring device, by suitable construction of motor switches and relays and the rest must be removed by numerical treatment of the digitized signal in a computer.

2.9.2 SIGNAL TRANSMISSION

The analytical instrument is often placed in a different room from the recorder or computer, or a recording apparatus is used simultaneously for several instruments. The signal can be transmitted between the measuring instrument and the recording apparatus or the computer in an analog or digital form.

Transmission of a digital signal requires that each measuring instrument be provided with an analog-to-digital converter or a common converter can be used for all the instruments in the room. A digitized signal can be transmitted through a multiconductor cable or through a single conductor after a parallel-to-series conversion. A disadvantage of the latter method is a substantially lower data rate and the necessity of using a more complex interface module. The data rate is given in Baudes (Bd), i.e. bits per second, and attains values as large as 100,000 Bd. An advantage of digital information transfer is virtually no sensitivity of the signal to interfering impulses and noise along the transmission path.

Analog signals are transmitted over short distances, using shielded cables. Then a good quality analog-digital converter can be shared by many measuring instruments.

2.9.3 SIGNAL CONVERSION

Analog information in the form of an electric voltage cannot be treated by a digital computer. Its conversion into digital form involves determination of how many times a selected unit is contained in the amplitude of the analog signal. The resolution of the converter is determined by the number of binary places obtainable at its output and the related unit selected.

The measuring instrument signal is usually recorded as a function of an independent quantity, such as time, wavelength or magnetic field intensity. To process the dependence, a digital value of the independent variable must be known for each value of the measured signal. The measured signal values are either digitized at regular intervals of the independent variable value, thus simultaneously determining the independent variable values for further numerical treatment, or analog-digital conversion is performed on both the dependent and the independent variable, at fixed time intervals, signalled by timer impulses. More complicated procedures are also employed, in which the frequency of data collection is controlled on the basis of real-time evaluation of the signal by an on-line computer.

The data rate affects the resultant precison and resolution of the whole measuring system, for example the peak separation and integration in gas chromatography. An increase in the frequency of data collection need not improve the precision of the information evaluation and makes the numerical treatment of the results more difficult and time-consuming. The limiting factor for increasing the data rate is the speed of the analog-digital converter or of the recording unit and thus with some analytical methods the data must be stored in a buffer memory and recorded on a suitable medium after the experiment. A similar situation occurs when using magnetic tape as the recording medium, where complete data files are stored on the tape.

When several measuring instruments are to be connected to a recording unit or a computer, a digital or analog multiplexer is used, considering the arrangement of the whole measuring system. In view of the high prices of good quality analog-digital converters, it is mostly advantageous to connect a number of measuring sites with a single converter, using an analog multiplexer.

2.9.4 THE FORM OF INPUT DATA

In off-line treatment of experimental data, the form and number of digits in the input numerical data must be adjusted to the requirements of the program, or the program instruction for entering the data must be modified to match the form of the data on the punched tape from the recording device. Below, brief rules of the FORTRAN programming language input instructions will be given.

These instructions control the transfer of numerical or alpha-numerical information from peripheries to the operation memory of the computer and can be divided into operation instructions and format instructions.

Operation instruction for entering the data, "READ (i, n) list", determines the number of the input device, i, which is to be used for entering the information and refers to instruction number n of the given FORMAT instruction, specifying the data form. The list given after the parenthesis consists of a sequence of elements that are entered in the order they are given. Each element is referred to the corresponding specification in the FORMAT instruction, which specifies the length and shape of the input zone for a given text number.

Example:

READ (1,20) A,XPR,B1,L
:
20 FORMAT (F8.2,3X,E10.4,F5.0,I3)

Survey of specifications used:

Type I is used for entering integral data and has form In, where n is the length of the field in which the number is located on the input medium, e.g. the number of columns of data on a punched card. If the number of digits is smaller than n they must be placed in the right-hand part of the zone.

Example: ..., I4, ... _ _ 23 ...

Type F specifies entering of a rational number and is used in the form Fn . m, where n is the length of the field and m is the number of digits to the right of the decimal point. In reading, the decimal point need not be given in the input data and the computer places it before the n-th digit from the right. If the decimal point is given, parameter m in the specification is ignored.

Example: ..., F8.4, ... _ −4.5391 ...

Type E specifies a rational number with a decadic exponent. It is used in the form, En . m, where parameters n and m have the same significance as in type F specification. The exponent part need not be given.

Example: ..., E11.3, ... _ _ 0.549E+02 ... i.e. 54.9

Type X has the form nX and specifies the number of signs that are to be ignored in reading

Type H is used for entering a text and has the form nH _ _ _ _ _, where n is the length of the text chain.

Example: ... 4HMEAN ...

Any specification can be used repeatedly, if a coefficient of multiplicity is given first, e.g. 6E15.4.

For entering a text, specification '....' can also be used, which is employed for transfer of a chain of signs enclosed by the apostrophes. *Example*: ... 'DETERMINATION OF IMPURITIES' ...

Sign "/" in the format specification induces taking of another card, with reading beginning on a new line. The same effect can also be attained by judiciously using parenthesis; however, proper understanding of this possibility requires a deeper study of the FORTRAN language.

304

3 General Use of Computers in Analytical Chemistry

3.1 Numerical Data Processing

The digitized output data signal of analytical instruments mostly consists of pairs of data representing a dependence between two variables, e.g. a titration curve (the dependence of pH or potential on titrant consumption), a polarographic curve (the current vs voltage dependence), a chromatogram (the time dependence of the voltage), etc. Only sometimes does the signal consists of isolated data, corresponding to e.g. the concentration of the test component in a flowing gaseous mixture. These data are obtained at preset time intervals, but their dependence on time is not monitored and evaluated.

A great variety of mathematical operations and procedures are employed in handling various types of experimental dependence. The most common of them will be briefly discussed below. For the relevant mathematical principles and definition of conditions for their validity, reliability and precision, the reader is referred to text-books on numerical mathematics.

3.1.1 HANDLING OF DIGITIZED INPUT DATA

Numerical data corresponding to the analytical signal consist of two components, namely, the analytical signal itself, transformed to a numerical form, and the noise arising during conversion of the signal to numerical form.

The sensitivity of the instrument is given by the ratio of the measured signal to the mean noise value. The noise can be characterized by its mean amplitude,

$$a = \sqrt{[(\sum_{i=1}^{n} (y_i^{(r)} - y_i^{(t)})^2)/n]} \tag{3.1}$$

where $y^{(t)}$ is the value of the analytical signal itself and $y^{(r)}$ is this signal subject to noise. Ideally, the noise should contain the whole spectrum of frequencies (white noise), but various noise sources with non-uniform spectra are encountered in practice, such as impulses from switching contacts, the noise of semiconductor elements and temperature-dependent noise sources.

Periodic noise, whose frequency is higher than that of the measured signal, can mostly be eliminated from the output signal, using two basic approaches, namely, RC filters for analog signals or computer treatment of digitized signals. Numerical procedures are more advantageous, because they enable not only programmed removal of interfering frequencies, but also many other corrections and modifications of the measured signal.

In numerical smoothing, combinations of numerical procedures for removal of noise, elimination of outlying points and modification of the signal to a form suitable for further treatment are generally used. It should be borne in mind that this combination must be selected depending on the character of the given signal, to obtain a marked improvement in the signal-to-noise ratio. A high degree of signal smoothing often distorts the signal and leads to information loss.

For the sake of simplicity it will be assumed that the measured signal is digitized at regular intervals of the independent variable time, wavelength, temperature, frequency. This procedure of digitization of analog signals is simplest and therefore also most frequently used.

In the method of the sliding mean, the signal is calculated at a certain point as the mean of a certain number of signal values at points symmetrically distributed around the point to be determined. This procedure is chiefly applicable in determining the signal baseline and in checking signal consistency. It leads not only to signal smoothing, but also to flattening, which is very unfavourable, especially in detection of sharp peaks.

The drawbacks of this procedure can be partially suppressed using a weighed mean

$$\bar{y}_i = \frac{\sum\limits_{j=-n}^{n} w_j y_{i+j}}{\sum\limits_{j=-n}^{n} w_j}$$

where the signal value at the i-th point is determined from the values at $2n + 1$ points symmetrically distributed around the test point assigning

306

certain weights (w_j) to these values. The weights must be found from the character of the signal and from experience.

The individual procedures may differ in the value of n, i.e. the number of points treated, in the values of the weights, symmetry of the smoothing function and in whether for $j < 0$ only the measured signal values, y_{i+j}, or also the smoothed values, $\bar{y}_{i+j}$, are considered. The shape of the smoothing function can be determined by the choice of weights. If the weights are zero for all points for $j > 0$, a one-sided smoothing function is obtained, i.e. only previous values of the measured signal are considered. By using one-sided exponential smoothing functions common RC analog filters are numerically imitated.

The above procedures pay little attention to the character and variability of the signal shape. This drawback is partially eliminated by using polynomial smoothing function, e.g.

$$y = a_0 + a_1x + a_2x^2 + \ldots + a_nx^n \tag{3.2}$$

The polynomial coefficients, $a_0 - a_n$, are determined by the least squares method (see Section 3.3.1), evaluating a certain number $(j \geq n + 1)$ of signal values. The corrected signal is then calculated from Eq. 3.2 at the central point of this interval. The method can further be improved by introducing weights. This method affects the character of the measured signal far less than the mean methods, but is sensitive to the selection of the polynomial order and to the number of points considered in calculating the coefficients in Eq. (3.2). The smoothing effect of the function increases with decreasing polynomial order and increasing number of points treated.

A special problem in signal smoothing is the removal of random, markedly outlying, signal values, which are called spikes, consisting often of only one or two outlying points and being produced by various impulses in the line voltage. When the above smoothing methods are used, spikes can be transferred to resemble narrow peaks and, on chromatograms, are then recorded and integrated. They are mostly identified using algorithms distinguishing very sharp changes in the unfiltered signal and rapid return to the original level. The signal at these outlying points is mostly replaced by values obtained by interpolation from the neighbouring values.

The signal-to-noise ratio is often improved by using the time-averaging process that utilizes the stochastic character of the noise and is based on summing the signal values obtained under analogous conditions. The signal

307

is obtained either by repeating the measurement, or by multiple parallel evaluation of the analytical information. By summing a great number, n, of digital values of the output signal, the value of coherent amplitudes increases n-times, whereas the mean noise amplitude is only $\sqrt{n}$-times greater; hence the sensitivity of the method is improved. This procedure has many limitations, based chiefly on difficulties in maintaining identical measuring conditions, but is often used successfully.

The signal can also be smoothed using Fourier analysis of the signal frequencies. Considering the character of changes of the measured quantity, the signal is expanded in a trigonometric series in a certain interval (Fourier expansion), the frequencies corresponding to noise are excluded and the smoothed signal is obtained after back-transformation. In practical application of this procedure, it is, however, often difficult to determine the number of coefficients in the Fourier transformation and unsuitable coefficient selection causes distortion of the signal. Another disadvantage of this method is its tediousness, even if a computer is used.

3.1.2 DESCRIPTION OF THE SIGNAL SHAPE

For further treatment of an analytical signal, especially for determination of important points, the signal must be approximated by a continuous function over the whole measuring range or at least at shorter intervals of the independent variable values. A functional dependence is easier to treat mathematically than a set of data; it enables interpolation, location of extremes and resolution of overlapping peaks. In curve fitting, a function must be found that best corresponds to the analytical signal shape. These questions have been solved and suitable approximate functional dependences are known for most characteristic analytical signal shapes.

Straight lines are used to describe conductometric and amperometric titrations, in determining the baseline in chromatograms and in approximation of parts of the potentiometric titration curve. In determining the end-point of a conductometric titration, the intercept of two straight lines that enclose a certain angle is determined. The reliability of the estimation of the equivalence point depends on the magnitude of this angle and on the calculated confidence intervals for the parameters of the regression straight lines, $y = a_0 + a_1 x_1$, that are obtained by the least squares method (see Section 3.3.2).

Parameters of a linear dependence are also determined in evaluation of spectral signals and in chromatography. Reconstruction of the signal baseline, approximated by a straight line or a series of segments, is the basis for calculation of the signal amplitude and the peak area. All inaccuracies in this procedure markedly affect the precision and reliability of the concentration data obtained.

Sigmoid curves usually originate from potentiometric titrations, calorimetric measurements, etc. An acid-base titration curve can be described, up to the equivalence point, by the Henderson-Hasselbalch equation,

$$pH = pK + \log \frac{[A]}{[HA]} \tag{3.3}$$

where $[A]$ and $[HA]$ are the concentrations of the titrated and untitrated acid, respectively, and $pK = -\log K$, where K is the dissociation constant. The potential dependence on the concentration in redox titrations is described by the Nernst equation,

$$E = E^0 - \frac{RT}{nF} \log \frac{[Red]}{[Ox]} \tag{3.4}$$

where E^0 is the normal potential and $[Red]$ and $[Ox]$ are the concentrations of the reduced and oxidized forms, respectively. These titration curves can often be described, up to the equivalence point, by the quite general relationship

$$z = z_0 - k \cdot \log \frac{y_{max} - y}{y} \tag{3.5}$$

The inflection point on the curve is considered as the equivalence point and the shape of the curve beyond this point is mostly unimportant.

Stepped curves are known, e.g. from classical polarography and can be described by the expression

$$y = \frac{y_{max}}{1 + \exp(a(z - z_0))} \tag{3.6}$$

where y_{max} is the value of the flat part and z_0 is the value of variable z for which $y = \frac{1}{2}y_{max}$, i.e. the half-wave potential in polarography. The relationship between a sigmoid curve and one step on a stepped curve is best seen when Eq. (3.6) is rewritten to express the dependence of z on y, i.e.

$$z = z_0 + \frac{1}{a} \ln \left(\frac{y_{max}}{y} - 1 \right)$$ (3.7)

A signal containing a peak is obtained e.g. in chromatography, in absorption molecular spectrometry, in nuclear magnetic resonance spectrometry, in photometering the photographic plate in emission spectrography, in IR, UV and visible spectroscopy, etc. Many functional relationships have been proposed and used in mathematical treatment of these signals. The two most important ones, the Gauss and the Lorentz relationships, will be discussed below.

It is assumed that the signal maximum value, y_{max}, is located at $z = z_0$. Then the *symmetrical Gauss relationship* is

$$y = y_{max} \exp \left(-(z - z_0)^2 / b^2 \right)$$ (3.8)

The *asymmetrical Gauss relationship* differs in the value of coefficient b for the left-hand and right-hand parts of the peak. The *Lorentz relationship* in the form

$$y = y_{max} / (1 + 2(z - z_0)^2 / b^2)$$ (3.9)

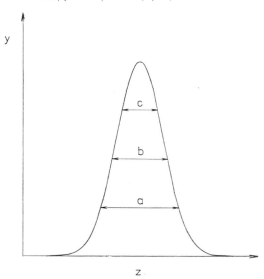

Fig. 54. Peak profile characterization.

310

is also often used. For description of unsymmetrical peaks, the function

$$y = y_{max} \exp\left(-(z - z_0)^2/cz\right) \tag{3.10}$$

is also used. In extreme cases the peak shape is *rectangular*,

$$y = \begin{cases} = y_{max} & \text{for} \quad (z_0 - b/2) \leqq z \leqq (z_0 + b/2) \\ = 0 \end{cases} \tag{3.11}$$

In practice, peaks often have somewhat different shapes from the Gaussian or Lorentzian profile. Their profile can, at least approximately, be characterized by the ratio, $q = a/c$, see Fig. 54. The q values for the rectangular, Gaussian and Lorentzian profiles are 1.000, 2.196 and 3.000, respectively.

A general curve can be described over the whole interval of independent variable x or over a part by the polynomial

$$y = a_0 + a_1 x + a_2 x^2 + \ldots + a_n x^n . \tag{3.12}$$

The problems of selecting a suitable form and order, the value of n of the approximating polynomial, the transformation of the coordinate system and the procedure for the calculation of the coefficients are discussed in detail in Section 3.3.2. A graphical plot of the match of the approximating function with the recorded signal values is often used to check the suitability of the approximation used. Some examples are given in Fig. 55.

Cyclically varying signals consisting of various peaks or their combinations can be described by the *Fourier series*. It can be expanded in the interval from $-L$ to L,

$$y = \frac{a_0}{2} + \sum_{n=1}^{\infty} \left(a_n \cos \frac{n\pi t}{L} + b_n \sin \frac{n\pi t}{L} \right) \tag{3.13}$$

where

$$a_n = \frac{1}{L} \int_{-L}^{L} y \cos \frac{n\pi t}{L} \, dx \tag{3.14}$$

$$b_n = \frac{1}{L} \int_{-L}^{L} y \sin \frac{n\pi t}{L} \, dx \quad \text{for} \quad n = 0, 1, 2, \ldots \tag{3.15}$$

The remaining problem is the finding of coefficients a_n and b_n. If the plot of the function is symmetrical, $b_n = 0$, and the series contains only the a_n terms with cosines, so that the determination of the coefficients is simpler.

311

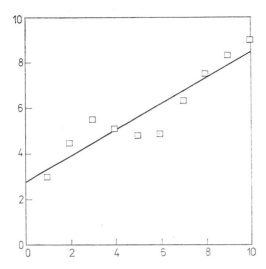

Fig. 55. Suitability of the approximation used.

(a) Ten points fitted by the straight line $y = a_0 + a_1 x$. Extrapolation to $x = 0$ yields $y_0 = 2.75$; the sum of the squares of the deviations $S = 4.000$.

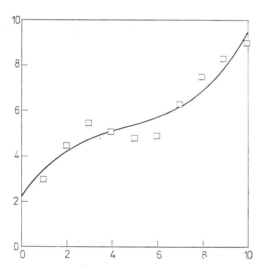

(b) The same points as in (a), fitted by a third degree polynomial, $y = a_0 + a_1 x + a_2 x^2 + a_3 x^3$; extrapolation yields $y_0 = 2.33$; $S = 2.28$.

312

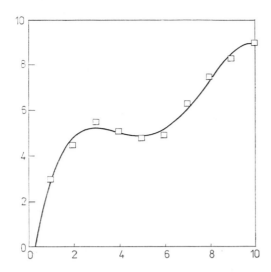

(c) The same points as in (a), fitted by a fourth degree polynomial, $y = a_0 + a_1 x + a_2 x^2 + a_3 x^3 + a_4 x^4$; extrapolation yields $y_0 = -1.33$; $S = 0.37$.

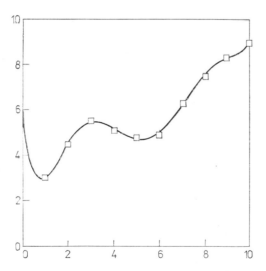

(d) The same points as in (a), fitted by a sixth degree polynomial, $y = a_0 + a_1 x + a_2 x^2 + \ldots + a_6 x^6$; $y_0 = 5.000$; $S = 0.038$.

313

3.1.3 DETERMINATION OF CURVE CHARACTERISTICS

The determination of certain important points on a curve is closely connected with evaluation of the analytical information itself from the signal obtained. An experienced analyst can, at a glance at an analog recording of the instrument output signal, determine the positions of the peaks and other important characteristics. In numerical computer treatment, sophisticated procedures and algorithms must be used, which are mostly highly demanding on computer time. In addition to simple jobs, such as determination of the intercept of straight lines in conductometric titrations or of the inflection point on a mathematically described titration curve, the chief application is in gas chromatography and NMR spectrometry.

In determining the inflection point on a titration or calorimetric curve, a numerical signal recording is sometimes used directly and the procedure is based on the fact that the second derivative (see Section 3.1.4) changes sign at the inflection point. The exact position of the inflection point is determined by interpolation. The signal shape in the vicinity of the inflection point can also be approximated by a suitable function $f(x)$ and the position of the inflection point can be determined by solving the equation

$$f''(x) = 0 \tag{3.16}$$

In determining the characteristics of a curve obtained in gas chromatography or NMR spectrometry, the problems to be solved involve the determination of the beginning and the end of a peak and of the maximum position, resolution of overlapping peaks and peak integration. The solution of these tasks directly affects the precision of the results. For the determination of the peak position and for separation of peaks, two different approaches are generally taken.

In the first method it is generally assumed that a relationship describing the peak shape is known. After locating the most pronounced maxima on the recording, the theoretical curve is calculated using the function describing the peak shape and the position of further peaks are found by comparing the theoretical and the experimental curves. This procedure is repeated until a good fit of the theoretical curve to the signal recording is obtained. This procedure is frequently used, e.g. in NMR spectrometry.

In the second method, the beginning and the maximum of the peak are located using derivatives, overlapping peaks are determined employing second derivatives and, after reconstruction of the baseline, the signal at the maximum is calculated. These data are used for approximation of the peak shape, separation of overlapping peaks and simulation of the theoretical signal shape. Final parameters are obtained by an iteration procedure and enable the determination of the peak areas and thus also of the qualitative and quantitative analytical data. When using this method, a smoothed signal is required, because differentiation is very sensitive to the presence of noise in the signal.

3.1.4 INTERPOLATION AND EXTRAPOLATION

The mathematical procedures are used for calculation of the signal at points at which the measurement was not carried out, interpolation being location of the signal inside an interval limited by the measured points; extrapolation involves location of the signal outside this interval.

Interpolation is employed in treatment of digitized recordings leading to exclusion of outlying values and replacement of a complex function by simpler interpolation polynomials.

A common relationship for *linear interpolation* has the form,

$$f(x) = f(x_0) + q \, \Delta^k f(x_0) \tag{3.17}$$

where $q = (x - x_0)/h$, $h = x_i - x_{i-1}$, $\Delta^k f(x_0)$ is the k-th difference of function f at point x_0. Eq. 3.17 can be rewritten in a more lucid form,

$$y = y_0 + (x - x_0) \frac{y_1 - y_0}{x_1 - x_0} \tag{3.18}$$

For *quadratic interpolation*, the Newton formulae,

$$f(x) = f(x_0) + q \, \Delta f(x_0) + \frac{q \cdot (q - 1)}{2} \, \Delta^2 f(x_0) \tag{3.19}$$

and

$$f(x) = f(x_n) + q \, \Delta f(x_{n-1}) + \frac{q \cdot (q - 1)}{2} \, \Delta^2 f(x_{n-2}) \tag{3.20}$$

315

are used, where $q = (x - x_n)/h$. For x close to point x_0, the first Newton interpolation formula is suitable, whereas for x close to point x_n the second formula is advantageous.

Differences of various orders are usually written in tables:

$$x \quad f(x) \quad \Delta f(x) \quad \Delta^2 f(x)$$

$$
\begin{array}{llll}
x_0 & y_0 \\
 & & \Delta y_0 \\
x_1 & y_1 & & \Delta^2 y_0 \\
 & & \Delta y_1 \\
x_2 & y_2
\end{array}
$$

Interpolation formulae with a higher precision and relationships for interpolation of values that are not equidistant can be found in larger monographs on numerical mathematics. The interpolation precision is affected by the shape of the curve; e.g. at a maximum or a minimum it is poorer than for monotonic (i.e. concave or convex) curves.

Extrapolation can be performed analogously to interpolation, but it is always a very unreliable operation and may yield imprecise values, especially for independent variable values far from the last measuring point; therefore, it is seldom used in practice where it can be avoided.

3.1.5 NUMERICAL DIFFERENTIATION AND INTEGRATION

Differentiation of the instrument output signal yields information on the rate of change and can be used, e.g. for controlling the frequency of signal recording in an on-line arrangement. When the signal is treated off-line, differentiation can be performed in two ways. The signal is either smoothed, its shape approximated by a functional relationship and differentiated using an analytically derived relationship or the derivative is obtained numerically directly from the stored values of the digitized signal. The latter procedure is not only simpler and thus also more rapid, but can also be used for signal smoothing and for identification of characteristic points on the recorded curve.

In deriving the relationships for numerical differentiation, it is assumed that the signal values recorded are equidistant $(x_{i+1} = x_i + h)$; the function is expanded in a Taylor series, higher derivatives are omitted and the general

316

relationship

$$y_i^{(k)} = \sum_{j=i-a}^{i+b} c_j y_j \qquad (3.21)$$

is obtained, in which c_j values are coefficients.

This formula expresses the k-th derivative as a linear combination of signal values y_j at a certain number of points $(a + b + 1)$. The following relationships are common:

$$y_1' = (y_2 - y_1)/h \qquad (3.22)$$

$$y_2' = (y_3 - y_1)/2h \qquad (3.23)$$

$$y_2' = (-2y_1 - 3y_2 + 6y_3 - y_4)/6h \qquad (3.24)$$

$$y_2'' = (y_1 - 2y_2 + y_3)/h^2 \qquad (3.25)$$

$$y_2''' = (-y_1 + 3y_2 - 3y_3 + y_4)/h^+ \qquad (3.26)$$

Formulae for higher derivatives and those with higher precision, based on a greater number of points, can be found in textbooks on numerical mathematics. However, in practical applications only the simplest formulae are mostly used and the required precision is attained by choosing a sufficiently small step in signal recording. The number of points considered in the calculation depends on the purpose for which the derivative is computed.

Relationships for numerical integration can be derived analogously. The trapezium rule,

$$I = \int_{x_0}^{x_n} f(x)\, dx = \frac{h}{2} (y_0 + 2y_1 + \ldots + 2y_{n-1} + y_n) \qquad (3.27)$$

and the Simpson relationship

$$I = \int_{x_0}^{x_n} f(x)\, dx = \frac{h}{3} (y_0 + 4y_1 + 2y_2 + 4y_3 + \ldots + 4y_{n-2} + 2y_{n-1} + y_n) \qquad (3.28)$$

where $n = 2m$ must be even, are most often used. When employing the trapezium rule, the signal values over the given interval are summed, taking the values at the limiting points with half weight. For integration of signals

317

with non-equidistant measuring points, the Gauss relationship is used; this formula can be found in the literature. In integration of analytical signals a high precision of the integration relationship is unnecessary, because the errors caused, e.g., by baseline fluctuations and peak overlapping in chromatography are larger than the error of the integration method.

3.2 Statistical Treatment of Analytical Data

Computers are now often used in mathematical-statistical treatment of analytical data, because this treatment involves many rather complex computations and several simple logic operations, it can mostly be carried out quite mechanically, sometimes has the character of mass data treatment and is time-consuming when a computer is not used. It is still often performed off-line, so that general multipurpose computers can be used. Modern on-line information transfer is applied more and more extensively in analytical practice. It is then necessary to use a specifically oriented computing system. The advantage of immediate transfer of information and often the possibility of automatic control of analytical experiments is partially offset by the use of a single-purpose, specific program or computer.

Some manufacturers of computing systems have met the requirements of statistical treatment of experimental data and supply relatively cheap minicomputers and programmable calculators dedicated to statistical computations, which are quite sufficient for common off-line statistical treatment of analytical results. If, however, a more extensive set of data is to be evaluated and statistical testing and some special data treatment procedures e.g. for spectra are to be carried out, a table-top computer with a printer and peripheral equipment for reading data from some permanent storage media are more suitable.

To treat a greater set of data or for frequent handling of smaller data sets using an off-line computer, a suitable computer program is essential. Program libraries mostly contain programs for all common procedures of numerical mathematics, mathematical statistics, etc., and many programs have been published. Still it is often necessary to prepare a special program for the given purpose and the given structure of the data treated. Therefore, a few notes are given below concerning the development of programs for statistical treatment of analytical results.

(1) The program must exclude logic errors during the computation and

must consider all possible computational variants. A greater number of alternatives must be considered especially with:

(a) Various procedures of data treatment in dependence on the size of the data set treated;

(b) Computation of the confidence interval; if parameter σ is known, $z(\alpha)$ is either computed from critical values of the normal distribution, or s is estimated and the interval calculated from critical values $t(v, \alpha)$. The computation is described by

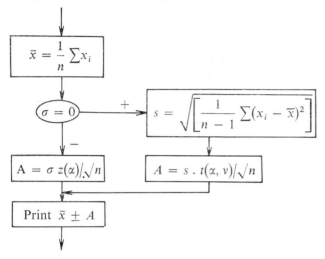

If the σ value is not known, it is set equal to zero (in practice, $\sigma = 0$ never occurs).

(2) In computation of the confidence interval (see 1.b) and in statistical testing, critical values are required, e.g. $z(\alpha)$, $t(v, \alpha)$, noncentral $t(v, \alpha, \delta)$, Grubbs criteria $T(\alpha, n)$ and others. Tables of critical values can be stored in the computer memory or can be calculated. However, the critical values are usually tabulated only for certain significance levels and only for certain numbers of degrees of freedom, so that they must often by interpolated during recall from the memory. Calculation of critical values as integration limits is tedious and hardly usable in practice. Various approximations are used more often; for some see the Appendices. Testing can also be carried out by obtaining the significance level for the calculated value of testing criterion Θ by solving $\int_{-\Theta}^{\Theta} g(x)\,dx = (1 - \alpha)$, which is obviously substantially faster than looking for Θ for a given $(1 - \alpha)$, and the decision

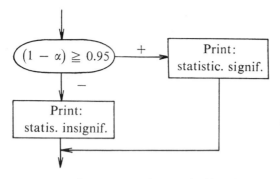

is taken. This procedure is practically advantageous only when the frequency function of testing characteristic $g(x)$ is readily integrable or its numerical integration is rapid and sufficiently precise (extreme precision is not required!). Various approximations are most suitable (see Appendix D, E).

(3) If some result may be outlying, e.g., the Grubbs test is used (see Section 1.5) according to the scheme

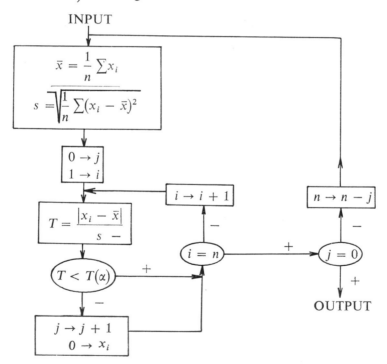

by means of which all results are tested, not only the highest and lowest ones; this procedure is often faster than ordering the results according to the magnitude and selecting the limiting ones.

The mathematical procedures for evaluation of parameters and statistical testing were variously modified to simplify and speed-up the calculation; when using computers, most of these modifications become pointless. Moreover, even some graphical procedures are now converted into numerical and carried out on computers.

In addition to characterization of results, i.e. finding of the most probable value after excluding outlying results and determining the precision of this value, e.g. in the form of the confidence interval, statistical treatment is also used to make information contained in a large data set "denser". However, this procedure of making information "denser" must allow for later extraction of information that seems irrelevant at the moment or whose existence is so far unknown.

3.3 Analysis of Experimental Dependences

3.3.1 INTRODUCTION

3.3.1.1 Dependence between Two Variables

Regression methods are used in evaluation of experimental dependences between two or more variables. Dependences between two variables can basically be of three kinds:

a) *Functional dependences*, $y = f(x)$, where a single value y_i corresponds to a certain argument value x_i;

b) *Regression dependence*, where a certain probability distribution of the values of random quantity y_i holds for a fixed argument value, x_i, and thus $E[y_i] = f(x_i)$;

c) *Correlation*, where both variables have random character and there is a certain relationship between them. If the relationship is strong, equation $E[y_i] = f(E[x_i])$ is valid.

Regression and correlation, i.e. dependences in which one or both variables are random quantities with a certain probability distribution, are termed *stochastic dependences*, in contrast to functional dependences in which the variables are fixed quantities.

321

Experimental dependences are virtually never functional dependences, because the variables are almost always determined with a certain error given by the limitations of the measuring instruments or human senses. Even in approximations of complex mathematical functional dependences, errors must mostly be considered, following from rounding-off the numerical values.

If nothing is known about experimental quantities assumed to have random character, then the first question asked is whether there is a relationship between these values and what the intensity of the relationship is. Evaluation of the dependence between two variables is called *data correlation*.

3.3.1.2 Data Correlation

A correlation is a stochastic dependence between two random variables. The degree of this dependence is characterized by the correlation coefficient, ϱ, whose exact value cannot, of course, be found and is estimated. For a linear dependence it can be estimated from the relationship

$$r = \frac{\Sigma(x_i - \bar{x})(y_i - \bar{y})}{\sqrt{[\Sigma(x_i - \bar{x})^2 (y_i - \bar{y})^2]}} = \frac{N\Sigma x_i y_i - (\Sigma x_i)(\Sigma y_i)}{\sqrt{\{[N\Sigma x_i^2 - (\Sigma x_i)^2] \cdot [N\Sigma y_i^2 - (\Sigma y_i)^2]\}}}$$

(3.29)

Correlation coefficient ϱ has values near -1 or 1 for a close dependence between x and y; positive values hold for a direct dependence, negative for an inverse dependence. When the two variables are completely independent, it is zero. The estimated value, r, attains values which are not statistically significantly different from -1 or $+1$ for a close direct or inverse dependence and from zero for independent variables.

Because all the sums necessary for estimation of the correlation coefficient are also used in calculation of the parameters of the linear regression dependence, the two calculations are often combined.

3.3.1.3 Regression Methods. The Least Squares Method

If the correlation coefficient between two variables is significantly different from zero, then the variables are related. Finding this relationship, i.e. finding the functional relationship of a stochastic or regression dependence between the mean values of the two variables, is not easy. Assuming that both variables y and x may be subject to errors, optimization methods (see Section 3.3.3) must be used to estimate the parameters of the appropriate functional relationship. Provided that normal distribution is present and the variances of the experimental values of x and y are the same, the parameters of the functional relationship must be found in such a way that the sum of the squares of the shortest distances of the points from the curve corresponding to the functional relationship is minimal. The shortest distance corresponds to a line perpendicular to the tangent to the curve. When the variances of values x and y are different (they can generally be different for various values of the variables), the conditions for the optimal state are even more complex.

Regression methods can be used when the dependence treated has regression character, i.e. when independent variable x is known precisely and only the dependent variable, y, is subject to error. Thus a functional relationship is sought between the mean values of variables y and x. However, regression methods are also used when the conditions for a regression dependence are not fully met, because regression calculations are much simpler than optimization calculations. Before a regression method is applied, it must be decided which variable will be selected as dependent and which as independent. Generally, the variable that is subject to a larger error is selected as dependent. If dependent and independent variables cannot be unambiguously distinguished, then, e.g., the variable that can be explicitly expressed is selected as dependent.

If the error of the dependent variable has normal distribution, the least squares method can be used. The parameters of the functional dependence must meet the condition that the sum of the squares of the differences between the experimental values and the values calculated from the regression equation is minimal. The sum of the squares of the deviations with normal distribution (2nd central moment) is proportional to the variance; hence this method involves variance minimization. Partial differentiation of the regression functional relationship with respect to all the parameters to be found yields a set of normal equations equal in number to the number of parameters, so that an unambiguous solution is possible using the methods of linear algebra.

In addition to the best known and most frequently used least squares method, there are other regression methods based on other regression criteria, such as the *method of minimization of the sum of the absolute deviations*, that assumes an exponential error distribution on both sides, or the *method of minimization of the maximal deviations* (Tchebyshev) assuming uniform rectangular error distribution.

Special attention should be paid to the *group method*, which utilizes the fact that the 1st moment of the normal distribution equals zero; therefore, the criterion of zero sum of the deviations is applied. Because this criterion would not yield unambiguous estimates of the parameters, the experimental points must be divided into the same number of groups as the number of regression parameters. The above rule is then applied to each of these groups and a set of normal equations is obtained. For linear regression, this method is identical with the old graphical centre of gravity method. Regression solution by the group method is numerically easier and faster than by the least squares method, but problems arise in separating the points into groups. Various approaches can be taken, e.g. groups can be formed by dividing a series ordered according to the magnitude of the independent variable, or by random selection; the groups can be approximately the same size, or middle groups can be preferred at the expense of the side groups. The results obtained by the group method generally do not differ significantly from the results obtained by the least squares method.

3.3.1.4 The Form of Functional Relationships

The assumed form of the regression dependence can be implicit with respect to the dependent variable,

$$f(x, y, a, b, ...) = 0 ; \quad a, b - \text{parameters}$$

and still the sum of the squares of the deviations of this variable is to be minimized. Looking for optimal parameters of such a functional relationship is part of the field of *optimization methods* (the Gauss-Newton method); to calculate the sum of the squares of the deviations (optimization criteria), the dependent variable value conforming to the implicit function, $f(y_i) = 0$, must be found for each value of the independent variable (see Appendix F).

If the assumed regression dependence is given by an explicit expression for the independent variable, $y = f(x, a, b, ...)$, the method of finding the optimal values of the parameters depends on the form of functional relationship f; the form with respect to variable x is unimportant, but if the function is non-linear with respect to the parameters, non-linear regression methods must be used, which are generally classified as optimization methods.

Sometimes, linearization can be achieved by transformation of the dependent variable. The chief assumption of regression methods, the normal error distribution, is then related to the transformation variable, which may lead to incorrect results, provided that the weighed least squares method is not used.

3.3.2 THE LINEAR REGRESSION METHOD

The method is based on the following assumptions:

i) the independent variable x is determined precisely, dependent variable y is determined with random error;

ii) the error of dependent variable y has normal distribution;

iii) the assumed function $y = f(x, a, b, ...)$ is linear with respect to the parameters.

The method can sometimes be used even if all these assumptions are not completely met, but the effect of deviations from these assumptions on the results must be considered during the interpretation of the results.

In the future minicomputers or table-top calculators will probably be available in every laboratory or will be coupled on-line with measuring instruments. Minicomputers can communicate with the user by a symbolic language, similar to large computers, but are handicapped by limited memory capacity and computation rate. This handicap can be partially alleviated by using macroinstructions (e.g. for matrix operations), saving the memory capacity and accelerating the computation.

However, from the point of view of programming strategy, the methods of numerical mathematics must be selected carefully, with maximum economy and effectiveness of the programs; for this reason, many algorithms published e.g. in Fortran cannot be used on minicomputers.

3.3.2.1 The Least Squares Method

The principle of the linear regression method, i.e. the least squares method, will be demonstrated on the regression relationship for a general

325

straight line,

$$y = f(x, a, b) = a + bx \tag{3.30}$$

If the number of pairs of experimental values x_i, y_i is N ($i = 1, 2, ..., N$) and the value of dependent variable y corresponding to independent variable x_i according to Eq. (3.30) is denoted as y_i^* and called the expected value, then the condition holds that

$$\sum_{i=1}^{N} (y_i - y_i^*)^2 = \min \tag{3.31}$$

Substituting the expected value y_i^* yields

$$\sum (y_i - a - bx_i)^2 = \min \tag{3.32}$$

The expression on the left-hand side can be considered as a function of two variables, unknowns a and b, which has the form of a paraboloid and which attains a minimal value for a certain combination of parameters a and b. For local maxima and minima it holds that differentiation with respect to the appropriate variable is zero. If Eq. (3.32) is differentiated partially with respect to a and b, two equations are obtained,

$$\sum 2(y_i - a - bx_i) = 0$$
$$\sum 2(y_i - a - bx_i) \cdot x_i = 0 \tag{3.33}$$

and their rearrangement ($\Sigma 1 = N$)

$$aN + b\Sigma x_i = \Sigma y_i$$
$$a\Sigma x_i + b\Sigma x_i^2 = \Sigma y_i x_i \tag{3.34}$$

leads to a set of normal equations that can be written in the matrix form,

$$\begin{vmatrix} N & \Sigma x_i \\ \Sigma x_i & \Sigma x_i^2 \end{vmatrix} \cdot \begin{vmatrix} a \\ b \end{vmatrix} = \begin{vmatrix} \Sigma y_i \\ \Sigma y_i x_i \end{vmatrix} \tag{3.35}$$

A symmetrical square matrix whose elements are formed by summation of only the independent variable values, multiplied by the vector of coefficients a and b, equals the vector whose elements are formed by summations of y and x.

The values of parameters a and b can be obtained by solving set of equations (3.35) by any method. The resultant relationships for parameters a

326

and b have the form,

$$b = \frac{N(\Sigma y_i x_i) - (\Sigma y_i)(\Sigma x_i)}{N(\Sigma x_i^2) - (\Sigma x_i)^2} \tag{3.36}$$

$$a = \frac{(\Sigma y_i)(\Sigma x_i^2) - (\Sigma y_i x_i)(\Sigma x_i)}{N(\Sigma x_i^2) - (\Sigma x_i)^2} \tag{3.37}$$

or

$$a = (\Sigma y_i - b\Sigma x_i)/N$$

When independent variable x changes in an equidistant manner, the summation values in general form Σx_i^n, can be expressed analytically. Placing the step $\Delta x = x_{i+1} - x_i$ before the summation

$$\sum_{i=1}^{N} x_i^n = (\Delta x)^n \cdot \sum_{i=1}^{N} k_i^n$$

is obtained where the k_i are integral numbers. If $k_1 > 1$, the summation can be replaced by the difference between two sums, each of which begins at unity:

$$\sum_{i=1}^{N} k_i^n = \sum_{i=1}^{k_N} i^n - \sum_{i=1}^{k_1-1} i^n$$

The two sums $\big($or one sum when $k_1 = 1\big)$ can then be replaced by analytical expressions $\big($see the formula for n up to 6$\big)$.

$$\Sigma i = N(N + 1)/2$$
$$\Sigma i^2 = N(N + 1)(2N + 1)/6$$
$$\Sigma i^3 = N^2(N + 1)^2/4$$
$$\Sigma i^4 = N(N + 1)(2N + 1)(3N^2 + 3N - 1)/30$$
$$\Sigma i^5 = N^2(N + 1)^2(2N^2 + 2N - 1)/12$$
$$\Sigma i^6 = N(N + 1)(2N + 1)(3N^4 + 6N^3 - 3N + 1)/42$$

Using these formulae, the expressions for e.g. estimation of the parameters of a general straight line are simplified $(k_1 = 1)$

$$a = \frac{2}{(\Delta x)N(N + 1)}\left[(\Sigma y)(2N + 1)(\Delta x) - 3(\Sigma xy)\right]$$

$$b = \frac{6}{(\Delta x)^2 \, N(N^2 - 1)} \, [2(\Sigma xy) - (\Delta x)(N + 1)(\Sigma y)]$$

so that only two sums, Σy and Σxy, and the Δx value are required for the calculation.

3.3.2.2 Statistical Characteristics

From the mathematical point of view, the optimal values of parameters a and b are determined unambiguously and precisely, but, from the statistical point of view they are only estimates of unknown parameters α and β of the objective functional relationship. These estimates are affected by the number and selection of experimental pairs x_i, y_i and by the precision of the determination of y_i. Therefore, regression calculations involve not only the determination of the optimal values of the parameters but also of the statistical characteristics of the estimates of the regression dependence coefficients.

(a) Usually the first quantity to be calculated is the sum of the squares of the deviations S_0, which was minimized by the above procedure. It can be calculated according to the definition, i.e. by summation of the squares of the differences between the experimental and expected values of dependent variable y or, if expected values y_i^* are not calculated, by modification of Eqs. (3.32) and (3.33) using the equation

$$S_0 = \Sigma(y_i - y_i^*)^2 = \Sigma y_i^2 - a\Sigma y_i - b\Sigma y_i x_i$$

(b) Another quantity is standard deviation S_1

$$S_1 = \sqrt{[S_0/(N - 2)]}$$

calculated as the square root of the sum of the squares of the deviations divided by the number of degrees of freedom. The number of degrees of freedom is the number of experimental points less the number of parameters (two for linear regression). This quantity is preferable to S_0 as it permits comparison of the results of linear regression for dependences of various amounts of experimental data or various numbers of parameters in the assumed functional relationships. It has the same dimension as quantity y.

(c) The square of the standard deviation, S_1^2, is the estimate of the variance of dependent variable y, more precisely of the errors of this variable whose normal distribution has been assumed.

(d) Standard deviations of the parameter estimates are calculated from the formulae

$$S(a) = S_1 \sqrt{\{(\Sigma x^2)/[N(\Sigma x^2) - (\Sigma x)^2]\}}$$

$$S(b) = S_1 \sqrt{\{N/[N(\Sigma x^2) - (\Sigma x)^2]\}}$$

(e) The confidence intervals of the parameter estimates are defined by the expressions

$$a \pm t_{\alpha,\nu} \cdot S(a); \quad b \pm t_{\alpha,\nu} \cdot S(b)$$

where $t_{\alpha,\nu}$ is the critical value of the Student t-distribution for $\nu = N - 2$ degrees of freedom and significance level α.

(f) The confidence interval of expected value y_0^*, calculated from regression relationship $y_0^* = a + bx_0$ for arbitrary x_0 is defined by the relationship

$$y_0^* \pm t_{\alpha,\nu} \cdot S_1 \sqrt{\{(\Sigma x^2 + Nx_0^2 - 2x_0\Sigma x)/[N(\Sigma x^2) - (\Sigma x)^2]\}}$$

If x_0 is varied continuously in the interval of experimental values x_i or over a larger interval, the confidence intervals thus defined form a "confidence band", which is narrowest in the middle of the experimental interval and widens on both sides. For $x_0 = 0$ it becomes the confidence interval of the estimate of parameter a. This confidence band lucidly illustrates possible errors in extrapolations using the regression relationship.

3.3.2.3 The Least Squares Methods for Some Types of Functional Relationships

The least squares method can be simply generalized for any type of polynomial but, especially for regression calculations on minicomputers when the memory capacity must be used economically, single purpose regression subroutines for certain concrete types of functional relationships are often useful. A survey of the formulae for calculation of parameter estimates and of statistical characteristics is given below.

Single-parameter Functional Relationships

Relationship:		$y = a$	$y = ax$	$y = ax^2$
Sums required:	B1	$= \Sigma 1 = N$	Σx^2	Σx^4
	B2	$= \Sigma y$	Σyx	Σyx^2
	B3	$= \Sigma y^2$	Σy^2	Σy^2

Parameter estimate: $a = $ B2/B1

$$S_0 = \text{B3} - a \cdot \text{B2}$$
$$S_1 = \sqrt{[S_0/(N-1)]}$$
$$S(a) = S_1/\sqrt{\text{B1}}$$

Two-parameter Functional Relationships

Relationship:		$y = a + bx$	$y = a + bx^2$	$y = ax + bx^2$
Sums required:	B1	$= \Sigma 1 = N$	$\Sigma 1 = N$	Σx^2
	B2	$= \Sigma x$	Σx^2	Σx^3
	B3	$= \Sigma x^2$	Σx^4	Σx^4
	B4	$= \Sigma y$	Σy	Σyx
	B5	$= \Sigma yx$	Σyx^2	Σyx^2
	B6	$= \Sigma y^2$	Σy^2	Σy^2

System determinant: $D = \text{B1} \cdot \text{B3} - \text{B2}^2$

Parameter estimate	a:	$a = (\text{B3} \cdot \text{B4} - \text{B2} \cdot \text{B5})/D$
	b:	$b = (\text{B1} \cdot \text{B5} - \text{B2} \cdot \text{B4})/D$
	$S_0 =$	$\text{B6} - a \cdot \text{B4} - b \cdot \text{B5}$
	$S_1 =$	$\sqrt{[S_0/(N-2)]}$
	$S(a) =$	$S_1 \cdot \sqrt{(\text{B3}/D)}$
	$S(b) =$	$S_1 \cdot \sqrt{(\text{B1}/D)}$

Three-parameter Functional Relationship $y = a + bx + cx^2$

Sums required:	B1	B2	B3	B4	B5	B6	B7	B8
	Σx	Σx^2	Σx^3	Σx^4	Σy	Σyx	Σyx^2	Σy^2

Auxiliary calculations: $D1 = \text{B2} \cdot \text{B4} - \text{B3}^2$; $D4 = N \cdot \text{B4} - \text{B2}^2$
$D2 = \text{B2} \cdot \text{B3} - \text{B1} \cdot \text{B4}$; $D5 = \text{B1} \cdot \text{B2} - N \cdot \text{B3}$
$D3 = \text{B1} \cdot \text{B3} - \text{B2}^2$; $D6 = N \cdot \text{B2} - \text{B1}^2$

System determinant: $D = N \cdot D1 + \text{B1} \cdot D2 + \text{B2} \cdot D3$

330

Parameter estimates: $a = (B5 . D1 + B6 . D2 + B7 . D3)/D$
$b = (B5 . D2 + B6 . D4 + B7 . D5)/D$
$c = (B5 . D3 + B6 . D5 + B7 . D6)/D$

Sums of squares of deviations: $S_0 = B8 - a . B5 - b . B6 - c . B7$

Standard deviation: $S_1 = \sqrt{[S_0/(N - 3)]}$

Standard deviation of estimated parameters:

$$S(a) = S_1 . \sqrt{(D1/D)}$$

$$S(b) = S_1 . \sqrt{(D4/D)}$$

$$S(c) = S_1 . \sqrt{(D6/D)}$$

For the regression equation of a general straight line $y = a + bx$, when all the sums required for the calculation of the correlation coefficient are available, the relationship

$$r = b . \sqrt{[D/(B1 . B6 - B4^2)]}$$

can be used to estimate its value.

3.3.2.4 Generalization of the Linear Regression Method

As already mentioned, one of the conditions for application of the linear regression method is the linearity of the functional relationship with respect to the parameters to be found. If this condition is not met, then apparently very different functional relationships, such as

a) $y = a_1 + a_2 x + a_3 x^2 + a_4 x^3$
b) $y = a_1 + a_2 x^2 + a_3 x^3 + a_4 x^5$
c) $y = a_1 \sin x + a_2 \cos x + a_3 \log x$
d) $y = a_1 x_1 + a_2 x_2 + a_3 x_3$

can be written generally as

$$y = \sum_{j=1}^{L} a_j . f_j = a_1 f_1 + a_2 f_2 + \ldots + a_L f_L$$

where L is the number of parameters and the individual partial functions f_j are

a) polynomial (regular): $f_j = x^{j-1}$ ($f_1 = 1$)
b) polynomial (irregular): $f_j = x^{e_j}$, $e_1, e_2, e_3, e_4 = 0, 2, 3, 5$
c) transcendental functions: $f_1 = \sin x$, $f_2 = \cos x$, $f_3 = \log x$
d) several independent variables: $f_j = x_j$

Fulfilling the condition of the minimum of the squares of the deviations,

$$\sum_{i=1}^{N} \left(y_i - \sum_{j=1}^{L} a_j \cdot f_{j,i} \right)^2 = \min$$

where $f_{j,i}$ is the value of the partial function for the j-th parameter and the i-th experimental point, then leads to a set of normal equations,

$$\sum_{k=1}^{L} \left[a_k \cdot \sum_{i=1}^{N} (f_{j,i} \cdot f_{k,i}) \right] = \sum_{i=1}^{N} (y_i \cdot f_{j,i}) \quad j = 1, 2, \ldots, L$$

3.3.2.5 Matrix Methods of Obtaining Normal Equations

A general system of normal equations can be written in the matrix form as

B $\times$ **A** = **C**

where **A** is the vector of parameters of dimension L, whose elements are identical with the required parameter estimates: $A_j = a_j$, **C** is the vector of the right-hand sides of the system of normal equations and also of dimension L, whose elements consist of summations

$$C_j = \sum_{i=1}^{N} (y_i \cdot f_{j,i})$$

and **B** is the square symmetrical matrix with dimension $L \times L$, whose elements are formed by the sums

$$B_{j,k} = \sum_{i=1}^{N} (f_{j,i} \cdot f_{k,i})$$

When filling in matrices **B** and **C** in the classical way, i.e. by actual calculation of the sums indicated, advantages connected with the symmetry of matrix **B** can be utilized (with polynomial regression it even suffices to calculate Σx_i^n up to the power equal to twice the highest power of x in the regression functional relationship to fill in matrix **B**), but the procedure involves tedious programming and computation.

The formation of matrices **B** and **C** is substantially simplified by application of matrix calculus. Some minicomputers employing the BASIC language permit matrix operations to be carried out using macroinstructions which facilitate programming and substantially accelerate the computation. The use of matrix operations can be recommended even for computers that do not contain these macroinstructions, by using appropriate subroutines.

Two methods of forming matrices B and C by matrix operations are available; their block schemes are given in Appendix C.

a) If a regular matrix F with dimensions $(L \times N)$ is generated, its elements consisting of the values of partial functions of the given functional relationship, x_i,

$$F_{j,i} = f_{j,i}$$

and transposed matrix G with dimension $(N \times L)$ is formed

$$G = \mathrm{trn}\,(F) \quad \text{or} \quad F^t$$

$$G_{i,j} = F_{j,i}$$

then matrix multiplication yields the required matrices, B and C

$$B = F \times G$$

$$C = F \times Y$$

where Y is a vector of dimension N containing the values of dependent variable y_i. This procedure is very rapid and economical in programming but places great demands on the memory capacity, because two matrices with dimensions $L \times N$ must be stored.

b) In the second procedure all the elements of matrices B and C are first set equal to zero. Then vector F_i with dimension L and with elements identical to the i-th column of matrix F in the previous procedure is generated in a loop passing through all the experimental points (for i from 1 to N). Transposition of vector F_i yields row vector G_i and multiplication of the two vectors yields square matrix E_i with the same dimension as matrix B,

$$E_i - F_i \times G_i$$

Gradual addition of matrix E_i to $B(B + E_i \rightarrow B)$ yields resultant matrix B:

$$B = \sum_{i=1}^{N} E_i$$

Scalar multiplication of vector F_i by y_i leads to a vector whose gradual addition forms matrix C:

$$(y_i) \cdot F_i \rightarrow F_i ; \quad C + F_i \rightarrow C$$

$$C = \Sigma y_i F_i$$

This procedure is somewhat slower than the previous one, but places lower demands on the memory capacity and thus enables treatment of extensive data sets. To calculate the sum of the squares of the deviations, it is necessary to pass through all the experimental points once more after solving the set of normal equations and to calculate the expected values of y.

The block schemes of the algorithms for the basic matrix operations are given in Appendix A.

3.3.2.6 Solution of a Set of Normal Equations

A set of normal equations written in matrix form as

$$B \times A = C$$

can be solved with respect to the required parameter vector A, e.g. by an elimination method. However, it is preferable to use a matrix method for the solution, i.e.

$$A = B^{-1} \times C$$

as the product of square matrix B^{-1}, inverse with respect to initial matrix B, and vector C. This procedure is more advantageous, because minicomputers permitting matrix operations carried out using macroinstructions contain an instruction for matrix inversion and because knowledge of the elements of the inverse matrix is important in calculation of the statistical characteristics.

Prior to actual matrix inversion by macroinstruction it should be tested whether the matrix is singular; this can be carried out using another macro-instruction calculation of the matrix determinant and testing it for difference from zero.

Many methods are available for matrix inversion; for computer treatment, elimination methods are most advantageous. Solution of a set of equations or matrix inversion carried out on a computer must always be considered as approximate methods, because rounding off errors can have a markedly adverse effect in view of the great number of arithmetical operations. Therefore, the classical Gauss elimination method was modified, yielding e.g. the *Gauss-Jordan elimination method* (the method of the maximum element) by means of which the precision was to be improved. Modified elimination methods place greater demands on programming, but do not completely remove the imprecision of inverted matrices. The results of linear regression can be refined better by consistent use of matrix operations.

Inverted matrix B^{-1} obtained by the classical Gauss elimination method (whose block scheme is given in Appendix B) either by means of macro-instructions on a minicomputer, or by a subroutine, should be considered as an approximate solution. Hence the parameter vector, obtained as the product with vector C,

$$A = B^{-1} \times C$$

must also be considered approximate. These parameter values are substituted into the functional relationship and the vectors of expected dependent variable values y_i^* are calculated for all independent variable values x_i ($i - 1, 2, ..., N$). The vector of deviations $\Delta y_i = y_i - y_i^*$ is obtained by subtracting vector y_i^* from vector y_i. Linear regression of Δy_i is carried out according to the same functional relationship as that used in the primary regression calculation,

$$\Delta y_i - \sum_{j=1}^{L} (\Delta a_j) \cdot f_{j,i} \qquad (3.38)$$

If parameters a_j have been determined precisely, deviations Δy_i are randomly distributed and a new regression calculation according to Eq. (3.38) yields zero vector corrections for parameters Δa_j. If, however, parameters a_j were not originally determined precisely (due to imprecise inversion of matrix B^{-1}), then the corrections to the parameters are non-zero and their addition leads to improvement in the parameter precision. This procedure is repeated until the corrections to all the parameters are sufficiently small. This is basically the principle of the Gauss-Newton method of non-linear regression applied to linear regression. From the point of view of programming, there are only a few instructions added, no field of variables is required, but the method is slower. However, a great advantage is a posteriori control of the regression results. When a set of incorrectly constructed normal equations is used, the classical procedure yields imprecise and often incorrect results; this method reveals divergence of the refining procedure, signalling an incorrectly chosen regression functional relationship.

3.3.2.7 Calculation of Statistical Characteristics by Matrix Procedures

With the first method of construction of a set of normal equations (see Section 3.3.2.5), the calculations following the solution of the set of equations can also be performed solely by matrix operations. By multiplying rectangular matrix G (transposed F) by parameter vector A, vector of dimension N is obtained for expected values of the dependent variable

$$Y^* = G \times A$$

Subtraction of the expected value vector from that of the experimental values yields the vector of deviations R:

$$R = Y - Y^*$$

Transposition of this vector gives a row vector of the same dimension, T,

$$T = \mathrm{trn}\,(R)$$

and multiplication (in the order T, R) yields a "matrix" with dimension (1×1), i.e. a single number $S = S_0$,

$$S = T \times R$$

which equals the sum of the squares of the deviations. Division of S_0 by the number of degrees of freedom and taking the square root yields the standard deviation,

$$S_1 = \sqrt{[S_0/(N - L)]}$$

When the second, more economical procedure for construction of a set of normal equations is used, all the experimental points must again be passed in a loop and row vector G_i must be formed directly for each of them. Multiplication with the parameter vector gives expected value y_i^* corresponding to x_i,

$$y_i^* = G_i \times A$$

In the same loop the squares of differences $y_i - y_i^*$ must be summed to obtain number S which was set equal to zero at the beginning

$$S + (y_i - y_i^*)^2 \to S$$

336

To complete the standard deviations of the parameter estimates, the diagonal elements of the inverse matrix must be known. Generally, the standard deviation of the j-th parameter in a functional relationship, a_j, is given by

$$S(a_j) = S_1 \cdot \sqrt{[(\mathbf{B}^{-1})_{jj}]}$$

i.e. by the product of the standard deviation and the square root of the j-th diagonal element of the inverse matrix.

In the previous section, inverse matrix $\mathbf{B}^{-1}$ was considered as only approximate, which is natural from the point of view of calculation of parameter estimates that should be as precise as possible. However, from the point of view of the calculation of standard deviations, its precision is mostly sufficient. If the standard deviations of the parameter estimates are also to be determined with the highest possible precision, the inverse matrix must be made more precise (see Appendix B).

To calculate the standard deviation of expected value y_0^* for an arbitrary value x_0, all elements of the inverse matrix must be known. If, analogously to the second procedure for matrix construction of a set of normal equations, the vector on the partial functions $\mathbf{F}_0$ corresponding to value x_0 is generated, then row vector $\mathbf{G}_0$ is obtained by transposition. The standard deviation of the expected value y_0^* is then given by the expression

$$S(y_0^*) = S_1 \cdot \sqrt{[\mathbf{G}_0 \times (\mathbf{B}^{-1}) \times \mathbf{F}_0]}$$

3.3.2.8 Statistical Significance of Parameters

The confidence interval for the parameters determined by the linear regression method is obtained by multiplying the standard deviations of the parameter estimates by the critical value of the Student t-distribution and adding the products to (and subtracting them from) the estimated value,

$$a_j \pm t_{\alpha,v} \cdot S(a_j)$$

where α is the selected significance level and v is the number of degrees of freedom $(v = N - L)$. The critical values of the t-distribution are tabulated for given significance levels and degrees of freedom. However, in computer treatment it is awkward to store extensive tables of critical values in the memory or to use less extensive tables with the necessity of interpolations.

The most advantageous approach is the use of a polynomial functional relationship approximating the critical values. The constants of the approximating polynomial for the critical values of the t-distribution are given in Appendix D.

Using the confidence interval thus obtained, statistical significance of the parameters and hence of the terms of the functional relationship can be tested. This is actually the principle of testing the zero hypothesis. If the limits of the confidence interval are located on both sides of zero (i.e. zero lies inside the confidence interval), or if it holds that $t_{\alpha,\nu} \cdot S(a_j) \geqq |a_j|$, then the parameter is statistically insignificant.

In empirical evaluation of experimental dependences, this testing has several consequences. If the fit of the curve of the chosen functional relationships to the experimental points is good (see the following section), then terms whose parameters are statistically insignificant are redundant. However, if the fit is poor, then the term is inadequate and must be replaced by another that fits the experimental dependence better.

3.3.2.9 Criteria of Fit

The main purpose of empirical evaluation of experimental dependences is attaining the best fit with the minimum number of parameters in the chosen empirical functional relationship and obtaining the best possible smoothing of data that are subject to random error.

The main problem in e.g. polynomial regression is selection of the degree of the polynomial. Too low a degree generally results in a poor fit but in good data smoothing. On the other hand, too high a degree usually leads to good fit but to poor smoothing because the curve has a tendency to "oscillate" between the points, i.e. to pass through points that are subject to error.

Theoretically, the selection of the degree of the polynomial or of the number of parameters in the functional relationship should follow the rule that the standard deviation of the fit, S_1, should be approximately the same as the error of measurement of dependent variable y, as a further improvement in the fit is pointless. A drawback of this rule lies in the fact that the error in the measurement of y is mostly unknown.

One of the criteria for testing the fit is testing for inflection points on the empirical curve in the given interval and comparison with the estimated number of inflection points on the experimental dependence. Location of inflection points involves solution of the equation, $f''(x) = 0$, for x. On the basis of this criterion, the degree of a polynomial that has more inflection points than the experimental dependence is unacceptably high.

Another criterion of the fit is evaluation of the confidence intervals of the parameters of the empirical functional relationship. In practice it often suffices to follow standard deviations of the parameter estimates (bearing in mind that the critical value of the t-distribution for $\alpha = 0.05$ and a large number of degrees of freedom is approximately 2). According to this criterion, the increase in the degree of the polynomial can be stopped when the parameter of a further added term in the functional relationship is statistically insignificant.

In polynomial regression, the inverse proportionality between the magnitude of the sum of the squares of the deviations and the degree of the polynomial is well known. The sum of the squares of the deviations decreases sharply at the beginning and then more slowly. Evaluation of the ratio of the sums of the squares of the deviations for polynomials with a lower and a higher degree and testing the significance of the difference in the sums of the squares of the deviations by the F-test can serve as another criterion of fit.

An anomalous situation may sometimes appear when the sum of the squares of the deviations is greater for a polynomial with a higher degree than for one with a lower degree. This is caused either by a rounding-off error during regression calculation or by incorrect use of the polynomial. If a fixed number of parameters is selected for the polynomial, the optimum combination of the exponents from the point of view of the fit is found, then a further polynomial term is added and the procedure is repeated, then an anomalous increase in the sum of the squares of the deviations is not encountered.

It can be recommended to use polynomials with some terms omitted, rather than ones with a regular series of exponents. The computing time is lengthened owing to looking for the optimum combination of exponents, but this drawback is outweighed by a better fit with a smaller number of parameters. When selecting integral or, at most, half-number exponents, the process of seeking the optimum combination of exponents can be algorithmized; when rational exponents are chosen, the problem becomes an optimization procedure.

A more exact criterion of fit need not be based on knowledge of the measuring error, but on the assumption that the error has normal distribution, i.e. is of random character. If the differences between the experimental and expected values of the dependent variable are plotted in dependence on the independent or dependent variable, it is readily found whether the deviations have random character. When the fit is poor, the deviations have a non-random, regular character, e.g. are sinusoidal.

Testing by the so-called sign test is based on a similar principle. The number of deviations with positive and negative signs are found; from tables of critical values, the number of "iterations" that can exist in a system ordered in a certain way at a given significance level is found (an iteration is a group of deviations with the same sign). Assuming that the iterations have normal distribution, the interval in which the number of "iterations" can appear can be determined exactly (at the given significance level). If the number of iterations is smaller, e.g. 2, then one half of the deviations has one sign and the other half has the opposite sign; this unambiguously indicates poor fit. Equally suspicious is the opposite case of a high number of iterations; deviations can regularly change sign.

3.3.2.10 The Weighed Least Squares Method

Sometimes, as will be shown below, it is advantageous to modify the least squares method and introduce the weighed least squares method in which the parameters of the functional relationship satisfying the condition

$$\sum_{i=1}^{N} w_i (y_i - y_i^*)^2 = \min$$

are sought; w_i are the weights that assume non-negative values. In this method, the individual values of the dependent variable contribute to a different degree to the minimized quantity.

Weights can be simply introduced formally into regression calculations. In all summations, each summation term is multiplied by an appropriate weight $(\Sigma w_i x_i^n$ and $\Sigma w_i y_i x_i^m)$. If the selected functional relationship contains an absolute term, then N appears in the set of normal equations obtained by the summation of unit values $(N = \Sigma 1)$. This term must now be replaced by the sum of the weights (Σw_i). In calculation of the statistical characteristics, when the number of degrees of freedom is calculated from N, the significance of N remains unchanged. In using the two matrix procedures for construction of a set of normal equations, the elements of matrix F or of vector F_i must be multiplied by the square root of the weights:

$$F_{j,i} = \sqrt{(w_i)} \cdot f_{j,i}$$

and vector Y must be replaced by vector R whose elements are formed by the products,

$$R_i = \sqrt{(w_i)} \cdot y_i$$

The block schemes of the two methods of regression calculations given in Appendix C use weights.

The most important and most problematic step in the weighed least squares method is the choice of weight values. Three examples are given below.

a) Exclusion of Outlying Points

If the weights of all points are the same and equal to unity $(w_i = 1)$, then the weighed least squares method is converted into the classical, non-weighed least squares method. If zero weight is chosen for some points, then these points are not considered at all in the regression calculation. In this way some points, which are e.g. subject to gross error, can be excluded from the regression calculation. In calculating statistical characteristics, the number of degrees of freedom must then be decreased by the number of points excluded, $v = N - L - W$, where W is the number of points excluded.

Outlying points can be excluded on the basis of subjective or objective criteria. In practice, subjective criteria are often chosen, after comparing numerical values of differences $y_i - y_i^*$ or their percent values, $(100 (y_i - y_i^*)/y_i)$, or after examining the computer graphical output.

An objective criterion for exclusion of outlying points is provided e.g. by the Grubbs test, in which the ratio of the absolute difference $y_i - y_i^*$ and the standard deviation S_1 is compared with the critical value of the Grubbs test for the chosen significance value; if the actual ratio is higher, the point is excluded.

b) Transformation of the Dependent Variable

Transformation of both variables, dependent and independent, is usually considered, but there is a qualitative difference between the two transformations. Transformation of the independent variable is only a formal operation that does not change the linearity of the functional relationship with respect to the parameters.

In contrast, transformation of the dependent variable often converts non-linear functional relationships into linear and enables regression calculation to be carried out by the linear regression method. As already mentioned, such transformations can lead to invalidity of basic assumptions involved in the least squares method and thus to gross errors.

For example, for a rigid functional relationship between the variables, two expressions are possible,

either $\quad y = a \cdot \exp(bx)$ $\hfill (3.39)$

or $\quad y' = \ln y = a' + bx \quad (a' = \ln a)$ $\hfill (3.40)$

However, with a real stochastic dependence, there is a principal difference between the two expressions. If dependent variable y values are measured with the same absolute error, then Eq. (3.39) is more correct, if they are measured with the same relative error (which often occurs), then Eq. (3.40) is more correct.

The weighed least squares method enables elimination of errors caused by the transformation and permits the use e.g. of Eq. (3.40) (straight line) even if the y values are measured with the same absolute error.

If the initial dependent variable y is subject to error ε, then after transformation of the functional relationship, $y' = g(y)$, the new variable y' is subject to an error determinable using the Taylor expansion and neglecting the terms with higher derivatives.

$$y' = g(y \pm \varepsilon) = g(y) \pm \varepsilon \cdot \frac{dg(y)}{dy}$$

This change in the magnitude of the error is eliminated by introducing the weights

$$w = \left| \frac{1}{\dfrac{d}{dy} g(y)} \right|$$

342

Correctly, the expected values of variable y^* should be substituted into functional relationship $\frac{dg(y)}{dy}$ in order that the weight values not be subject to a secondary error. However, these values are not available before carrying out regression calculation; therefore, experimental values y are used for calculation of the weights, provided that they are not subject to a large error. It is also possible to recalculate the weights after regression calculation, using the expected values, and repeat the calculation.

The following table summarizes the weight functional relationships for the most frequent transformations:

$g(y)$	$\dfrac{dg(y)}{dy}$	w
$1/y$	$-1/y^2$	y^2
y^2	$2y$	$1/y$
$\sqrt{y}$	$\frac{1}{2}y^{-1/2}$	$\sqrt{y}$
$\exp(y)$	$\exp(y)$	$\exp(-y)$
$\ln(y)$	$1/y$	y

c) Linear Regression of Points with a Given Standard Deviation

One of the main conditions for the use of the least squares method is normal distribution of errors. The same distribution, i.e. errors with the same variance, is implicitly assumed, leading to the classical least squares method with the same (unit) weights.

If each value of the dependent variable y is determined with exactly defined standard deviation S_i, then the weights of these points equal

$$w_i = S_i^{-2}$$

It is frequently necessary to carry out gradually applied linear regression. If the dependent variable is a function of several independent variables, the regression dependence on one variable can be studied first, then on another variable, etc., using the weighed least squares method.

Sometimes, when it is not desired to assign different weights to the parameter estimates obtained, the effect of the standard deviations of the individual points on the resultant sum of the squares of the deviations is of interest. It can readily be derived that

$$S_0 = S_0' + \sum_{i=1}^{N} S_i^2$$

where S_0' is the sum of the squares of the deviations not considering the S_i values.

3.3.2.11 Multiple Linear Regression

The above method of gradual linear regression cannot be generally recommended; complex regression treatment of the dependence on all the variables is more suitable and has been made possible by development in computing techniques.

As shown in Section 3.3.2.4, a classical example of multiple linear regression, i.e. the functional relationship, $y = a_1 x_1 + a_2 x_2 + \ldots + a_L x_L$, where the x_j are various independent variables, was generalized from the point of view of the linearity of the relationship with respect to the parameters. This generalization assumes, of course, real independence of variables x_j. To check this, the data correlation method must be used. Any stronger dependence among variables x_j produces singularity of matrix $\boldsymbol{B}$ during regression calculation, leading to insolvability of the set of normal equations. These considerations should precede experimental planning, to ensure successful data treatment.

Special attention should be paid to functional dependences on two variables, each of which is raised to several powers, i.e. to two-dimensional polynomial regression. Two-dimensional polynomials can be lucidly represented by rectangular, square or triangular parameter matrices but for regression calculations they must be considered as linear (with a one-dimensional parameter field).

3.3.3 OPTIMIZATION (NON-LINEAR REGRESSION)

3.3.3.1 Introduction

Optimization methods are used to find the optimal values of parameters satisfying a given optimality condition. This condition is usually related to the purpose function. There are methods for location of a maximum, minimum or for a zero purpose function.

344

Here methods for minimization of the purpose function will be of principal interest. If the purpose function is the sum of the squares of the deviations between the experimental and expected values of the dependent variable calculated from any functional relationship, the method is called "non-linear regression".

If only dependent variable y is a random quantity, then the optimization criterion is the condition of minimum sum of the squares of the deviations,

$$\sum_{i=1}^{N} (y_i - y_i^*)^2 = \min$$

or minimum sum of the weighed squares of the deviations,

$$\sum_{i=1}^{N} w_i(y_i - y_i^*)^2 = \min$$

between the expected and experimental values of y, where the expected values y_i^* are calculated either from the non-linear functional relationship,

$$y_i^* = f(x_i, a_1, a_2, ..., a_L)$$

or from a functional relationship, from which y_i^* cannot be explicitly expressed,

$$f(x_i, y_i^*, a_1, a_2, ..., a_L) = 0$$

To calculate the y_i^* values from this functional relationship, iteration methods for finding the roots of an equation are most often used (see Appendix F).

When both variables, x and y, are random quantities, formulation of the optimization criterion is not unambiguous. If the errors in the two variables are comparable, the criterion of minimum sum of the squares of the shortest distances of the experimental points from the functional relationship curve can be used,

$$\sum_{i=1}^{N} d_i^2 = \min$$

Distances d_i from a point to the intercept of the normal (the line perpendicular to the tangent) with the curve can be determined approximately from

$$d_i = \frac{(\Delta x_i) \cdot (\Delta y_i)}{\sqrt{[(\Delta x_i)^2 + (\Delta y_i)^2]}}$$

where $\Delta x_i = x_i - x_i^*$ and $\Delta y_i = y_i - y_i^*$ are the differences between the experimental and expected values. The purpose function thus constructed is also applicable for complex functional relationship curves (containing maxima, minima, etc.).

Another simple criterion is given by

$$\sum_{i=1}^{N} (x_i - x_i^*)^2 + \sum_{i=1}^{N} (y_i - y_i^*)^2 = \min$$

but its use is limited to simpler monotonic curves.

If the errors in the two variables are not comparable, then the criterion

$$\sum_{i=1}^{N} v_i(x_i - x_i^*)^2 + \sum_{i=1}^{N} w_i(y_i - y_i^*)^2 = \min$$

can be used, where v_i and w_i are weights defined by

$$v_i = (S_x)_i^{-2}$$

$$w_i = (S_y)_i^{-2}$$

The situation can be further complicated by the necessity of an iteration solution for x_i^* or y_i^* or both from the implicit functional relationship.

3.3.3.2 Properties of Optimization Methods

There are many optimization methods. They can be classified e.g. according to

   i) rapidity of calculation,

   ii) necessity of good initial estimate of the parameters,

   iii) necessity of knowledge of the partial derivatives of the functional relationship with respect to the individual parameters.

An ideal method that would not require a good initial estimate and would lead rapidly to the final result does not exist. In solving optimization problems, the two methods are often combined, one slower, but independent of the initial estimate, the other faster, applied after sufficient refinement of the parameters to be optimized.

Emphasis is placed on small computers in this chapter, hence on rapid methods, even if a good initial estimate is required. The third criterion in classification of optimization methods, the necessity of knowing the derivatives, is connected with no negative effects, because analytical differentiations can, in principle, always be performed.

Optimization problems are generally complicated by ambiguity of the solution, because other local minima appear. However, these complications are usually not encountered in non-linear regression.

3.3.3.3 Linearization

The linearization method is not, strictly speaking, part of the optimization methods and is dealt with here because it yields good initial estimates of the parameters of non-linear functional relationships.

A special case of linearization was discussed in Section 3.3.2.10. Some non-linear functional relationships can be linearized with respect to the parameters by suitable transformation of the dependent variable and the parameter estimates can be determined by linear regression methods. Good initial estimates of the parameters can thus be obtained for optimization methods with or without the use of weights.

Generally, other functional relationships, where transformation of the dependent variable is not useful, can be linearized. This is especially true with fractional functions; the dependent and independent variables must not be rigidly distinguished and x', y', and the products of the original variables x and y are used as new variables. For example, the fractional function

$$y = a + b/(x - c)$$

can be linearized in the form

$$xy = (b - ac) + ax + cy$$

and thus the regression relationship

$$y' = a_0 + a_1 x_1' + a_2 x_2'$$

where $y' = xy$, $x_1' = x$ and $x_2' = y$, is solved by the linear regression method. The parameters obtained (after recalculation to the original form) are, of course, inaccurate statistically, but are mostly numerically close to the accurate values that are readily refined by an optimization method.

It often occurs in these linearizations that the linearized functional relationship has a greater number of parameters than the original one. This causes a certain ambiguity on recalculation to the parameters of the original relationship, which, however, rarely leads to greater complications.

3.3.3.4 The Gauss-Newton Method

This method is of the greatest importance in the present context, because it is the fastest of all optimization methods and the formal procedure is the same as that for the linear regression method.

This method is based on the fact that any function can be expanded in a Taylor series. Assume a function of variable x and L parameters $a_1 - a_L$, $f(x, a_1, ..., a_L)$. The values of the function for parameters a_j $(j = 1, 2, ..., L)$ close to parameters a_j^0 can be expressed by the expansion

$$f(x, a_1 ... a_L) = f(x, a_1^0 ... a_L^0) + \sum_{j=1}^{L} \Delta a_j f'_{a_j}(x, a_1^0 ... a_L^0)$$

where $\Delta a_j = a_j - a_j^0$, $f'_{a_j} = (\partial/\partial a_j) f(x, a_1^0 ... a_L^0)$ and higher derivatives are neglected.

If the initial estimate of parameters a_j is used for calculation of the expected values of dependent variable y_i^* and it is assumed that deviations $\Delta y_i = y_i - y_i^*$ are caused by inaccuracies in the estimation of the parameters, corrections to the parameters are obtained by solving the linear regression equation

$$\Delta y_i = \sum_{j=1}^{L} \Delta a_j \cdot f'_{a_j}(x_i, a_1 ... a_L)$$

Addition of corrections Δa_j to the original parameters a_j yields new parameter estimates. Using the new parameter estimates, new expected values y_i^* and derivatives f'_{a_j} are calculated and the procedure is repeated.

If this iteration process converges correctly, the parameter estimates approach limiting values and the sum of the squares of deviations $\Sigma(\Delta y_i)^2$ approaches a minimum value. A criterion for termination of the iteration process can be the constancy of the parameter estimates or the constancy

348

of the sum of the squares of the deviations, expressed by the chosen number of decimal places. If the initial parameter estimates are incorrect (too far from the correct values), the iteration process diverges. The algorithm for the Gauss-Newton optimization method is substantially simplified when the matrix operations are used. In contrast to the linear regression method, initial matrix **F** must contain the values of the partial derivatives of the given functional relationship with respect to the individual parameters and individual values of x_i,

$$F_{j,i} = \frac{\partial}{\partial a_j} f(x_i, a_1 \ldots a_L)$$

A block scheme of this optimization method is given in the Appendix G. It is important (and not readily evident) that this method also yields all the data required for calculation of the standard deviations of both the parameters and the y values calculated from a non-linear functional relationship. These data are again contained in the elements of the inverse matrix $\mathbf{B}^{-1}$.

3.3.3.5 Modifications of the Gauss-Newton Method

a) Implicit Functional Relationship

If the functional relationship is implicit with respect to dependent variable y,

$$f(x, y, a_1 \ldots a_L) = 0$$

then the Gauss-Newton method must be modified, so that parameter corrections are obtained by solving the linear regression relationship,

$$\Delta y_i = \sum_{j=1}^{L} \Delta a_j \cdot F_{j,i}; \quad F_{j,i} = \frac{\left(\dfrac{\partial f}{\partial a_j}\right)_i}{\left(\dfrac{\partial f}{\partial y}\right)_i}$$

Hence matrix **F** is filled with the ratios of the partial derivatives with respect to the parameters and those with respect to y.

b) Stochastic Dependence

If both variables are subject to error, one criterion of the optimum state is the condition of minimum sum of the squares of the distances of the points from a curve. The parameter corrections are then determined by solving the regression relationship

$$\frac{(\Delta x_i)(\Delta y_i)}{\sqrt{[(\Delta x_i)^2 + (\Delta y_i)^2]}} = \sum_{j=1}^{L} \Delta a_j \cdot F_{j,i}; \quad F_{j,i} = \frac{\left(\dfrac{\partial f}{\partial a_j}\right)_i}{\sqrt{\left[\left(\dfrac{\partial f}{\partial x}\right)_i^2 + \left(\dfrac{\partial f}{\partial y}\right)_i^2\right]}}$$

When a simpler condition is used for the optimal state, then the regression relationship

$$\sqrt{[(\Delta x_i)^2 + (\Delta y_i)^2]} = \sum_{j=1}^{L} \Delta a_j \cdot F_{j,i}; \quad F_{j,i} = \frac{\left(\dfrac{\partial f}{\partial a_j}\right)_i}{\sqrt{\left[\left(\dfrac{\partial f}{\partial x}\right)_i + \left(\dfrac{\partial f}{\partial y}\right)_i\right]}}$$

must be solved.

c) Marquardt Modification of the Gauss-Newton Method

The principal drawback of the Gauss-Newton method, the necessity of having good initial parameter estimates, can be removed using this modification.

If a unit matrix multiplied by a number λ from the interval $(0, \infty)$ is added to matrix **B** forming a set of normal equations, then the procedure converges even for distant estimates, though more slowly. With large λ, the method is converted into the gradient method; with $\lambda = 0$ it becomes the classical Gauss-Newton method. There are certain rules for selection of the magnitude of number λ which can be included in the algorithm, but when using minicomputers it is most suitable to vary the λ value manually during the computation.

This modification is considered in the block scheme of the Gauss-Newton method, but zero is substituted for λ in the beginning. For this modification it is useful to store the vector of the parameters, corresponding to the lowest sum of the squares of the deviations obtained, in the memory.

3.3.3.6 Survey of Further Optimization Methods

Other optimization methods can be used to carry out non-linear regression, as a special case of optimization. These methods are classified according to the number of optimized parameters into single- and multi-parameter or, in dependence on the requirement of knowledge of derivatives of the purpose function, into differentiation and non-differentiation.

An important group of optimization methods are methods optimizing a single parameter, such as the *direct search method*. This method requires only calculation of the values of the purpose function, but not of their derivatives. In searching for the minimum of the purpose function (i.e. the minimum sum of the squares of the deviations between the calculated and measured values) the value of the optimized parameter changes in an equidistant manner. In this way, passage through a minimum is indicated; the magnitude and sign of the increment to the optimized parameter are then changed (e.g. halved) and the procedure is repeated. Another procedure is fitting a parabola to three points in the vicinity of the minimum and calculation of the coordinates of this parabola. The procedure is repeated closer to the minimum, until the change in the value of the parameter or of the purpose function between the two successive computing cycles is not less than a preset value. Satisfaction of both criteria simultaneously is often required, to prevent oscillation of the calculation around very flat or very steep minima of the purpose function. The method based on this principle is identical with the pit-mapping method for a single-dimensional problem.

The direct search method is extended for two parameters by using the "*probe algorithm*" *method*, in which the parameters are alternately optimized by the above method. This method is not as effective as e.g. the *Rosenbroke method* in which the two parameters are simultaneously optimized in a direction selected so that the search for the minimum is as effective as possible. Another well-known method is the *simplex method* which, for two parameters, requires an initial calculation of 3 values of the purpose function at points forming a triangle. Then, instead of the point with the highest value of the purpose function, a new value of the purpose function is calculated on the opposite side of the line connecting the points with lower values of the purpose function, i.e. the triangle is rotated around one side. Rotation and gradual decreasing of the size of the triangle (simplex) leads to location of the purpose function minimum.

All these methods can be generalized for more than two parameters but the computation is very tedious. From the point of view of computing economy, the pit-mapping method, especially Sillén's **LETAGROP**, is most suitable for optimization of a greater number of parameters. In this method, it is necessary to compute $(n + 1)(n + 2)/2$ values of the purpose function in each step for n optimized parameters, which just suffices for determination of the parameters of an n-dimensional paraboloid whose minimum is to be found. The procedure is repeated until the required precision is attained.

Among the methods requiring knowledge of the derivatives of the purpose function with respect to the individual parameters, there are *gradient methods*, the best known being the "greatest slope" and the Powell methods. The Gauss-Newton method, discussed in detail in the previous section, also belongs here.

Important time-saving in optimization computations can be achieved by dividing the parameters in a non-linear functional relationship into non-linear and linear and optimizing only the non-linear parameters. The appropriate value of the purpose function and the values of the linear parameters are obtained by the linear regression method. Another possibility of time-saving is offered by discarding the optimum state condition, the minimum of the sum of the squares of the deviations, and using a group method (zero sum of the deviations). The results must, of course, be refined using a least squares method.

3.4 Simulation of Experimental Data

In practice it is often required to obtain a set of numerical data very similar to the data obtained experimentally and to use this data for testing certain theoretical problems. An advantage of this approach, in addition to elimination of tedious and often costly experimental work, is the possibility of obtaining quasi-experimental data satisfying the required mathematical model and subject to defined errors with chosen values of the parameters and their distribution. The sets of data thus formed can be used in, e.g.

1. searching for a suitable algorithm for non-linear regression for e.g. the test problem,

2. solution of regression problems for error distributions other than linear,

3. formation of new tests and calculation of the appropriate critical values.

Computer simulation of experimental data subject to error consists of two steps:

— calculation of the expected value of a quantity on the basis of a mathematical model of the behaviour of the system,

— calculation of an error that is added to the expected value.

The first step involves no problems and thus the whole procedure is reduced to calculation of the errors as random quantities with given characteristics and distribution. For this purpose, the random number concept must be introduced.

A random number is a random quantity with a uniform distribution in the interval (0, 1). Random numbers can only be obtained by a random process, e.g. by amplification of electronic noise and its digitization and storage in the computer memory (random number generators). However, this procedure is rarely used in practice, because it is unduly complicated and pseudo-random numbers are satisfactory for the above purposes; these can be obtained directly in the computer using special algorithms. Because the computation of a pseudo-random number is always based on the preceding pseudo-random number, a disadvantage and typical peculiarity is the fact that repeated use of the algorithm yields the same series of random numbers. Therefore, the first random number must be selected before the generation and it must be ensured that it is always different (e.g. on the basis of the data). Another unpleasant property of pseudo-random numbers is their periodic character; after a certain number of generated random numbers the series is repeated. In selecting a generator of pseudo-random numbers an attempt should be made to find the longest possible period.

There are various algorithms for generating random numbers with uniform distribution. Such generators are part of the software in some minicomputers and pocket calculators. An example of such an algorithm is given below:

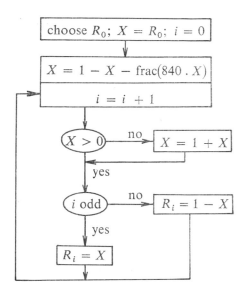

The quality of a random number generator is evaluated by statistical testing of the generated series of random numbers. They are, however, most often tested using their moments. It can be proved that the n-th moment of random numbers from interval $(0, 1)$ approaches the value, $1/(n + 1)$. In testing, the 1st and 2nd moments are mostly calculated and are compared with the values 0.5 and 0.333, or the 2nd central moment is calculated, which theoretically equals $1/12$.

Random numbers with uniform distribution have, however, limited applicability. In most cases, random numbers with a certain, exactly defined distribution are required. One of the ways of obtaining them is using the formula,

$$R = \int_{-\infty}^{r} f(x)\,dx$$

where R is a random number with a uniform distribution in the interval $(0, 1)$ and r is a random number with the distribution (i.e. the probability density), $f(x)$. This procedure is quite general and enables obtaining of random numbers with any distribution.

When random numbers with normal, most frequently required, distribution are needed, a simple method can be used, based on the "central limiting lemma". According to this lemma, the sum of random quantities with any distribution yields a random quantity with a distribution that limits to the number of numbers summed to normal distribution. Thus sums of random numbers with a uniform distribution can also yield random numbers with normal distribution. In practice, six addends often suffice. The mean value of such sums is $n/2 = 3$ and the variance is $\sqrt{(n/12)} = 1/\sqrt{2}$. Random number generators with normal distribution, used in computers, generate normalized random numbers with mean value 0 and variance 1 that can be obtained from the formula,

$$r = \sqrt{2} \left(\sum_{i=1}^{6} R_j - 3 \right)$$

Simulation of e.g. a regression dependence then involves three steps. The number of experimental points and the interval for generation of the values of independent variable x are selected first, then the expected values of the dependent variable are generated according to the given functional relationship and finally the errors formed by product σr_i are added to the y_i values. With simulation of a general stochastic dependence, both values y_i and x_i can be subject to error.

3.5 Other General Uses of Computers in Analytical Practice

The use of computers in analytical work enables introduction of some methods that provide additional information on an analytical result and methods of obtaining results, by means of which the properties of the results and analytical methods can be improved in a preset direction. Some of these methods are rather complicated and often involve several steps. Their description is outside the scope of this chapter; therefore, only two will be mentioned, namely, the method of optimization of analytical procedures and the promising pattern recognition methods, which are described further in the specialized literature.

3.5.1 OPTIMIZATION OF ANALYTICAL PROCEDURES

In analytical practice, a new method must often be developed or an established method selected and the most suitable conditions must be found for its application to a concrete analytical problem. The search for the most suitable conditions for application of an analytical method is generally termed optimization; this procedure must not, however, be confused with "optimization" of coefficient values, e.g. in non-linear regression. The logic basis is similar but each procedure serves a different purpose and employs a somewhat different method; optimization of coefficients is purely mathematical while that of an analytical procedure is predominantly an experimental method whose results are treated mathematically.

If an analytical method is considered as a system that is to provide a certain amount of information, the operation of this system must be optimized from the point of view of the properties of the information required. In optimization the response function must predominantly be found that characterizes the properties of the analytical information, i.e. the analysis result. The factors must further be selected that affect the result or the course of the process and finally the optimum conditions are sought. The third and sometimes also the second step of an optimization procedure are generally carried out using computing techniques.

The selection of a suitable response function is very important for practical applicability of the optimization results; it should be borne in mind that only the conditions characterized by a given response function are optimized and therefore the conditions optimal for a certain response function can be markedly suboptimal for another response function. Simple quantities are sometimes selected as response functions, such as the analytical signal intensity, or signal-to-noise ratio. In other cases, it is required that the response function characterize several properties simultaneously and then quantities such as the equivocation, information performance, rentability, etc. are chosen. When optimal conditions are to be found for simultaneous determination of several components, then compromise conditions are sought using a single response function or, on the other hand, the determination of major components is optimized separately using an response function other than that for independent determination of minor and trace components.

Before looking for the optimal conditions, under which the response function attains the most suitable (usually extreme) value, the factors that statistically significantly affect its value must be found. These can be quantitative factors (pH, temperature, concentration, etc.) or qualitative factors (kind of instrument, procedure, etc.). Quantiative factors must be studied at at least two levels, i.e. at two different values of each factor. The statistical significance of the effect of factors on the value of the response function is

356

usually based on the fact that they express the response function value as a linear combination of the effect of the individual factors and random measuring errors. The overall variability of the response function is then divided into independent components corresponding to the individual factors and their statistical significance is tested with respect to the residual variability corresponding to random errors, using the F-test. This variance analysis is described on p. 97 and further pages in Vol. IA. The effect of the factors can be found by a complete factor experiment, when measurement must be performed for N factors at two levels and for n-fold repetition of each experiment, including all combinations of the factors and their levels; this involves $n \times 2^N$ measurements. In a shortened factor experiment, the measurement is carried out for only certain combinations. The evaluation of the factor experiment that includes the significance of the individual factors is facilitated using computation techniques.

Finding of the optimal combinations of the factor values is a further step in the optimization of an analytical procedure. Various methods can be used, the most common being the "greatest slope" and "simplex" methods. In both cases, the principle is controlled movement of an experimentally determined point in an N-dimensional factor space toward the optimum in the direction of the greatest gradient of the objective function value. The two procedures and programs suitable for analytical practice have been published. Details on optimization of analytical methods can be found in the monograph by D. L. Massart, A. Dijkstra and L. Kaufman [9].

3.5.2 PATTERN RECOGNITION

Methods of pattern recognition have successfully been used recently for evaluation and interpretation of analytical information, for classification and identification purposes and for choice of an optimal analytical method or procedure [10, 11]. The greatest advantage of the pattern recognition methods is the possibility of using a great number of various parameters, those that have the character of quantitative, numerically expressed quantities and qualitative parameters, provided that they can be assigned numerical descriptions. The correctness of the final decision is not adversely affected even if parameters that are later found to be quite irrelevant are used. Practical use of pattern recognition requires a computer with a sufficient memory capacity.

Pattern recognition belongs among procedures called artificial intelligence and includes methods enabling analysis of multivariant experimental data and classification of objects characterized by these data into classes according to the similarity of properties studied. Moreover, these methods can sometimes be used for testing a working hypothesis. The pattern recognition algorithm usually involves four parts: (1) Preprocessing in which the original data are converted into a form suitable for the computer and transformations, autoscaling, descriptor assignment, etc. are carried out. (2) Feature selection is the determination of the minimum number of variables relevant for the given classification and is carried out in various ways. One technique for feature selection is searching for an optimum set of features for representation of a dissimilarity between classes using the entropy concept (see 1.4.1). Two step feature selection was studied by Štrouf and Fusek [12] who introduced the concept of "intrinsic dimensionality" for the least number of features relevant for the given classification. (3) Classification of objects into classes according to feature similarity can also be carried out in various ways; some are based on fitting to various empirical functions, others on assignment of the minimum distance from the given object in the N-dimensional space to the pattern, etc. (4) Recording and representation of the results of classification are the last part of the whole pattern recognition process.

Analytical results are generally very suitable for treatment by a pattern recognition method, chiefly because modern analytical methods enable ready obtaining of a sufficient amount of data from spectra, chromatograms and other recordings, results, etc. with a satisfactory precision. The use of pattern recognition methods in analytical practice is now rather wide, especially in treating complex analytical signals in mass and molecular spectroscopy, in structural analysis carried out by a combination of various methods, etc. They have a special importance for identification of organic substances and can also be used for selection of a suitable analytical method or of an optimal procedure, e.g. a suitable combination of mobile and stationary phase in chromatography. Some practical applications of pattern recognition methods are closely related to the use of information theory in analytical chemistry and are thus mentioned in the monograph by K. Eckschlager and V. Štěpánek [2]. For a survey of the use of pattern recognition methods, see also [10, 11, 13].

4 Special Use of Computers in Analytical Chemistry

4.1 Gas and Liquid Chromatography

In the last 25 years, gas and liquid chromatography have found more widespread application than practically any other modern analytical method. With an increasing number of active chromatographs (in 1972 the number of working gas chromatographs in the world was estimated at over 100,000 [14]), efforts also increased to replace the relatively laborious, manual, quantitative evaluation of chromatograms. Development proceeded from the mechanical disc integrator of detection response through electronic integrators with hardware logic to mini and microcomputers.

Minicomputers have been used as intelligent integrators of the chromatographic signal since the end of the sixties. The systems for processing chromatograms were either substystems of large program systems serving large analytical laboratories (e.g. the Mülheim [15] system, the Dupont [16] system) or worked independently. These specialized systems became available commercially at the beginning of the seventies and were offered by practically all the bigger producers of chromatographic instrumentation (Hewlett-Packard [17], Varian [18], Perkin-Elmer [19], Siemens [20]). It was expedient to compensate the relatively high price of the processor and peripheries of these systems by increased capacity; the systems were capable of simultaneously processing the output data of as many as 40 chromatographs.

Large systems were not suitable for smaller laboratories, and the latter still remained dependent on electronic integrators, whose outputs could be processed in an off-line arrangement by any computer facility.

Significant changes were brought about by the introduction of the microcomputer technique at the beginning of the seventies. The large-scale integration (LSI) technology made it possible to construct a relatively powerful

359

computer with a 10K byte memory, including the peripheral interfaces, on a single printed-circuit board with dimensions of 30×20 cm, all for a fraction of the price of a minicomputer.

The first commercial microcomputer-based integrators were the Autolab-Spectra Physics System IV, 1971, and the Minigrator from the same firm, with Rockwell processors. The small capacity of the RAM (random-access memory) in both systems did not provide for good baseline correction. More elaborate systems for a single channel are, for instance, the Varian CDS 101 and CDS 111 integrators with the Intel 8008 processor.

The function of a microcomputer need not be confined to mere processing of the output data from the chromatograph; the capacity of the microcomputer can also be utilized to digitally control the analytical instrument. Then the microcomputer constitutes an intelligent centre of the instrument and the individual functional units of the chromatograph, such as the column oven, sample-inlet port, detectors and flow/pressure control constitute peripheries of the central processor. An instrument designed in this way is characterized by extraordinarily high reproducibility of the working conditions. The individual working conditions are set by a functional keyboard; the chromatogram, together with the appropriate qualitative and quantitative data, is recorded by a digital plotter. The Hewlett-Packard 5830A gas chromatograph was the first instrument of this generation, followed by the contemporary 5880A model with a 16 bit SOS-MOS processor with a 160 ns cycle.

Next, the logic structure of an intelligent chromatographic integrator with a microcomputer will be considered. Though the instruments produced by different manufacturers differ from each other in details, there are certain functional blocks which are common to all chromatographic integrators.

The first block is formed by the data acquisition system; an essential component of its electronics is a voltage-to-digital converter. A continuous voltage signal comes to the input of the system from the chromatographic detector and a sequence of converter readings is produced at the outlet.

A further block consists of a digital filter; a typical filter used in chromatographic systems consists primarily of an algorithm of the creation of moving sums from the individual converter readings (bunching). The programmed control of the number of addends makes it possible for the filter to adapt itself to varying peak width in an isothermal chromatogram. In addition to the sequence of converter readings produced by the data acquisition system, the parameters set by the operator can also be found at the input of the block. At the output are usually smoothed values of the signal and its first (sometimes also second) derivative.

The most important component of the integrator is the peak detector. The latter is usually composed of a constant part and a series of switchable branches which logically track the individual parts of the eluted peak. The input data are smoothed values from the output of the digital filter and the parameters set by the operator; the output is usually a table containing information on each peak.

The last block corresponds to final calculations and print-out. The program carries out the final baseline corrections and prints out the retention times and corrected peak areas. With more elaborate instruments, this block also performs complete quantitative processing of the chromatogram according to a chosen method of quantitative analysis. The input data are tables of peaks and a table of standards and the output is a formatted report.

4.1.1 DATA ACQUISITION SYSTEM

A basic task of the hardware (sometimes also the software) of the data acquisition system is the extraction of useful information contained in the continuous analog signal of the chromatogram and its conversion into tabular form suitable for processing by a computer.

The transition from a continuous signal to a table of discrete data is always associated with some loss of information. Every table has only a finite step length and the precision of the individual tabulated values is limited. The magnitude of the maximum step (sampling interval) and the minimum precision of the tabulated data are determined both by the properties of the analog signal being converted and by the manner and purpose of the processing of the digitized data as well as the required precision of the final results. Thus, for a given voltage signal of the chromatographic detector, the minimum sampling rate will differ according to whether high-frequency noise produced by the detector is to be studied or whether the areas of individual peaks are to be determined. In the first case, the spectral composition of the noise would certainly play a decisive role and the sampling rate would have to be determined in compliance with the Nyquist theorem.

The question of maximum sampling interval, or minimum sampling rate, with regard to quantitative evaluation of chromatograms, has received considerable attention [21−24]. The results obtained by various authors vary within rather broad limits, 20−100 samples per peak. With commercial systems, about 10−25 samples per half-peak width are common, which corresponds to approximately 20−50 samples per peak. The sampling interval will be discussed later in connection with digital filtering.

It is necessary to determine the precision of representing the individual digital samples of a chromatogram. Classical chromatographs are fitted with the output voltage attenuator, which makes it possible to vary the range of the signal within the limits of 1−1024 in eleven binary orders of magnitude. A good-quality line recorder with a precision and repeatability of 0.3 and 0.1% of the full scale deflection, respectively, can be connected to the divider. Then it is meaningful to distinguish deflections as small as 1/1000 of the full scale on the recorder. A binary representation of 1000 levels can be provided by 10 bits (1024 levels). Hence, if a good-quality classical chromatographic recording system is to be simulated, it is necessary to have a 10 bit converter fitted with an autoranging amplifier of 11 ranges. In fact, e.g. the Mülheim system has an eleven-range amplifier, a ten-range one can be found in the IBM 1130 system [25], the commercial PRAG 320 system from Siemens also uses a ten-range amplifier with 11-bit conversion precision. However, it seems to be more advantageous to limit the amplifier ranges and increase the conversion precision accordingly. Thus, the Varian CDS 101 and CDS 111 systems employ 4-stage amplifiers but the converter is a 12-bit one.

Of a number of techniques providing for analog-to-digital conversion [26], practically only two are employed in chromatographic systems: the first is based on the voltage-frequency converter and the second is the dual-slope technique.

The voltage-frequency converter, familiar from electronic integrators, provides an output signal with frequency proportional to the input voltage. On combining this converter with a counter, a clock and a control logic, an analog-to-digital converter is obtained with dynamics determined, at a given maximum frequency, by the sampling interval. Thus, a converter with a maximum frequency of 1 MHz at a sampling interval of 100 ms yields a maximum of 10^5 discrete values (approximately 17 bits). Because of its exclusively integration principle of operation, the converter is very resistant to noise. In chromatographic systems, it is used in the products of Autolab-Spectra Physics (System IV, Minigrator) and Perkin-Elmer.

An analog-to-digital converter based on the dual-slope technique works in two phases. In the first phase, duration of which is constant with a given converter, the voltage signal is integrated. In the second phase a standard voltage of opposite polarity is supplied to the integrator input and the integration is continued as long as the output is zero. The duration of the second phase, measured by a counter, is proportional to the value of the voltage supplied. Common-mode rejection is accomplished by synchronizing the beginning of the first, measuring phase with the mains frequency. Converters of this type work in Varian and Siemens chromatographic systems.

A schematic diagram of a typical data-acquisition system used in chroma-

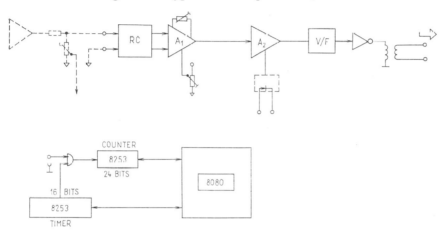

Fig. 56. Data-acquisition system.

tography is given in Fig. 56. After analog filtering, amplifier $A1$ adjusts the level of the detector signal to a normalized value of $0-10$ V. The normalized signal is led to amplifier $A2$, whose gain can be adjusted in four stages on command from the computer. After amplification follows an analog-digital converter whose analog part is galvanically separated from the digital part (counter, control), so that the entire analog part of the system is floating and can be connected to the chromatograph. The digital part of the converter has common ground with the interface and microcomputer. Galvanic separation is accomplished by either ferrite transformers or optocouplers.

The gain switching of amplifier $A2$ can be performed either by the converter logic or directly by the processor, as shown in Fig. 56. In the first case, the digital information transmitted to the interface consists of information on the magnitude of the signal within the limits of the given order (e.g. 12 bits) and setting of the range (e.g. 2 bits). When the range switching is provided directly by the processor, data on the range need not be transmitted.

4.1.2 DIGITAL FILTERING

The data acquisition system described in the previous chapter changes a continuous voltage signal from the chromatographic detector into a sequence of time-equispaced data points. The filtering block, in which the above data are preprocessed for the purposes of the peak detector, provides essentially two functions: performing a time-normalization of the digitized chromatogram and improving the signal-to-noise ratio.

The basic normalization technique, employed in chromatographic systems since the beginning of the seventies, is called bunching. This creates a new sequence of partial sums of data points from an equispaced sequence of data points; the number of addends in the individual partial sums is adjusted by a program during recording of the chromatogram so that each peak is represented by an approximately equal number of partial sums, i.e. bunches. If, for instance, a narrow peak at the beginning of the chromatogram is represented by a total of 40 data points, a broad peak at the end of the chromatogram, determined by 400 data points, is represented for the peak detector by 40 partial sums of 10 data points. The sum of the original 400 values is the same as the sum of the 40 partial sums, i.e. the operation preserves the total integral. It remains to determine the rules according to which the number of addends in the given region of the chromatogram can be set or changed. The initial number of addends is closely related to the width of the first significant peak in the chromatogram. Assuming that about 20 bunches correspond to the half-width of the first peak, and if the half-width of the first peak and the sampling interval of the data-acquisition system are W (sec) and τ (sec), respectively, then the number of addends in one such bunch is determined by the closest whole number value or, better, by the closest binary order of the expression, $W/20\tau$.

During the rest of the chromatogram, the number of addends must be changed by a program. In practice, it is usual to increase the number of addends by binary multiples. The command to pass over to a higher order

364

(doubling the number of addends) comes from the logic of the peak detector either by virtue of a parameter set by the operator or automatically. In the first case, the operator sets the time interval after which the number of addends is doubled on the keyboard. In the second case, the peak detector measures the true number of bunches by which the last peak (and/or its first half) was defined; after a certain, predetermined number of addends has been exceeded, the number is automatically doubled.

After the normalization procedure, digital filters usually follow. In the course of filtering, the processed data (bunches) are replaced by estimated values that differ only very little from the data measured, but ensure a smoother course of the chromatogram. The estimate is based on knowledge of the course being smoothed in the neighbourhood of the sample.

Linear digital filters can be described by the relationship

$$y_i^* = \frac{1}{2k + 1} \sum_{j=-k}^{k} A_j y_{i+j} \tag{4.1}$$

where $2k + 1 = n$ is the filter span and the A_j's are coefficients characteristic of the given filter. The A_j values can be looked upon as weight contributions of the individual values of the input function y_i to the smoothed value y_i^*. The coefficients of a number of filters have been summarized by Savitzky and Golay [27].

In chromatographic applications two types of filter are mostly used: moving average and least-squares curve fitting. With the moving average, the A_j coefficients equal unity for all j within the interval $(-k, k)$. The resultant filter can be expressed by the recursive formula

$$y_i^* = y_{i-1}^* + \frac{1}{n} \left(y_{i+k} - y_{i-k-1} \right) \tag{4.2}$$

The least-squares curve fitting replaces the input values y_i by values of the interpolation polynomial of the m-th degree, fitted to neighbouring points $(i - k, y_{i-k}), ..., (i, y_i), ..., (i + k, y_{i+k})$ by the least-squares method. A mathematical formulation of the problem and its solution yield coefficients A_j whose values depend on the degree of the polynomial, m, and the span of the filter $n = 2k + 1$ [28, 29]. Milne [29] has published tables of coefficients A_j for $n \leq 21$ and for $m = 2, 3$. Similar tables can also be found in the work by Savitzky and Golay [27].

Special attention will be paid to differentiating filters. A linear filter will be considered first:

$$y_i^* = \frac{1}{n} \sum_{j=-k}^{k} A_j y_{i+j} = \sum_{j=-k}^{k} a_j y_{i-j} \tag{4.3}$$

where $a_j = A_{-j}/n$. Suppose that a sequence of first derivatives, y_i', corresponding to the original signal y_i are supplied to the input of the filter; this data can be processed by filter (4.3) and the results will be smoothed values of the first derivative:

$$y_i'^* = \sum_{j=-k}^{k} a_j y_{i-j}' \tag{4.4}$$

The filtering represents a numerical convolution of the function being smoothed, $y(t)$, with a filtering function $a(t)$; it holds that

$$\sum_{j=-k}^{k} a_j y_{i-j}' = \sum_{j=-k}^{k} a_j' y_{i-j} \tag{4.5}$$

and the application of linear filter (4.4) with filtering coefficients a_j and span $2k + 1 = n$ to the first derivative of the signal is equivalent to application of a linear filter with coefficients a_j' to the original signal. The filtering coefficients are discrete values of the first derivative of the filtering function $a(t)$. Filters yielding higher derivatives can be derived in a similar way.

In chromatographic systems, the individual filtering procedures are combined; after the normalization procedure (bunching) the moving average and differentiating filter usually follow. For the purposes of peak detection, the derivatives should be in standard form with dimension $\mu V/s$. If the data from the converter are supplied to the system in μV every $\tau(s)$, the tabulated values of A_j and/or a_j' must be divided by the product τB^2, where B is the number of addends in the partial sums of the bunching procedure. The actual form of the filter differs from system to system. Thus, the Varian commercial program combines bunching with the moving average (5 points) and differentiating filter ($k = 5$, $m = 2$). Fozard et al. [30] recommend $k = 12$, $m = 3$. The Siemens commercial Prag program sets $k = 3$ and $m = 2$ and the derivatives are filtered by an exponential filter. Charrier et al. [31] calculated a moving average from 7 points for data on the baseline and a least-square curve fitting filter with $k = 3$ for data outside the base; the derivatives are calculated directly from smoothed data,

366

i.e. as the differences, without using differentiating filters. Hegedus and Peterson [32] chose $m = 3$ and $k = 4$. The small Autolab [33] computing integrators calculate the first derivative with $m = 2$ and $k = 2$ after bunching and a 5-point moving average.

4.1.3 PEAK DETECTOR

The most important part of the integrator is the peak detector, which strongly affects the intelligence of the entire system for processing chromatograms. With most modern integrators, the peak detection is based on tracking the smoothed first, sometimes also the second, derivative of the signal. In the course of the detection of a peak or a group of peaks, the values of the derivative are compared with predetermined, threshold values by employing decision blocks which ensure branching of the program. One of the possible stages in the program corresponds to each program branch.

In order to detect an ideal single peak on a horizontal baseline, at least three different program states are necessary. The basic, and, at the same time, also the initial state of the program is the **BASE** state; the appropriate program branch processes the signal on the baseline. The detector remains in the **BASE** state as long as the smoothed value of the derivative does not exceed the positive threshold value. As soon as an overlap is detected, the decision block switches over the state of the program to state **FRONT**, and the corresponding branch tracks the ascendent part of the peak as long as the first negative value of the derivative appears. Then the decision block switches over the state of the program to **BACK** and the detector tracks the descending part of the peak as long as the derivative exceds the negative threshold value and the program returns back to the **BASE** state. If the decision block in branch **BACK** is supplemented to test the overlap of the positive threshold in addition to the negative one, then this is simple detection logic for a group of incompletely resolved peaks, which are separated by a perpendicular line by the detector. If the positive threshold is exceeded, the logic of branch **BACK** returns the state of the program directly to state **FRONT**.

The above logic is not sufficient to detect an ideal peak or a group of peaks on a non-horizontal, drifting baseline. If peak areas are to be corrected with respect to a non-horizontal baseline, the coordinates of at least one point at the end of the group of peaks must be determined. To accomplish this, the condition of a single overlap of the negative threshold is obviously too weak and must be made more stringent. Usually it is required that the signal derivative vary within the positive and negative thresholds at least for a certain time, corresponding to the peak width, after the elapsing of which the actual point on the chromatogram can be regarded as a baseline point. To determine the base another branch, PKEND, is thus needed which is situated after branch BACK and is activated at the moment when the derivative exceeds the negative threshold. On passing over from state BACK to state PKEND a time window fulfilling the condition for passing over to BASE in delimited. Branch PKEND has three outputs: return to BACK, passing over to FRONT and passing over to BASE. If the detector detects exceeding of the positive threshold in the PKEND branch before the time window has run out, it changes the state over to FRONT; analogously, if the signal derivative decreases below the negative threshold, the state returns to BACK. If neither of these two cases occurs, the detector changes the state of the program to BASE at the moment the time window runs out.

If the peaks or groups of peaks are not ideal, the detection logic has to be extended further. The detector must eliminate nonchromatographic shifts of the baseline, spikes and other nonidealities occuring in the course of the chromatogram. The requirement of tangent separation calls for a further extension of the logic. With modern chromatographic integrators, the detectors generally have ten or more logic branches.

In the detection of peaks, a significant role is played by the signal derivative threshold values. The threshold values are set in the form of parameters of the type SLOPE SENSITIVITY, usually in $\mu V/s$, from the integrator keyboard by the operator. More elaborated systems compound the threshold values from two components; the maximum permissible baseline drift and the noise component. In order to make a quantitative estimation of noise, either the absolute values of the smoothed data differences or directly the absolute values of the smoothed signal derivatives are averaged. In every case, such a system requires a short preparation period, during which it determines the actual noise level, before the processing proper of the chromatogram commences. In some cases, the extent to which the existing noise is to be considered is left to the operator's discretion by an adjustable parameter of the type SIGNAL/NOISE.

The performance part of the individual branches of the peak detection program requires closer consideration. At the moment of exceeding the threshold, branch BASE stores the coordinates of the beginning of the peak in the buffer memory and starts integration. The integration usually consists of simply adding the partial sums. Because of the finite threshold value, the starts of the peaks are recognized with a certain delay by the logic. The computer memory permits the error so produced to be eliminated by the base-integral technique which assumes the baseline to be integrated in the BASE branch actualized. In branch FRONT integration continues; on transition from FRONT to BACK the coordinates of the peak maximum are stored. Branch PKEND records the coordinates of the end of the peak, together with information on the method of termination and the total integral. The coordinates of the beginnings, maxima and ends of peaks, together with the uncorrected peak areas, constitute one item in the table of peaks. To save the memory, the data are reduced before recording them in the table. For instance, the time coordinates of the beginnings and ends can be stored relatively with respect to the time coordinate of the maximum (retention time); on subtracting the voltage coordinate of the beginning of the peak, starting form the base, from the partial sums (the formation of an auxiliary horizontal baseline) the necessity to record the voltage co-ordinates of the peak beginnings is eliminated, etc. The storing of information about the termination of the peak on its PKEND-BASE transition, which provides for reconstruction of the baseline and accurate correction of the peak areas, is of crucial importance.

A common part precedes each of the variable branches of the peak detector. This ensures reading of the converter data, organizing the time basis of the chromatogram, performing the normalization and filtering of the signal, and tracking the actuality of the time-dependent parameters of processing the chromatogram.

4.1.4 RECONSTRUCTION OF THE BASELINE AND CORRECTION OF PEAK AREAS

Whereas the above program parts of processing chromatograms (data acquisiton, filtering, peak detection) must proceed in real time, i.e. simul-taneously with the chromatographic analysis taking place, the corrections of peak areas and the quantitative processing of the corrected areas belong in the category of programs which are indepedent of real time.

369

The corrections of the peak areas, inclusive of the reconstruction of the baseline, constitute a program part that processes the data stored during the chromatographic analysis in the table of peaks. In compliance with the philosophy of creating the table, the tabulated chromatogram is divided into peak groups which begin and end at the baseline. Each individual group then has its own baseline formed by a connecting line between the beginning and end of the group. The corrections of the peak areas within a group consist of subtracting the trapezoidal areas bound by the reconstructed baseline, auxiliary horizontal baseline used in the integration, and perpendicular lines drawn between the beginning and end of each peak from the uncorrected areas stored in the table of peaks.

A special strategy is required by tangent correction, which precedes the procedure described above. The peaks localized by the peak detector on a solvent tail are provided by the detector with an indication of tangent separation in the table of peaks. The rules according to which the ends of such peaks are determined differ from those used with peaks on the baseline and/or peaks separated by the perpendicular line. In correcting peak areas on a solvent tail, it is above all necessary to construct an auxiliary baseline connecting the minimum at the end of the solvent peak with the end of the last peak on the tail. The areas of minor peaks are corrected with respect to this auxiliary baseline, and the area below the auxiliary baseline is added to the solvent peak area. The areas of solvent peaks are then corrected, together with other peaks, with respect to the baseline of the entire chromatogram.

4.2 Molecular Spectroscopy

Molecular spectroscopy consists of a group of spectroscopic methods that study the interactions of electromagnetic waves with molecules of substances. According to the character of the electromagnetic radiation, its energy level and the physical basis of the interaction, molecular spectroscopy in the ultraviolet and visible regions, in which quanta of radiating energy interact with valence electrons of molecules, and infrared spectroscopy based on absorption of quanta of thermal energy leading to an increase in the internal energy corresponding to vibrational and rotational motion of interacting molecules, can be distinguished. High-frequency spectroscopic methods (EPR − electron paramagnetic resonance and NMR − nuclear

magnetic resonance) are also included in the group of molecular spectroscopic methods. They are based on energy transitions of electronic or nuclear
magnetic moments after absorption of high-energy quanta of ultrashort
radiowaves.

Methods of molecular spectroscopy are now an effective aid to chemists
in both research and control analytical laboratories. By using suitable
combinations of these methods, almost all problems of structural analysis
can be solved, various kinetic problems and processes can be followed and
the methods can be used for determination of substances in simple and
multicomponent mixtures. They are optimally used in combination or
coupled with some separation methods.

With increasing demands of theoretical and applied research, emphasis is
being shifted from simple molecules and systems to progressively more
complex ones; with greater precision, sensitivity and selectivity of the
methods required, with simultaneous rapidity and simplicity of these
methods, molecular spectroscopic instrumentation is becoming more
sophisticated. The necessity for automatic control of technological processes
and of the environment (e.g. infrared analyzers of gaseous pollutants in the
atmosphere) must also not be overlooked.

These stringent demands can be met only by progress in instrumentation
and by use of computing techniques. Some problems cannot be solved at
all without application of computers. In this short review, it is impossible
to exhaust all the application possibilities of computers in molecular
spectroscopy and the discussion of selected examples cannot be very
detailed.

Computing techniques can be used at all stages of molecular spectroscopic experiments. The applications can be divided into two categories —
(1) applications leading to improvement in the spectral experiment, including
its evaluation and data processing; (2) — applications to solution of theoretical problems of molecular spectroscopy. The two categories can further
be divided into a number of partial problems.

The first category:

— control of the spectrometer operation in order to obtain experimental
values of the highest possible precision, corrected for errors caused by optical,
mechanical and electronic parts of the instrument,

— special experimental techniques — combination of a physical experiment with mathematical treatment and measurement under extreme experimental conditions;

— evaluation of experimental spectra and digitization of spectra for further numerical treatment,

— cataloguing of spectra,

— analytical applications — optimizing of experimental conditions, calculation of the error of the determination, statistical treatment of the results.

The second category:

— solution of theoretical problems of structural analysis,

— study of equilibrium and kinetic processes,

— theoretical calculations in chemical physics in the field of molecular spectroscopy (this field partially overlaps that of quantum chemistry).

4.2.1 EXAMPLES OF THE USE OF COMPUTERS IN SPECTROSCOPIC EXPERIMENTS

Analyst-spectroscopists try to obtain reproducible experimental values with the highest possible precision. In recording spectra, the signal-to-noise ratio is of prime importance. With increasing sensitivity of the recording, the noise also increases and the spectrum becomes irregular. Smoothing of a spectral recording, considering the selection of the most probable values, can be carried out numerically. The experimental recording is separated into individual points and a polynomial is fitted to a certain number of points. The number of points depends on the degree of the polynomial and the ratio of these two values must be selected correctly; the number of points should be sufficiently large and the polynomial degree relatively low. In this way, all the points in the spectrum are treated. The calculated values represent the most probable shape of the spectral dependence, corrected for the instrumental noise.

Another example is the transformation of spectra to a linear scale. The precision of the wavenumber scale is important for the reproducibility of infrared spectra. Any spectrometer exhibits a certain non-linearity of the wavenumber scale, depending on the instrument quality and caused by turning of the grating or by opening and closing of the slit. This non-linearity can be corrected numerically with respect to wavenumber standards; the spectrometer operation can be programmed and controlled by a computer to obtain the correct linear wavenumber scale of the spectrum. This correction is indispensable with high-resolution spectra or when non-linear prism spectra are converted into a linear recording. An analogous application is computer control of constant homogeneity of the mangetic field in NMR spectroscopic experiments.

The main problem in the treatment of data from absorption molecular spectroscopy is overlapping of the absorption bands in the experimental spectra. This problem, caused by imperfect spectrometer resolution, cannot be solved in any other way than numerically by calculation of theoretical spectra and separation of overlapping absorption bands. Basically, it is necessary to separate the experimental envelope curve of the absorption spectrum into individual absorption bands, in agreement with the physical character of the test substance and under certain assumptions (the number of absorption bands, symmetric or asymmetric band shape) and to recompose the original experimental spectrum from these bands. A general program for band separation has been developed, in which the absolute number of separated bands depends only on the computer capacity [34−36].

Computing techniques are a good aid in measurements on systems under extreme experimental conditions. An example is the spectroscopic measurement of very low concentrations of test substances or of very weak interactions between electromagnetic radiation and a test substance. These measurements are constantly gaining importance, because trace amounts of substances must be determined in complicated systems. The spectroscopic signal (the measured impulse) is then similar to the instrument noise. In normal spectrum recording, the signal is lost in random noise; spectrum accumulation can then be used, enabling amplification of the measured signal, preserving a relatively low level of the instrument random noise (Fig. 57).

A small or medium computer with a sufficient memory capacity is coupled with the spectrometer. The measured signal and random noise are stored in the memory on repeated passage through the spectrum, where the signal has a single sign and the random noise has alternating signs (provided that fluctuations in the positive and negative directions are equally probable), so that the intensities of the measured signal are added, whereas the random noise is eliminated. The final spectrum recording, taken from the computer memory, contains the measured signal amplified with respect to the noise in the ratio $n/\sqrt{n}$, for n passages through the experimental spectrum.

Spectrum accumulation is a general method, but is used chiefly in NMR spectroscopy. A disadvantage is its tediousness, which places great demands on the stability of the instrument main parameters. In an effort to remove this disadvantage, a new method, *Fourier Transformation NMR Spectroscopy* (FT NMR) has been developed [37, 38].

FT NMR uses short-time pulses of a high-frequency magnetic field for

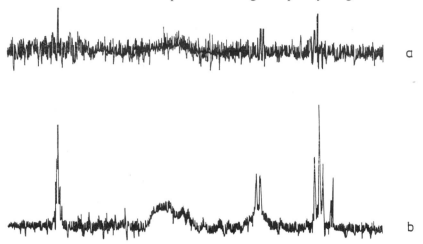

Fig. 57. Spectrum accumulation:
(a) normal recording of an NMR spectrum,
(b) recording of an accumulated NMR spectrum with n scans.

excitation of resonance in the spin system. In contrast to classical NMR, where resonance of the individual spins is excited gradually by smooth variation of the magnetic field, in FT NMR the resonances of all spins in the system are excited simultaneously. In recording of such a spectrum, the responses of all the resonance signals are recorded simultaneously; the resultant interference of their intensities over the whole frequency range is obtained. The recorded interferogram (the recording of decaying magnetization of the measured spin system excited by high-frequency pulses) has a time-dependence, not a frequency dependence as in classical NMR spectrometry (see Fig. 58). Therefore, it must be converted into a frequency dependence and the NMR spectrum must be represented in the same form

374

as that obtained by classical NMR spectrometry. This operation can be performed mathematically by a digital computer using Fourier transformation, accumulating interferograms in the computer memory. The main advantage of the method is a time saving with a substantial increase in the measuring sensitivity by improving the signal-to-noise ratio. In practice it is used for measurements on systems with a low content of nuclei with non-zero magnetic moments − e.g. ^{13}C NMR spectra.

In addition to NMR spectrometry, this method has also been applied

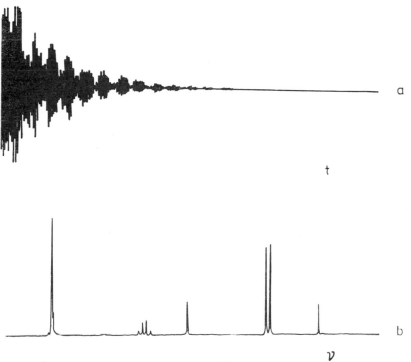

Fig. 58. ^{1}H Fourier-transform spectroscopy (^{1}H FT NMR):
(a) interferogram of the measured spin system,
(b) interferogram converted into a frequency dependence using the Fourier transformation (FT NMR spectrum).

in infrared spectroscopy. An example is a combination of an infrared spectrometer with a gas chromatograph. The principle of this combination is gradual recording of infrared spectra of individual substances leaving the chromatographic column. Therefore, the infrared spectrometer must have a high recording speed, a high sensitivity and must rapidly handle the experimental data. These requirements can only be met by an FT infrared spectrometer that operates on the same principle as FT NMR spectrometers. Evaluation and numerical treatment of the IR interferogram yield infrared spectra of the individual components of the test mixture; in this way the gas chromatograph — infrared spectrometer system becomes a very selective analyzer.

At present, computers are progressively more extensively used in analytical or statistical treatment of experimental data. In spectroscopy, they are chiefly employed for transformation into a form suitable for further treatment.

The process itself can be divided into two steps. The first step of a relatively lower level is automated recording of corrected experimental values in digital form (digitization of spectra). Classical analog recorders must be replaced by digital ones, connected with an on- or off-line computer. The classical spectrum recording is thus converted into a set of numerical values that serves as input data for further treatment.

An example is on-line coupling of a computer with an NMR spectrometer that permits obtaining of accurate values of chemical shifts, interaction constants and integral intensities of the resonance signals in an experimental NMR spectrum. Depending on the complexity of the problem and the spectrum, optimal numerical values of spectral parameters can be obtained by judicious modification of the program. For example, in evaluating the integral intensities of resonance signals, the signals can automatically be integrated only from a preset intensity value.

If a spectrum can be converted into digital form, it must also be possible to reconvert it into the original graphical form (i.e. decode the spectrum). This process can also be carried out by a computer.

4.2.2 INTERPRETATION OF MOLECULAR SPECTRA USING COMPUTERS

Methods of molecular spectroscopy in all spectral regions are an indispensable tool in identification and structural analysis of organic and inorganic substances. The problems of spectrum interpretation are, therefore, constantly studied. Computers are widely used for this purpose, because of the great number and variability of molecules studied and ever increasing demands on the precision and unambiguity of the determination of molecular structures.

The molecular spectrum is basically a mirror reflection of the molecule, in which all, or at least most, molecular parameters are coded. The task of spectrum interpretation is decoding of these parameters and obtaining the molecule structure by their synthesis. With simple molecules this can be done subjectively, without using other techniques, on the basis of experience expressed frequently in terms of purely empirical rules. It often suffices to compare the experimental spectrum parameters with tabulated values or with standard spectra. However, this procedure is inadequate for treatment of more complex molecules or when increased demands are placed on the interpretation process.

In thorough interpretation of molecular spectra, great requirements are placed on the erudition of the interpreter and on the interpretation time and thus computers are employed.

Any computerized spectrum interpretation must be based on analysis of experimental spectra, i.e. obtaining of experimental spectral values that form a set of input data for numerical identification or solution of the structure of the test molecule. The quality of this analysis depends on the quality of the spectrometer. It is pointless to apply a rigorous numerical method to a spectrum with poor quality, or, on the other hand, to use an approximate interpretation method with a perfectly resolved spectrum. Therefore, the spectrum level must correspond to the level of the computing technique to prevent a decrease in the level of the structural information contained in the experimental spectrum and to obtain as clear a solution of the molecular structure as possible.

An ideal method would involve fully automated spectrum interpretation, carried out solely by a computer, starting with obtaining of the spectral parameters (e.g. characteristic wavenumbers, integral intensities, absorption band half-widths etc.) by digitizing the corrected experimental spectrum. The computer program must further contain instructions based on the general laws of the given spectral method, on subjective assumptions and on requirements concerning the level of the solution of the structural problem [39].

General formulation of the algorithm for interpretation of a certain kind of spectrum is complex and difficult; the present programs are limited to certain classes of simpler compounds that permit unambiguous solution. However, simpler methods can also be used successfully for unambiguous interpretation and hence fully automated programs are not very advantageous [40].

Another way of finding the test molecule structure is gradual semiempirical spectrum interpretation. On the basis of partial structures confirmed by the experimental spectrum, e.g. in determination of the characteristic groups of atoms from an infrared spectrum and using general correlation rules, a probable structure can be proposed for the molecule. Structural parameters are calculated for the proposed structure and are compared with the experimental ones. If they agree, the structure is correct; otherwise, correction must be made to the structure proposed, the spectral values calculated again and again compared with the experimental values. This iteration is repeated until the calculated and experimental values agree satisfactorily, the degree of agreement being the degree of probability of the determined molecular structure.

Methods that are used much more often in practice are based on direct comparison of the test substance spectrum with that of a tabulated standard and require very extensive sets of standard spectra, suitably coded for computer handling. The experimental spectrum must then be coded in the same way, enabling the computer to scan the set of standard spectra and choose an identical spectrum. In optimal cases, a final and unambiguous solution is thus obtained. This ideal solution is, however, rather rare because it requires complete collection of standard spectra for all existing molecules and satisfaction of experimental conditions necessary for correct comparison. The program is thus adjusted so that standard spectra similar to the spectrum of the test substance are selected and ordered according to the degree of similarity. In this way, the set of spectra is narrowed and can be treated by a refined program, supplemented by further data.

One more problem is involved in this interpretation procedure, namely that the experimental conditions for obtaining the spectra of the test substance and the standard, involving the instrumental parameters and the sample preparation, be identical. For coding of the spectrum for computer treament, those spectral quantities that are affected least by the instrumental parameters must therefore be used.

Successful application of these comparison methods depends on meeting several basic requirements. The chosen spectral quantities

378

a) enable recognition of structural similarities of the molecules compared,

b) quantitatively determine the degree of similarity,

c) be as independent as possible of the instrumental and experimental measuring conditions (including the sample preparation).

These conditions are met by spectral quantities that remain constant during a change in the experimental conditions provided that the molecules have identical structures and are sensitive even to the smallest differences in the molecular structure. To choose molecules with similar structures from a collection of standard spectra, less specific parameters are suitable, whereas for differentiation among individuals in a certain class of similar compounds, specific spectral quantities are more suitable. An optimum procedure combines both these types of parameters.

For the above reasons, selection of suitable spectral quantities as input data for an interpretation program is very important. A great advantage of this comparison procedure is the possibility of determining the structures of molecules whose spectra are not contained in standard spectra collections.

4.2.3 CATALOGUING OF SPECTRA

Computerized interpretation of spectra is closely connected with the problem of storage, spectra cataloguing and organization of program libraries. A primary condition is the existence of extensive collections of spectra in all types of molecular spectroscopy. The spectra must be coded in the form of selected spectral quantities in a certain input data system for the computer and assigned with a catalogue code according to the principle features of the molecular structure. The test substance must be coded in the same way.

The actual interpretation consists of three operations, the coding of the test substance, comparison with the structures stored in the standard spectra, i.e. location of all similar structures and determination of the degree of similarity, and conversion of the results into a brief and lucid form.

The second operation requires most of the machine time, because all the stored spectra must be compared with the test substance spectrum. Therefore, efforts are made to simplify the relevant algorithm and thus to save time. It is impossible to limit the number of standard spectra if the structure determination is to be reliable. However, a certain possibility of simplification of the relevant algorithm exists during determination of the degree of

379

similarity, which is carried out simultaneously with comparison of individual pairs of spectra (of test and standard substances). A certain degree of similarity can be preset, above which the structures are relevant to the determination of the test substance structure and below which they can be neglected. In this way, the search is simplified and shortened.

From the user's point of view, the quality of a collection of standard spectra and of an algorithm is determined by the number of spectra collected and by the selection and weight of the spectral quantities characterizing the molecular structure. The principal criterion is the rapidity and completeness of the selection of structures identical with that of the test substance.

The data obtained by a single spectroscopic method are often insufficient for unambiguous determination of a molecular structure and several molecular spectroscopic methods must be combined. Infrared or Raman spectroscopy is often combined with NMR spectroscopy and sometimes simultaneously with NMR and mass spectroscopy. Computers are very useful here, but the algorithm must be modified so that the data from all the methods are treated simultaneously and the results are unified on a common basis.

However, the experience and intuition of the chemist-structural analyst cannot be replaced by the most perfect computer, in spite of all the advantages of automated interpretation of molecular spectra.

4.2.4 SIMULATION OF SPECTRA

Molecular structures can also be determined by using the opposite procedure to that described above. The interpreter proposes a structure for the test substance, calculates its theoretical spectrum and compares it with the experimental spectrum. This procedure is often applied in NMR spectroscopy.

The first step is the selection of suitable spectral parameters (chemical shifts, spin interaction constants, integral intensities) for an n-spin system representing the test molecule. The second step is the simulation of the theoretical NMR spectrum for this system. The third and last step is comparison of the simulated spectrum with the experimental spectrum, by an iteration process until the required degree of identity of the two spectra is attained (see Fig. 59).

The iterative procedure mostly operates with quantitatively expressed frequencies and ignores the curve shape and intensity. The simulated spectrum is then only symbolically depicted. However, the final calculated spectrum can be plotted and replaced for the experimental one.

Another field of computer application is evaluation of the analytical results of spectroscopic methods. In quantitative analysis, the requirements

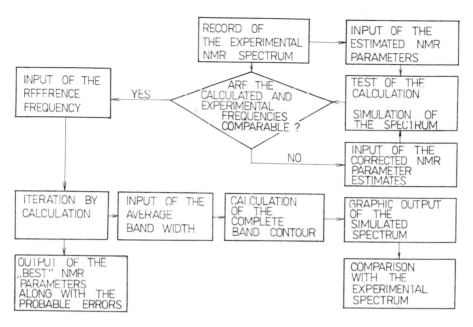

Fig. 59. Iterative procedure for the determination of the molecular structure by comparing a simulated spectrum with the experimental spectrum.

placed on spectroscopic methods and hence on computers are analogous to those placed on any analytical method, from the point of view of optimization of experimental conditions, calculation of the determination error or statistical treatment of the analysis results. The application of computers in this field is quite general and is discussed in other chapters of this book.

4.2.5 EXAMPLES OF THE USE OF COMPUTERS IN SOLVING THEORETICAL PROBLEMS OF MOLECULAR SPECTROSCOPY

Structural Analysis

Computers are most extensively used in solving various problems of structural analysis [41]. It is outside the scope of this chapter to review this field exhaustively; only a few typical examples will be given.

By direct correlation of experiments with the theory of infrared spectra, the structure of substances can be solved from the point of view of the force field operative in the given vibration, as well as considering the change in the potential energy of the set of particles (atoms) in the molecule during the vibration. The study of the bond order, the effect of polar structures and substitutents on the force field of the molecule and the computation of a model spectrum and of refined force constants of the substance are connected with this problem. A number of programs in common languages have been developed for these purposes [42].

Another example is the solution of the spatial arrangement of complex adducts of test substance molecules with shifting or relaxation agents in NMR spectrometry, which considerably affects the values of the induced shifts. Here the distance of the central paramagnetic ion of the reagent from the test nucleus and the space angle, formed by the line connecting the reagent paramagnetic ion with the test substance nucleus and the line connecting the reagent paramagnetic ion with the coordinated polar group of the test substance (coordination axis) in the complex adduct formed, must be calculated. Knowledge of the spatial arrangement and structure of the adduct is necessary for assignment of the experimental values to the individual protons of the test substance molecule.

One of the important problems of high-resolution NMR spectroscopy is obtaining precise values of chemical shifts and interaction constants from experimental NMR spectra (NMR spectrum analysis). The theory of analysis of high-resolution NMR spectra seems to be a successfully solved problem [43, 44]. However, practical application of the theory of analysis of high-resolution NMR spectra encounters many serious difficulties. These obstacles can be overcome either by using more sophisticated instrumentation or by applying better computing techniques. In the analysis of an NMR spectrum two principal problems must be solved, namely, assignment of the experimental frequencies of resonance signals to permitted transitions and the solution of the inverse spin secular problem, to which the calculation of the NMR parameters generally leads.

The assignment of the experimental frequencies is usually incomplete, because of possible ambiguous results and inevitable experimental errors, and therefore several NMR techniques must be combined. The other problem, the inverse spin secular problem, can be solved only when a sufficient amount of information is available. This solution is associated with further difficulties connected with the convergence properties of the numerical procedures used, the mathematical ambiguity of the solutions obtained, the large volume of the computation operations, etc. The test molecule can be described as a system of interacting spins and a theoretical spectrum (i.e. the number, position and relative intensity of the resonance signals) can be computed for this system (see Fig. 60). Programs have been developed for solution of a general n-spin system; the number of spins treated is limited by the capacity of the computer used.

It should, however, be pointed out that care must be exercised to ensure that the results of the mathematical solution are consistent with the physical basis of the problem. Mathematical formulation of most spectroscopic problems usually does not ensure physical correctness of the results obtained and thus any result of mathematical treatment must be checked from the physical point of view.

Another approach can be taken in structural analysis. If physical or analytical data are available on the test substance molecule, a set of all possible structures of the molecule can be computed. Computing techniques save much time in proposing possible structures and ensure that all the real possibilities are considered. The most probable structure is then selected and experimentally verified. This method has also been applied in solving the structures of large, complex molecules, such as those of peptides. The peptide molecule is chemically separated into a mixture of component amino acids, the mixture is chromatographed to separate them and the individual amino acids are identified in infrared and NMR spectrograms. The results obtained are used as the input data for computation of the most probable structure of the original peptide molecule.

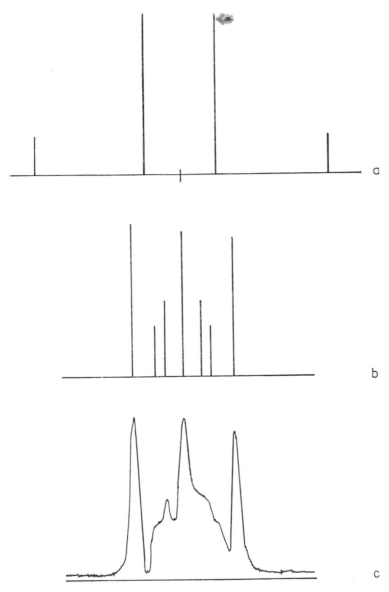

Fig. 60. Analysis of an NMR spectrum:

(a) calculated spectrum — AB system, J_{AB} O,

(b) calculated spectrum — $AA'XX'$ system; part XX',

(c) experimental spectrum of the same system to be compared with (b).

384

This is another field in which computing techniques are applied in spectroscopy [45]. Equilibrium constants of dissociation and association processes (homo- and heteromolecular association, intra- and inter-molecular hydrogen bonds, complexation equilibria, etc.) are chiefly calculated. Many programs have been developed at various solution levels. The choice and/or modification of a program must consider both the physical character of the equilibrium process and the experimental possibilities, suitability and number of input experimental data and the requirements on the precision of the results. The solution of functional dependences of spectral parameters and the correlation of theoretical and experimental dependences are closely connected with these problems.

An example is the study of equilibrium conditions in the association of phenol molecules by hydrogen bonding, using NMR spectroscopy. The shift in the resonance signal of the proton in the phenol $-OH$ group caused by molecular association exhibits characteristic concentration and tempe-rature dependences. From the concentration dependence of the association shift of the resonance signal, the type and stoichiometry of the associated complexes can be found and the equilibrium association constant calculated. Thermodynamic quantities that quantitatively describe the hydrogen bond involved can then be obtained from the temperature dependence of the association constant.

In most real cases, several parallel processes are involved rather than a single process; therefore, computers must be used. For the phenol as-sociation, theoretical dependences have been computed for di-, tri- and tetra-mers and for a series of equilibrium association constant values in a preset interval. The constant value that best described the experimental dependence was then considered as the resultant equilibrium constant.

Instrumental Aspects of Application of Computing Techniques in Spectroscopy

This question can be considered from the point of view of spectrometer-computer coupling and from that of the computer capacity. The two points of view are closely related and must be evaluated simultaneously.

385

Certain basic rules can be formulated:

1) Computers with small memory capacity are mostly used as dedicated machines and are coupled on-line with the spectrometer.

2) Computers with medium and large memory capacities are either connected on-line with several spectrometers or are used for solution of more complex theoretical problems.

An interface is required for on-line coupling of a computer with a spectrometer. This is basically an analog-to-digital converter transforming the sensor output signal into the input data for the computer. An example of on-line use of a computer is the control of the spectrometer operation (e.g. spectrum linearization). The computer must return the treated data to the spectrometer and control the spectrum recording through a digital-to-analog converter. Another example is replacement of spectral recording by direct numerical expression of the spectral parameters required, after correction of the spectrometer function.

In using an off-line computer, the spectrometer must be provided with an auxiliary apparatus recording the spectrometer output signal in the form required by the computer — e.g. on a punched tape or as a magnetic recording. Modern commercial spectrometers are adapted for on-line connection with a dedicated computer or for on-line or off-line coupling with a multipurpose computer, through a suitable interface.

4.3 Use of Digital Computers in Emission Atomic Spectrochemistry

Emission spectrochemistry can be divided into two groups, that in the optical region and that in the X-ray region. In recording X-ray spectra, direct recording techniques are now almost exclusively used. However, in recording optical spectra, photographic recording is still widely used and photoelectric recording predominates only with multichannel analyzers for industrial control.

The use of analytical computers is especially important for photographic recording, because the evaluation of such spectra requires complicated conversion of the density values into the intensity values and further interpretation is also difficult. It is advantageous to divide the whole complex of optimization problems in experimental techniques [46] into partial problems, independent from the point of view of applied cybernetics, that can be treated by independent programs and subroutines.

The basic problem is always obtaining the values of the density S, or transparencies, T (or the a_i values). The experimental values must be transformed using transformation parameters obtained on the basis of suitably chosen experimental data [47, 48]. The density values must be transformed to obtain the logarithms of the relative intesities ($Y = \log I$), or the relative intensity, I. Only these transformed values can be used for further optimization computations and for determination of concentrations.

Optimization calculations in emission spectrochemistry are auxiliary computing procedures that are often repeated several times during development of an analytical method. In this group belong the control of elimination of matrix effects [49, 50], mathematical treatment of evaporation processes [51, 52] and combined correlation-regression analysis for selecting an optimal pair of spectral lines [53−56]. The problems of measurement of the plasma temperature, the electron pressure in the plasma and the calculation of the number of emitting particles in the plasma are closely connected with this group. Calculation of these values is often combined with monitoring of the radial distribution of these parameters, which also requires use of computers.

Evaluation procedures in emission atomic spectroscopy are aimed at constructing analytical calibration straight lines and obtaining the detection limit and guaranteed purity limit values [57] with the appropriate standard deviations. Computing procedures for collective obtaining of analytical concentrations c_X and their relative precisions $s_{c_X,r}$ and arrangement of these values in final analytical reports also belong here.

The key problem in emission spectroscopy with photographic recording is the determination of systematic errors of simple transformation parameters [48] and construction of statistically treated analytical calibration straight lines [58].

4.3.1 THEORETICAL PRINCIPLES OF DENSITOMETRY AND OF TRANSFORMATION OF DENSITY VALUES

Densitometric measurement determines the intensity of light passed through an exposed site of a photographic emulsion, a_i, relative to an unexposed site, a_0. It holds that $a_i \leqq a_0$. Density of an exposed site on an emulsion is given by

$$S = -\log T = \log a_0 - \log a_i \qquad (4.6)$$

The S values are not proportional to the intensity values, I, leading to blackening, and the $S = f(I)$ dependence has two inflection points (Fig. 61), which makes the application of a polynomial to description of this relationship very difficult. Therefore, the $S = f(\log I)$ dependence, proposed by

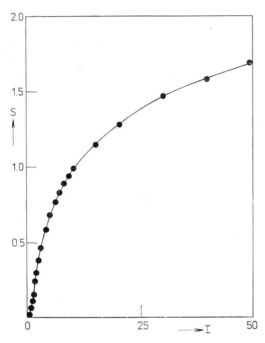

Fig. 61. The dependence of the change in the density values on the intensity values. ORWO-WU-3 emulsion, ORWO F-43 developer, $\lambda = 310$ nm.

Hurther and Dreifield and called the *photographic emulsion calibration curve* (the characteristic density curve), has been used since the beginning of densitometry. This curve is separated by point S_L into a proportional (linear) region and a curved (underexposed) part (Fig. 62). During the development of densitometry, transformation operations were sought for linearization of the photographic emulsion calibration curve. At present, the most satisfactory solutions are the Kaiser [59] general P-transformation

388

and the Török-Zimmer [60] *l*-transformation. These empirical equations require determination of the appropriate transformation parameters (constants), on whose accuracy depends the accuracy of the transformation and hence of the whole complex of optimization and evaluation operations. The general *P*-transformation is expressed by

$$P = S - \varkappa D_S \tag{4.7}$$

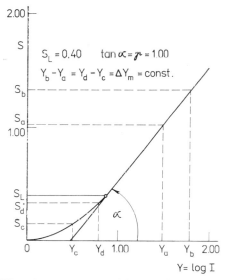

Fig. 62. The photographic emulsion calibration curve. ORWO WU-3 emulsion, ORWO F-43 developer, $\lambda = 310$ nm.

where S is the experimental density value, $\varkappa$ is the transformation parameter and D_S is derived from the density value,

$$D_S = S - W \tag{4.8a}$$

$$W = \log\left[(a_0 - a_i)/a_i\right] \tag{4.8b}$$

The W value is termed the Gaussian subtraction logarithm and was introduced into spectrochemical practice by Seidel [60]. The P value must further be divided by the slope of the transformed straight line, γ_P. The *l*-transformation is described by

$$l = s - d(k - s) \tag{4.9}$$

389

where $s = S/\gamma$, k is the transformation parameter, γ is the slope of the photographic emulsion calibration curve and d is a quantity derived from the reduced blackening, s.

$$d = s - W_s \tag{4.10a}$$

$$W_s = \log\left[(a_0 - a_{i,s})/a_{i,s}\right] \tag{4.10b}$$

$$a_{i,s} = 10^{(a_i - s.\gamma)} \tag{4.10c}$$

In this transformation, the slope of the proportional part of the photographic emulsion calibration curve must be determined first and the experimental density values, S, divided by it, obtaining reduced density values. The importance of this operation lies in the fact that the proportional part of the photographic emulsion calibration curve constructed from the s values always has a slope equal to unity.

For obtaining the transformation parameters, a judiciously prepared set of experimental data is required, which enables graphical construction of a preliminary curve or of the calibration curve and can be used as an input matrix in the computation. The set of experimental data is obtained either by using a rotating chopper or, more often, by using a stepped filter. A two-step filter with a constant of ΔY_m,

$$\Delta Y_m = \log T_b - \log T_a \tag{4.11}$$

where $T_b > T_a$, T_b and T_a are the transparencies of the two filter windows, is most effective. The ΔY_m value is a function of the wavelength and the filter must be recalibrated from time to time. When using a stepped filter (or a chopper) for calibration of a photographic emulsion, several spectra must be recorded with progressively decreasing total light intensity (by changing the size of the slit in the aperture) and maintaining the same exposure time. In the double spectra obtained, a series of density value pairs $\{S_{b,i}, S_{a,i}\}$, where $i = 1, 2, \ldots, N$ and $N \in \langle 20, 100 \rangle$, are measured and a preliminary or calibration curve is graphically constructed. From the point of view of application of digital computers, it is preferable to employ the Churchill [62] preliminary curve, modified by Plško [61], representing the $S_{b,i} = f(S_{a,i})$ dependence. This combined dependence (Fig. 63) consists of a linear part (proportionality between $S_{b,i}$ and $S_{a,i}$) and a non-linear part, the inequality, $S_{b,i} > S_{a,i}$, being generally valid. The difference in the density values,

$$eS_i = S_{b,i} - S_{a,i} \tag{4.12}$$

is, within experimental error, constant in the proportional part of the curve, and its value gradually decreases in the underexposed part, where point S_L, separating the two parts, is designated. The eS_i value from the

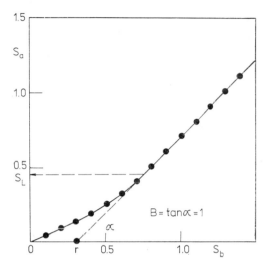

Fig. 63. Modified Churchill preliminary curve.

linear part of the preliminary curve permits determination of the slope value,

$$\gamma = eS/\Delta Y_m \tag{4.13a}$$

but a more correct value is obtained from averaged results,

$$\gamma = \frac{1}{N} \sum_{1}^{N} \frac{eS_i}{\Delta Y_m} \tag{4.13b}$$

The slope value, γ, can also be calculated from parameter r of the linear part of the preliminary curve (Fig. 63), defined by

$$S_{a,i} = B_i \cdot S_{b,i} - r_i \tag{4.14a}$$

The slope is then calculated using Eq. (4.14b) that follows from geometric considerations.

$$\gamma = r/\Delta Y_m \tag{4.14b}$$

The value of the transformation constant k can be calculated from Eq. (4.14c) that has been empirically verified [60]

$$k \equiv s_l = S_L/\gamma \tag{4.14c}$$

However, a transformation of the density is correct [63] only when the sum of the differences between Y_i or l_i and the filter constant ΔY_m is either zero or very close to zero,

$$\frac{1}{N} \sum_1^N (\Delta Y_i - \Delta Y_m) \simeq 0 \tag{4.15}$$

where $\Delta Y_i = Y_{b,i} - Y_{a,i}$ or $\Delta l_i = l_{b,i} - l_{a,i}$. It has been demonstrated [60] that $\Delta Y_i \equiv \Delta l_i$. Adequate relationships have been developed [59, 60] for calculation of transformation "constants" $\varkappa$ and k, but it has been found that these values are better obtained by using an iteration procedure [47, 48, 60], gradually refining the transformation parameter values and using Eq. (4.15) as a verification criterion.

4.3.2 THEORETICAL PRINCIPLES OF CONSTRUCTION OF THE ANALYTICAL CALIBRATION STRAIGHT LINE AND ITS STATISTICAL EVALUATION

Analytical calibration straight lines for emission spectrochemical methods are based on the Lomakin and Scheibe concepts, assuming the validity of the simple empirical relationship,

$$I_X = A_X \cdot c_X^{B_X} \tag{4.16}$$

where I_X is the relative intensity of a spectral line, A_X and B_X are the coefficients of empirical equation (4.16) connected with the vaporization and excitation processes of the test substance and c_X is the test substance concentration. After introduction of the reference element method, Malpica modified Eq. (4.16) and proposed using the relationship, $I_X/I_R = f(c_X)$, enabling exploitation of the compensating effect of fluctuations in the intensities of the analytical line, I_X and the reference line, I_R. A relationship analogous to Eq. (4.16) holds for the intensity of the reference element line,

$$I_R = A_R \cdot c_R^{B_R} \tag{4.17}$$

Dividing Eq. (4.16) by Eq. (4.17) and taking logarithms, the final equation is obtained

$$\Delta Y = A_{X,R} + B_X \cdot \log c_X \tag{4.18a}$$

where $\Delta Y = Y_X - Y_R = \log (I_X/I_R)$ and value $A_{X,R}$ is given by

$$A_{X,R} = \log a_X - \log a_R - B_R \cdot \log c_R \tag{4.18b}$$

This is a constant term determined by the vaporization properties of the test and reference elements and by the specific properties of the reference element. Because the reference element concentration must be constant, the $A_{X,R}$ value is also constant. Eq. (4.18a) is a straight line equation and thus the principal task of analytical calibration is determination of the $A_{X,R}$ and B_X values and statistical evaluation of the process of analytical calibration, including verification of the linearity.

In analytical calibration, concentration interval $\langle c_{X,\min}, c_{X,\max} \rangle$ must first be defined, within which the $A_{X,R}$ and B_X parameters ensure linearity for the experimental $|\Delta Y|$ values. Experimentally, by recording standard spectra with known values of the reference concentrations, $c_{X,j}$, an input data set is obtained,

$$\{ S_{X,L+U}, S_{X,U}, X_{R,L+U}, S_{R,U}, c_{X,j} \}$$

The S values represent the density, referred to the analytical and reference lines by subscripts X and R, respectively. Subscript $L + U$ denotes uncorrected values, i.e. line + background, subscript U denotes the background values. All density values must further be assigned index pairs, i, j, where $i - 1, ..., K, K \in \langle 4, 15 \rangle$ and $j - 1, ..., M, M \in \langle 4, 20 \rangle$. The K value is the number of repeated exposures for a single concentration value and the M value is the number of concentrations $c_{X,j}$ used for the calibration. The K value should be constant for all M values and M should equal at least four for one concentration order. Such a complete matrix is not always necessary; sometimes only certain values are corrected, using limited forms of the basic matrix. In obtaining the $\Delta Y_{i,j}$ values, the experimental density or transparence values form the basis and are first transformed to the P or l values, or are considered as Y values ($Y = \log I$).

Because the Y values are the logarithms of the relative intensities, they cannot be directly used for correction for the background intensity, and antilogs must be taken. The final formation of ΔY values requires correction for the background intensity [64]. In the first stage, this correction is neglected and ΔY is calculated from

$$\Delta Y = Y_{X,L+U} - Y_{R,L+U} \tag{4.19a}$$

where $Y_{X,L+U}$ is the logarithm of the analytical line intensity including the background $(L + U)$ and $Y_{R,L+U}$ is that for the reference line. Most often, only the analytical line intensity, $I_{X,L+U}$, is corrected and ΔY is obtained from

$$\Delta Y = \log\left(I_{X,L+U} - I_{X,U}\right) - Y_{R,L+U} \tag{4.19b}$$

If the reference line intensity, $I_{R,L+U}$, must also be corrected, then the ΔY value is given by

$$\Delta Y = \log\left(I_{X,L+U} - I_{X,U}\right) - \log\left(I_{R,L+U} - I_{R,U}\right) \tag{4.19c}$$

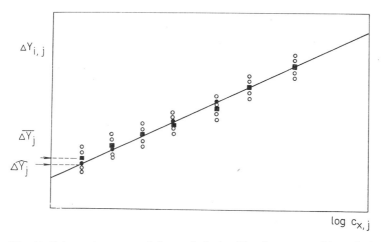

Fig. 64. Schematic course of the analytical calibration curve with typical values indicated.

In special cases only the reference line intensity is corrected and ΔY is calculated from the modified relationship,

$$\Delta Y = Y_{X,L+U} - \log\left(I_{R,L+U} - I_{R,U}\right) \tag{4.19d}$$

394

The final form of the input matrix is obtained from the ΔY values and logs of the concentrations, $\{\Delta Y_{i,j}, \log c_{x,j}\}$ and is used for computation of an analytical calibration straight line (Fig. 64).

The principal parameters of this straight line, $A_{X,R}$ and B_X, are obtained by minimizing the $F(A_{X,R}; B_X)$ function by the least squares method [58, 65],

$$F(A_{X,R}; B_X) = \sum_1^K \sum_1^M [\Delta Y_{i,j} - A_{X,R} - B_X \cdot \log c_{X,j}]^2 \equiv \text{MIN} \tag{4.20}$$

This complex calculation then also yields the standard deviations $s_{A_{X,R}}$, s_{B_X}, $s_{\Delta Y_j}$ and $\overline{s_{\Delta Y}}$ in accordance with Prokofiev's proposition [66] and the relative precision of the concentration determination, $s_{c_{X,R}}$ (%). Finally, correlation coefficient r, its standard deviation s_r and correlation determinant R (%) are obtained. The appropriate mathematical relationships are given by Eqs. (4.21a) to (4.24c):

$$A_{X,R} = \frac{[\sum_1^M \log c_{X,j} \cdot \sum_1^K \sum_1^M \log c_{X,j} \cdot \Delta Y_{i,j}]}{K[M \sum_1^M (\log c_{X,j})^2 - (\sum_1^M \log c_{X,j})^2]} -$$

$$-\frac{[\sum_1^K \sum_1^M \Delta Y_{i,j} \cdot \sum_1^M (\log c_{X,j})^2]}{K[M \sum_1^M (\log c_{X,j})^2 - (\sum_1^M \log c_{X,j})^2]} \tag{4.21a}$$

$$B_X = \frac{\left[\frac{1}{M} \sum_1^K \sum_1^M \Delta Y_{i,j} \cdot \sum_1^M \log c_{X,j}\right]}{K\left[\frac{1}{M} \sum_1^M (\log c_{X,j})^2 - (\sum_1^M \log c_{X,j})^2\right]} -$$

$$-\frac{[\sum_1^K \sum_1^M \Delta Y_{i,j} \cdot \log c_{X,j}]}{K\left[\frac{1}{M} \sum_1^M (\log c_{X,j})^2 - (\sum_1^M \log c_{X,j})^2\right]} \tag{4.21b}$$

$$s_{\Delta Y_j} = \sqrt{\left[\frac{1}{K-1}(\Delta Y_{i,j} - \overline{\Delta Y_j})^2\right]} \qquad (4.22a)$$

$$\overline{\Delta Y_j} = \frac{1}{K}\sum_1^K \Delta Y_{i,j} \qquad (4.22b)$$

$$\overline{s_{\Delta Y}} = \sqrt{\left[\frac{1}{K-1}\sum_1^K \frac{1}{M-1}\sum_1^M (\Delta Y_{i,j} - \overline{\Delta Y_j})^2\right]} \qquad (4.22c)$$

$$s_{c_{X,r}} = 230 \cdot \overline{s_{\Delta Y}} \cdot \frac{1}{B_X} \qquad (4.22d)$$

$$s_{A_{X,R}} = \overline{s_{\Delta Y}} \cdot \sqrt{\left\{\frac{\dfrac{1}{M}\sum_1^M (\log c_{X,j})^2}{K\left[\sum_1^M (\log c_{X,j} - \log \overline{c_X})^2\right]}\right\}} \qquad (4.23a)$$

$$\log \overline{c_X} = \frac{1}{M}\sum_1^M \log c_{X,j} \qquad (4.23b)$$

$$s_{B_X} = \overline{s_{\Delta Y}} \sqrt{\left\{\frac{1}{K\left[\sum_1^M (\log c_{X,j} - \log c_X)^2\right]}\right\}} \qquad (4.23c)$$

$$r = \sqrt{\left\{\frac{\sum_1^K \sum_1^M (\log c_{X,j} - \log \overline{c_X}) \cdot (\Delta Y_{i,j} - \overline{\Delta Y_j})}{K\left[\sum_1^M (\log c_{X,j} - \log \overline{c_X})^2 \cdot \sum_1^K \sum_1^M (\Delta Y_{i,j} - \overline{\Delta Y_j})^2\right]}\right\}} \qquad (4.24a)$$

$$s_r = \frac{1 - r^2}{\sqrt{(K \cdot M)}} \qquad (4.24b)$$

$$R = r^2 \cdot 100 \qquad (4.24c)$$

The calculated values have statistical character because they result from a stochastic dependence of the $\Delta Y_{i,j}$ and $\log c_{X,j}$ values. Therefore, some calculated values must be evaluated, using statistical tests [67], from the spectrochemical point of view. First of all, significant rejection of the $B_X = 1$

hypothesis must be evaluated, using the equation,

$$t_{c,B_X=1} = \frac{|(B_X - 1)|}{s_{B_X}} \tag{4.25a}$$

and comparing the calculated t_c value with the tabulated value, t_{tab}, of the Student test for the number of degrees of freedom, $f = (K \cdot M) - 1$, and a selected significance level, S (%), or calculating the t_{tab} value [67]. If the test shows that $B_X = 1$,

$$t_{c,B_X=1} \leqq t_{tab} \tag{4.25b}$$

the hypothesis cannot be rejected, but if

$$t_{c,B_X=1} > t_{tab} \tag{4.25c}$$

then $B_X \neq 1$ and the hypothesis is rejected. Spectrochemically, $B_X = 1$ is optimal, when $B_X < 1$ the results are still satisfactory and the case when $B_X > 1$ is least suitable.

The correlation coefficient is tested for significant rejection of the hypothesis, $r = 0$, using a further calculated value,

$$t_{c,r=0} = \frac{r}{\sqrt{(1 - r^2)}} \cdot \sqrt{(K \cdot M - 2)} \tag{4.26}$$

The inequality, $t_c \leq t_{tab}$, does not permit rejection of the hypothesis, but for $t_c > t_{tab}$ it is significantly rejected.

Unfortunately, testing of the correlation coefficient does not permit differentiation of linear sets from markedly non-linear sets, even if rigorous criteria are used. The fit of the $\Delta Y_{i,j}$ values to the theoretical $\widehat{\Delta Y}_j$ values must therefore also be tested [48]. The ΔY_j values are calculated from

$$\widehat{\Delta Y}_j = A_{X,R} + B_X \cdot \log c_{X,j} \tag{4.27}$$

and two quadratic deviations are obtained; using the equations

$$(s_1)^2 = \frac{K \sum_{1}^{M} (\Delta Y_j - \widehat{\Delta Y}_j)^2}{M - 2} \tag{4.28a}$$

397

$$f_1 = M - 2 \tag{4.28b}$$

$$(s_2)^2 = \frac{\dfrac{1}{M} \sum\limits_{1}^{K} \sum\limits_{1}^{M} (\Delta Y_{i,j} - \overline{\Delta Y_j})^2}{K - 1} \tag{4.28c}$$

$$f_2 = M(K - 1) \tag{4.28d}$$

The $t_{c,LIN}$ value is obtained from these deviations by dividing the higher values by the lower, i.e. $t_{c,LIN} > 1$; however, it usually holds that $(s_1)^2 \geqq \geqq (s_2)^2$. The theoretical t_{tab} value is obtained from the F-distribution tables for two degrees of freedom and a chosen significance level, S (%), or is calculated [67, 68]. The tests are evaluated analogously as in testing of the B_X value.

4.3.3 PROGRAMMED SOLUTION OF SPECTROCHEMICAL EVALUATION PROCEDURES

Effective programmed spectrochemical calculations require the use of a computer. Coupling of the densitometer or its detector through an analog-to-digital converter with a computer or at least with a printer or tape-puncher [69, 70] is extremely useful and excludes the human factor from the measurement. All programs must then contain a subroutine for conversion of the a_i values to S values, or, if the input data are the density values, S, for transformation to Y or I values. It should be pointed out that graphical determination or manual calculation of these values is either insufficiently precise or extremely time-consuming. The transformation subroutine can be based on the general P-transformation or on the l-transformation; the results of the two transformations are equally reliable. The chief requirement in using any transformation procedure is obtaining accurate transformation parameters without systematic error. The main parameter in both the transformation operations is the slope of the linear part of the photographic emulsion calibration curve, γ, using which the transformation constants can be determined.

4.3.3.1 Calculation of Transformation Parameters for the *l*-Transformation

The block scheme for the computation of *l*-transformation parameters (Fig. 65) contains two independent subroutines, one for computation of the parameters of a modified preliminary line and the other for the *l*-transformation, denoted as TRANSL.

The computation begins with ordering of the set of blackening values $\{S_{b,i}, S_{a,i}\}$ according to decreasing $S_{b,i}$. Then the subroutine for computation of the parameters of the modified preliminary line is started, based on least-squares minimization of the function.

$$\frac{1}{N} \sum_1^M [B_i \cdot S_{b,i} - r_i - S_{a,i}]^2 = \min \tag{4.29}$$

To suppress errors of measurement of the highest density values, first B_i and r_i values are calculated from the initial five pairs $(i = 5)$ of blackening values. The differences in the densities eS_i, are also obtained (Table I). The value of i is gradually increased up to N and all auxiliary parameters are always calculated. The $S_{a,i}$ value at which the first $B_i < 1$ value is obtained is considered as preliminary separating point $S_{L,v}$. The r_i value corresponding to this $S_{L,v}$ value is used for calculation of the preliminary γ_v value according to Eq. (4.14b), the preliminary value of transformation constant k_v being given by

$$k_v \equiv s_{l,v} = S_{b,v}/\gamma_v \tag{4.30}$$

The scheme of the selection procedure is given in Table I as operations 1, 2 and 3.

Subroutine TRANSL then calculates the transformed matrix of the $\{l_{b,i}, l_{a,i}\}$ values from the preliminary γ_v and k_v values, by solving Eq. (4.9) in the following series of operations (Fig. 66). Density values S are divided by the slope, giving the reduced density values. For further transformation, only density values smaller than transformation constant k or than s_l are used; the values equal to or larger than s_l need not be transformed, because it holds [60] that $s = l$. For values of s smaller than s_l, corresponding values of d are calculated using Eqs (4.10a) to (4.10c) and are recalculated to l according to Eq. 4.9.

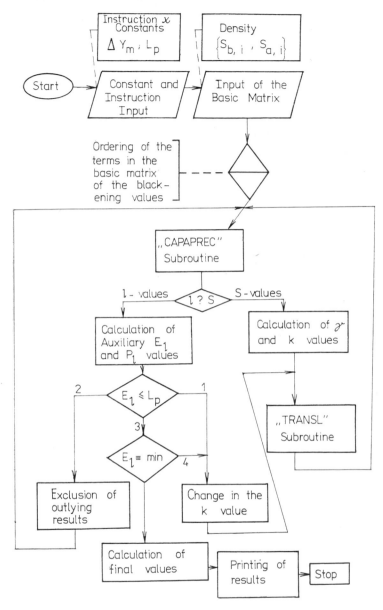

Fig. 65. Block scheme of calculation of the values of the transformation parameters of the *l*-transformation.

TABLE I

The field of blackening value pairs and the corresponding values for the modified preliminary curve

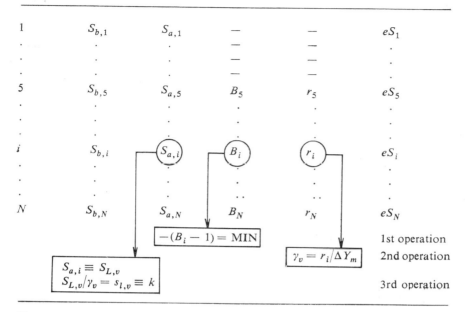

1	$S_{b,1}$	$S_{a,1}$	—	—	eS_1
.	.	.	—	—	.
.	.	.	—	—	.
.	.	.	—	—	.
5	$S_{b,5}$	$S_{a,5}$	B_5	r_5	eS_5
.	.	.	.	.	.
.	.	.	.	.	.
i	$S_{b,i}$	$S_{a,i}$	B_i	r_i	eS_i
.	.	.	.	.	.
.	.	.	.	.	.
N	$S_{b,N}$	$S_{a,N}$	B_N	r_N	eS_N

$$-(B_i - 1) = \text{MIN}$$ 1st operation

$$\gamma_v = r_i/\Delta Y_m$$ 2nd operation

$$S_{a,i} \equiv S_{L,v}$$
$$S_{L,v}/\gamma_v = s_{l,v} \equiv k$$ 3rd operation

Note:

The calculation of B and r values starts only for the whole set of density values 1 to 5.

Transformed matrix $\{l_{b,i}, l_{a,i}\}$ is treated by the CAPAPREC subroutine to obtain the preliminary line parameters (Table II) and auxiliary values B_i and r_i. Finally, differences in the l values, $el_e = l_{b,i} - l_{a,i}$, arithmetic mean $\overline{el}$ and standard deviation $s_{e,l}$ are calculated

$$\overline{el} - \frac{1}{N} \sum_1^N el_i \tag{4.31a}$$

$$s_{el} = \sqrt{\left[\frac{1}{N-1} \sum_1^N (el_i - \overline{el})^2 \right]} \tag{4.31b}$$

Testing of the accuracy and correction of the preliminary transformation parameters requires calculation of two auxiliary values, E_l and P_l:

$$P_l = \frac{1}{N} \sum_1^N \left(el_i / \Delta Y_m\right) \tag{4.32a}$$

$$E_l = \frac{1}{N} \sum_1^N \left(el_i - \Delta Y_m\right) \tag{4.32b}$$

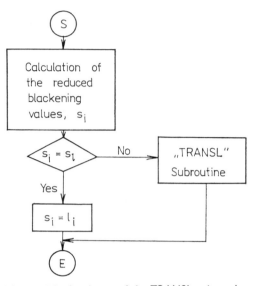

Fig. 66. Block scheme of the TRANSL subroutine.

If $P_l = 1$ and $E_l = 0$ for preliminary γ_v and k_v values, then these values can be considered as final. However, this case rarely occurs and E_l is usually greater than zero. To specify the precision limit of the iteration, a limiting value of L_p is chosen from 0.01 to 0.001 at the beginning of the calculation, expressing the degree of agreement between el_i and the filter constant and thus also (see Eq. 4.15) the accuracy of determination of the transformation parameters. Therefore, if $E_l > L_p$, iteration cycles are performed. (See Fig. 65, direction l). The variation in the k_v value is determined by the absolute value of auxiliary parameters E_l and P_l. A new slope value is calculated by the first operation given in Table II. Change in transformation constant k in dependence on the E_l and P_l values is shown in Table III. To compute the P_l and E_l values, only the above pairs of l values or of

402

TABLE II

The field of the pairs of transformed l values and the corresponding values for the modified preliminary curve

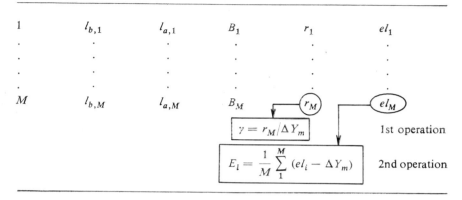

1	$l_{b,1}$	$l_{a,1}$	B_1	r_1	el_1
.	.	.	.	.	.
.	.	.	.	.	.
.	.	.	.	.	.
M	$l_{b,M}$	$l_{a,M}$	B_M	r_M	el_M

$$\gamma = r_M / \Delta Y_m \qquad \text{1st operation}$$

$$E_l = \frac{1}{M} \sum_1^M (el_i - \Delta Y_m) \qquad \text{2nd operation}$$

TABLE III

The change in the value of transformation constant k during iteration

Condition	$F_l \leq 0.01$		$E_l > 0.01$	
Step in k_v change	0.01		0.02	
Condition	$P_l < 1$	$P_l > 1$	$P_l < 1$	$P_l > 1$
Direction of k_v change	$-$	$+$	$-$	$+$

differences el_i are used, for which the $l_{b,i}$ value is smaller than s_l. The whole procedure is repeated with the new transformation constant until the new set, $\{l_{b,i}, l_{a,i}\}$, is obtained which is subjected to the above complex testing. The iteration cycles are continued until the condition, $E_l \leq L_p$ is satisfied.

This transformation constant value still cannot be considered final because the l values can be subject to gross random errors. Outlying l values are eliminated by defining an interval of validity of the el_i values,

$$\langle \overline{el} - x \cdot s_{el}; \; \overline{el} + x \cdot s_{el} \rangle \tag{4.33}$$

The x value is selected at the beginning of the computation. With $x = 1.55$, about 12 % of the pairs in the set $\{l_{b,i}, l_{a,i}\}$ are excluded. The use of lower (e.g. 1.0) or higher (e.g. 2.0) values of x did not lead to good results.

After exclusion of the outlying values from the defined interval, the computation continues along branch 3 in Fig. 68, during which the number of l pairs decreases from the original N to M; i.e. refinement of the transformation parameter values continues in successive steps (branch 4), but the E_l value is no longer compared with the given L_p value, but with the previous E_l value, until the smallest value, $E_l \equiv$ MIN, is obtained (direction 5). On attainment of minimal E_l, the final slope value is computed from the last transformed set (1st operation, Table II) and final E_l (2nd operation, Table II), from all the el_i values. The relative precision $s_{l,r}$ is calculated from

$$s_{l,r} = \frac{s_{el}}{\overline{el}} \cdot 100 \tag{4.34}$$

The computation ends with print-out of the result. The program, written in the FORTRAN IV language, is denoted as DEPALT G/K-M-78.

4.3.3.2 Computation of the General P-Transformation Parameters

The procedure given below yields more reliable general P-transformation parameters than the conventional method [71]. The values of the slope and parameter k that were determined previously are used, but a new limiting value of precision L_p is chosen between 0.05 and 0.001. It is advantageous to employ reduced set of values $S_{b,i}$, $S_{a,i}$, where $i = 1, ..., M$ and $M < N$, for the calculation. A block scheme of the DEPAP GP/KAPA-M-78 program is given in Fig. 67. The theoretical value of $S_{L,T}$,

$$S_{L,T} = k \cdot \gamma \tag{4.35}$$

is calculated first. The initial set of blackening values is divided into two subsets according to the criteria:

1st subset: $S_{a,i} \geqq S_{L,T} + 0.05$ \hfill (4.36a)

2nd subset: $S_{b,i} \leqq S_{L,T} - 0.05$ \hfill (4.36b)

The mean value of the transformation constant of the general P-transformation, $\varkappa$, is calculated from the two subsets by subroutine KAPA,

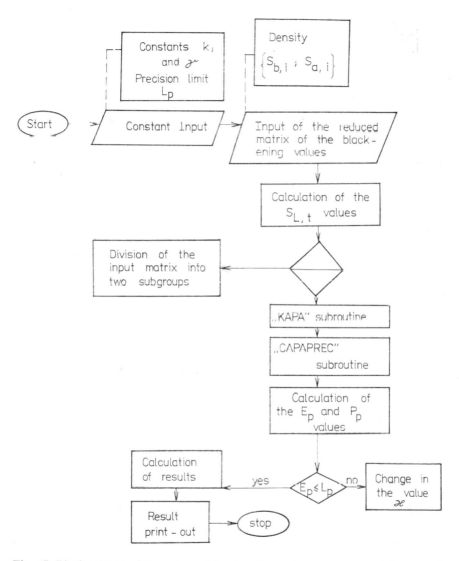

Fig. 67. Block scheme of the values of the transformation parameters of the general P-transformation.

using the summation relationship

$$\varkappa = \frac{\left[\dfrac{1}{K}\displaystyle\sum_1^K eS_q\right]_\mathrm{I} - \left[\dfrac{1}{Q}\displaystyle\sum_1^Q eS_j\right]_\mathrm{II}}{\left[\dfrac{1}{K}\displaystyle\sum_1^K eD_q\right]_\mathrm{I} - \left[\dfrac{1}{K}\displaystyle\sum_1^Q e\,D_j\right]_\mathrm{II}} \tag{4.37a}$$

where

$$[eS_q]_\mathrm{I} = S_{b,q} - S_{a,q} \qquad q = 1, \ldots, K \tag{4.37b}$$

$$[eD_q]_\mathrm{I} = D_{b,q} - D_{a,q} \tag{4.37c}$$

$$[eS_j]_\mathrm{II} = S_{b,j} - S_{a,j} \qquad j = 1. \ldots, Q \tag{4.37d}$$

$$[eD_j]_\mathrm{II} = D_{b,j} - D_{a,j} \tag{4.37e}$$

The parameters of the modified preliminary curve and auxiliary values of P_l and E_l are computed by the CAPAPREC subroutine. If $E_l > L_p$, the transformation constant is refined by the above iteration procedure, the iterations are ended when $E_l \leqq L_p$, the final value of γ_P is calculated,

$$\gamma_P = r_M/\Delta Y_m \tag{4.38}$$

and the results are printed. This program is also written in the FORTRAN IV language.

4.3.3.3 Programmed Calculation of the Analytical Calibration Straight Line

The block scheme for the calculation of the parameters of the analytical calibration straight line (Fig. 68) enables modification of the program as required. On application of instruction $L = 2$, ΔY values obtained, e.g. by a preceding calculation, can be introduced directly into the computation procedure. When $L < 2$, the $S_{X,L+U}$ and $S_{X,U}$ values are gradually read and if simultaneously $L \leqq 1$ the $S_{R,L+U}$ and $S_{R,U}$ values are also read and subroutine INPUT 0 is called up (Fig. 69). In this way, not only the density values, but also transparencies a_i can be read. On instruction $L = 0$ subroutine INPUT 0 recalculates transparencies a_i to the density values, S, and treats them further. First, subroutine TRANSL transforms all the

406

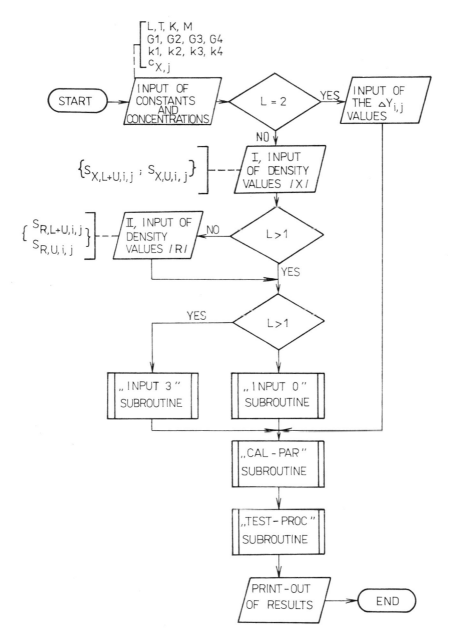

Fig. 68. Block scheme of the **CANCAL ACL-MO-78** program.

density values to l values and further treatment depends on the method for correction for the background. When instruction $T = 0$ is used at the input, the correction for the background is not carried out, but the ΔY values are computed directly. When $T = 1$, the analytical line value (X) is first corrected by subroutine CORREC; for $T = 2$, both lines are corrected and for

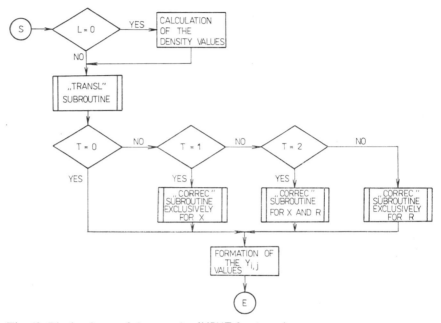

Fig. 69. Block scheme of the complex INPUT 0 subroutine.

$T = 3$ only reference line (R) is corrected. Subroutine INPUT 0 ends by finding the ΔY values for all four variants.

If, however, $L > 1$, i.e. $L = 2$ or $L = 3$, subroutine INPUT 3 is used (Fig. 70). Only the density values $S_{X,L+U}$ and $S_{X,U}$ are then introduced, transformed to the l values and the ΔY values are obtained from the l values; for $L = 3$, the ΔY values are obtained without correction,

$$\Delta Y = Y_{X,L+U} - Y_{X,U} \tag{4.39a}$$

When instruction $L = 4$, correction is made for the background with the analytical line and the ΔY value is obtained,

$$\Delta Y = \text{CORREC } Y_{X,L+U} - Y_{X,U} \tag{4.39b}$$

These two variants are used rarely when it is difficult to find a reference element for the given test element. (Fig. 71).

After obtaining the ΔY values, the parameters of the analytical calibration straight line (Eqs. (4.21a−4.24c)) are calculated by subroutine CALPAR and the testing values (see Eqs. (4.25a), (4.26), (4.27), (4.28a) and (4.34c))

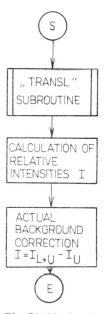

Fig. 70. Block scheme of the CORREC subroutine.

and testing parameters t_{fab} for three significance levels ($S = 95$, 99 and 99.9 %) are computed by subroutine TEST-PROC. The last computation employs the DOUBLE PRECISION method. The t_c and t_{tab} values are compared, the fulfilment of a hypothesis is denoted by + and letter T (true value) and the rejection by − and letter F (false value). The CANCAL ACL-MO-78 program is written in the FORTRAN IV language and ends with printing the results.

4.3.4 CALCULATION OF ANALYTICAL CONCENTRATIONS

The calculation of analytical concentration values is also programmed. The complete input set of the density or transparence values is defined by

$$\{S_{X,L+U,i,j}, S_{X,U,i,j}, S_{R,L+U,i,j}, S_{R,U,i,j}\} \qquad (4.39c)$$

where $i = 1, ..., N$, $N \in \langle 2, 7 \rangle$ specifies the number of repeated exposures of a sample and $j = 1, ..., M$, $M \in \langle 4, 100 \rangle$ gives the number of samples. For the computation, the transformation parameters for the analytical and reference lines and for the background must be given, the basic parameter values for the analytical calibration straight lines must be known and when the overall precision of the determination is to be found [72], the standard deviations, $s_{Ax}, {}_R s_{Bx}$ and $s_{\Delta Y}$ must also be known. Instruction L and T must

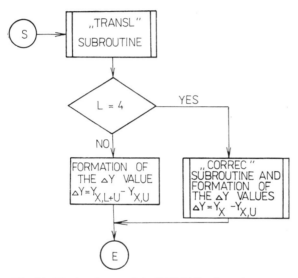

Fig. 71. Block scheme of the INPUT 3 subroutine.

be specified for control of the computing procedure. The computation scheme for the l-transformation is given in Fig. 72. The first step of the computation process is obtaining the density values (operation 1), mostly separately on a densitometer. At present, densitometers with dedicated computers are available that carry out the whole computation until the concentration values are obtained [69, 70]. In the second step, density values

410

S are transformed to the l values by subroutine TRANSL (operation 2). Transformed values of l are obtained in operation 3 (subroutine CORREC can also be used, according to the value of instruction T) and finally the ΔY values are computed (operation 4). The concentration values are calculated from the ΔY values,

$$c_{X,i} = 10^{(\Delta Y_i - A_{X,R})1/B_X} \qquad (4.40a)$$

Arithmetic mean c_X, standard deviation s_{c_X} and sometimes also relative precision of the determination $s_{c_X,r}$ are calculated from $c_{X,i}$ values.

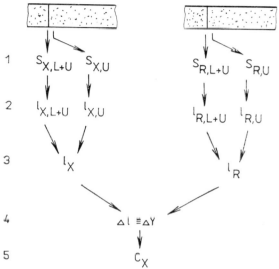

Fig. 72. Scheme of the calculation of the analytical concentration when using the l-transformation.

Sometimes it is also useful to find the effect of the precision of the analytical calibration on the relative precision of the determination, by gradual summation of the standard deviations,

$$s_{c_X,r}^* = 230 \frac{1}{B_X} \sqrt{[(s_{A_X,R})^2 + (s_{\Delta Y})^2 + (s_{B_X})^2 \cdot (\log c_X)^2]} \qquad (4.40b)$$

The $s_{c_X,r}^*$ value thus obtained is, however, always greater than the $s_{c_X,r}$ value calculated from repeated calculations of the $c_{X,i}$ values alone.

The scheme of the computation of the analytical concentrations using the general P-transformation is analogous to that given in Fig. 72, with the

411

difference that the density values are divided by the slope in the l-transformation, whereas in the P-transformation the transformed values of P are divided by the γ_P value.

4.4 The Use of Computers in Gamma-Ray Spectrometry and Activation Analysis

Gamma-ray spectrometry has experienced great development over the last $10-15$ years, chiefly because of the general use of semiconductor detectors [(GeLi) and high-purity Ge and Si(Li)] and because of intensive application of computers in the recording and especially evaluation of spectra. The progress in gamma-ray spectrometry has also favourably affected activation analysis, in which gamma-ray spectrometry is applied predominantly in fast and effective instrumental procedures.

4.4.1 INSTRUMENTATION FOR GAMMA-RAY SPECTROMETRY AND EVALUATION OF THE SPECTRA

A gamma-ray spectrometer consists of the following principal parts: a detector (semiconductor or scintillation), linear electronics (a preamplifier, an amplifier and shaping circuits), an analog-to-digital converter and a multi-channel memory for spectrum storage. A gamma-ray spectrum is then obtained as a set of digitized data contained in the individual channels. A typical gamma-ray spectrum, accumulated with a semiconductor detector, contains 4096 digital values (4096 channels). Multichannel spectrometers permit fast collection of data in a very short time. Gamma-ray spectra have a complex character and, owing to the complicated process of absorption of γ-radiation in matter, consist not only of the full-absorption lines (photopeaks) in which the required information on the energy and intensity of the test radionuclides is contained, but also of other peaks and components produced by various processes, such as Compton scattering, pair-production, scattering on the surrounding materials, etc. Typical gamma-ray spectra obtained with a semi-conductor Ge(Li) detector and a scintillation NaI(Tl) detector are given in Fig. 73. Scintillation detectors are used when high resolution is not required (medicine, industry, activation analysis with radiochemical separation, etc.). The poorer resolution of these detectors is partly compensated by their higher efficiency compared with semiconductor detectors. Only gamma-ray spectrometry with semiconductor detectors is discussed below;

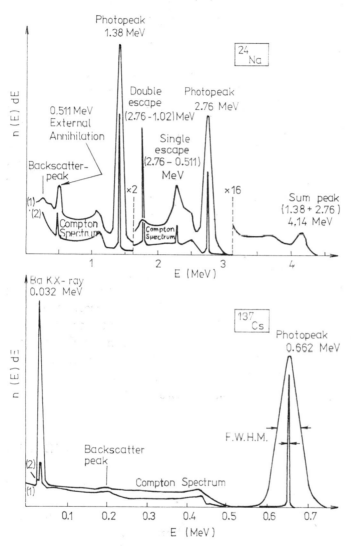

Fig. 73. Gamma-ray spectrum of ^{24}Na and ^{137}Cs, measured with NaI(Tl) (*1*) and with Ge(Li) (*2*) detector [74].

details on gamma-ray spectrometry with both detector types can be found in Crouthamel's monograph [73].

Computers have found extensive use in routine analytical applications of gamma-ray spectrometry (activation analysis, determination of fissile materials and fission products, etc). Two general methods of application of computers are available in modern gamma-ray spectrometry.

In off-line spectrum handling, spectra are obtained from one or more independent multichannel analyzers, transferred through a suitable medium (punched or magnetic tape, disc, etc.) to the memory of a large computer and handled by relatively complex programs usually written in a high-level language (Fortran, Algol).

The on-line method uses minicomputers (or microcomputers) that are part of spectrometers. Either part of the minicomputer memory is used for recording of the spectra, or the system is formed by direct coupling of a multichannel analyzer with a minicomputer. Using special measuring and handling programs, usually written in the minicomputer assembler, the spectra are evaluated and the final analytical results obtained immediately after the end of the measurement.

An advantage of the off-line method is the large memory capacity and the high speed of large computers, enabling the use of very complex handling programs. A disadvantage is the necessity of having access to a large computer and of transfer of the data from the analyzer to the computer. An advantage of the on-line method is rapid obtaining of the results, whereas a drawback is the necessity of using relatively simple programs, owing to the limited memory capacity of minicomputers. A block diagram of this system for simultaneous measurement with two Ge(Li) detectors is shown in Fig. 74.

4.4.2 PROGRAMS FOR HANDLING OF GAMMA-RAY SPECTRA

An experimental gamma-ray spectrum contains information on the presence of a certain nuclide and on its amount. The position of a line in the spectrum gives (through a calibration curve) the energy of the γ-transition, i.e. indicates the presence of a particular radionuclide. The intensity (area) of the line (photopeak) corresponds to the amount of radionuclide. Computers can be used in handling gamma-ray spectra, both for reduction of the data and, in the second stage, for interpretation of the spectra and for calculation of the concentrations of the elements (nuclides) in the given analytical method (activation analysis, determination of fission products, etc.). Most programs carry out the first stage of the computation, i.e. the reduction of the data, but programs for the second stage of the computation have also appeared recently and have been routinely used.

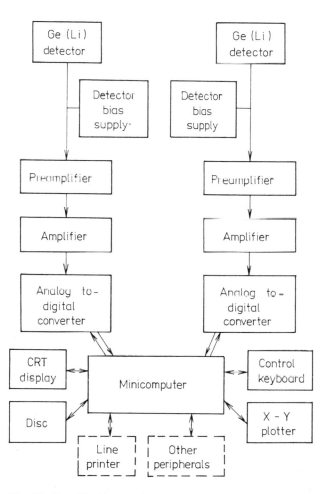

Fig. 74. Simplified block diagram of computer-based multichannel analyzer for gamma-ray spectrometry with two Ge(Li) detectors.

4.4.2.1 Data Reduction

Most programs for reduction of spectral data contain the following principal steps (procedures for data input, calibration and program organization are not included here):

a) smoothing of experimental data,
b) location of photopeaks in the spectrum,
c) determination of the background shape in the vicinity of a peak or of a group of peaks (multiplets),
d) precise determination of the peak positions and areas,
e) determination of the energies of γ-lines and the corresponding intensities.

Some programs, especially when gamma-ray spectrometers with minicomputers are used, omit some of these steps and some have a somewhat different operation order. The individual steps in the programs in the form commonly used with large computers are discussed below.

a) Spectrum Smoothing

Gamma-ray spectra, in contrast to other types of spectra, are subject to statistical fluctuations following from the statistical character of radioactive decay and of the photon detection process. Therefore, the experimental gamma-ray spectrum must usually be smoothed before application of procedures for peak location. Spectrum smoothing is discussed in detail in the work by Savitzky and Golay [27]. Some programs do not smooth the data before peak location but directly compute smoothed first or second derivatives of the spectra during the peak detection procedure.

The convolution technique, well suited for statistically scattered data, is most often used for smoothing. The smoothing method is usually based on the moving average [74]. If the data in the j-th point (channel) are smoothed using points from $j - m$ to $j + m$, the relationship

$$A_{j,s} = \frac{\sum_{i=-m}^{+m} C_i A_{j+i}}{N} \tag{4.41}$$

is used, where $A_{j,s}$ is the content of the smoothed channel and N is the normalizing factor. The weighting factors C_i are selected only once and specify the significance of point j. The following point, $j + 1$, is smoothed by omitting point $j - m$ and adding point $j + m + 1$. This procedure is repeated for all successive points in the spectrum. Factors C_i must be chosen so that a maximum smoothing effect is obtained with minimum distortion of the spectrum. Depending on selection of C_i, rectangular, exponential or other type of convolution can be attained; the number of points in the interval smoothed is simultaneously chosen (usually 5 to 7 points).

A far more exact procedure can be achieved by using the least squares method. In this technique one tries to fit the best curve through the block of points under consideration (5, 7 or more points). This curve can be a polynomial of the second, third or even of a higher order. In this way one obtains a set of $n + 1$ equations with $n + 1$ coefficients as unknowns. This procedure leads to relations of the type 4.41 for equidistant points (channels). The number of points used in the convolution technique should never be larger than the width of spectrum photopeaks. Usually, a 7-point least squares smoothing with a quadratic or cubic function is highly satisfactory [74].

b) Peak Detection in the Spectra

Most suitable methods for effective photopeak location are based on convolution techniques, in which the experimental spectrum is transformed into another spectrum which is more suitable for peak detection. It is usually relatively easy to distinguish the shape of an actual peak from slowly varying features in the spectrum, such as the Compton continuum and other scattering effects, and from purely statistical fluctuations between the individual channels in the spectrum (these fluctuations are often removed by smoothing). One of the most common methods is differentiation of the spectrum; as the spectrum is a set of values in discrete channels, differences are always determined that actually replace the derivatives. Savitzky and Golay [27] locate peaks using the smoothed first derivative. The first derivative of the spectrum changes sign with a negative slope exactly at the centre of each maximum in the spectrum. This also holds for other spectral shapes which might lead to identification of false peaks. Additional tests are therefore introduced to determine the statistical significance of the peak and to find whether the width of the peak corresponds to a real peak.

Many peak detection procedures are based on the second derivative of the spectrum, which indicates the centre of the peak by a narrow deep minimum, whose depth is a measure of the peak intensity, i.e. the peak statistical significance. The shape of the second derivative is a sensitive indicator of the spectral line distortion caused by a non-linear background and by overlapping peaks. The absence of a peak is indicated by zero second derivative, because the continuum in the spectrum is sufficiently linear within a narrow interval. Both derivatives of the spectrum can be calculated by the least squares method.

Mariscotti [75] proposed using the second difference of the spectrum. This method is simpler than the convolution procedure, but requires multiple smoothing of the spectrum before the peak-finding procedure itself. Examples of programs employing the first derivative (difference) technique are GASPAN [76] and GAMAN [77]; SAMPO [78] and HYPERMET [79] are typical examples of programs employing the second derivative (difference method).

Another principal method of peak detection in the spectrum is based on the cross-correlation technique that resembles method for signal separation from the noise used in radiocommunication (Black [80]). Op de Beck [81] proposed a cross-correlation technique with a square wave.

c) *Determination of Background Shape under the Peak*

In handling a spectrum, not only the shape of the background in the vicinity of the peak, but also that of the continuum under the peak or group of peaks must be found. With simple isolated peaks, the background can readily be approximated by a straight line. With multiplets this procedure is usually insufficient and a parabola or a higher order polynomial must be fitted to the background; see, for example, the SAMPO program.

The background shape is sometimes determined independently before the peak-fitting; otherwise, the background response can be included in the peak-fitting procedure. Gunnink [82] employs a discontinuous step-function in the GAMANAL program for determination of the background shape.

d) *Determination of Peak Position and Area*

Different procedures must be used for single peaks and for multiplets.

418

Single Peaks

The peak detection procedure (the first or second derivative) usually specifies the peak position within one channel and mostly specifies the positions of two or more partially resolved peaks (see Fig. 75). Using the second derivative method, even the position of two unresolved peaks can be

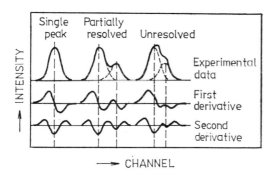

Fig. 75. Peak location by the first and second derivative method [82].

determined under certain conditions. The precise position of single peaks can then be found by fitting a parabola, Gaussian curve, etc. to the experimental points.

The area of a single peak can be determined by simple integration of the contents of the channels forming the peak, followed by subtraction of the background continuum under the peak. There are many simple digital methods for determining the peak areas directly from the experimental data, differing in the selection of the integration limits, data weighting and in the method of determining the background under the peak, which have been carefully compared by Baedecker [83] and Hertogen [84]. The method of calculation of the total peak area (TPA), which integrates the peak and subtracts a linear background is employed very frequently. The Wasson method is a modification of the TPA method; it differs from it in the selection of the background straight line. The Covell method chooses narrower integration limits than the TPA method. The Sterlinsky method and a combination of the Wasson and Sterlinsky method employ data weighting. The Quittner method fits a non-linear polynomial to the background. These simple methods for calculation of peak areas are mostly

419

used with analyzers coupled to small computers. Larger computers make it possible to employ more complex methods, also well suited for quantitative evaluation of multiplets.

Multiplets

Multiplets can be divided into two principal groups, partially resolved (the tops of two peaks are separated by a valley) and unresolved (a single maximum without a minimum). For a schematic representation of the two types, see Fig. 75.

A) Partially Resolved Multiplets — can be found by the above mentioned peak-finding procedures, but the determination of the areas of the individual peaks without peak-shape fitting procedures is difficult. A method based on the proportionality of the net area to the ratio of the heights of the measured peaks can sometimes be used, but an error is caused by the contribution of the impulses from one peak to the measured height of another peak.

B) Unresolved Multiplets — can be treated by fitting procedures on large computers. A suitable function is usually fitted to the peak shape and the continuum under the peak or the multiplet is approximated simultaneously. The peak shape can very often be represented by a Gaussian curve. More complex programs, however, also consider deviation from the Gaussian shape toward both peak limits (tailing effect), caused by the properties of the detectors and the linear electronics in spectrometers. Composite functions are used for fitting in these programs. For example, the SAMPO program [78] employs a function consisting of a Gaussian curve (the central part of the peak) and two exponentials to express distortion at the two sides of the peak. The parameters of the peak shape are usually determined from analysis of single peaks in the same spectrum (the peak shape calibration, which holds for the given experimental conditions). In evaluation of spectra the multiplet regions are fitted with such a composite function, using the parameters obtained in the peak shape calibration and approximating the background by a polynomial. The SAMPO program minimizes χ^2

$$\chi^2 = \sum_{i=k-l}^{k+m} (n_i - b_i - \sum_{j=1}^{np} f_{i_j})^2 / n_i \tag{4.42}$$

420

where i is the channel number, n_i is the number of counts in the i-th channel, k is the reference channel for the background polynomial, l and m specify the fitting interval, $b_i = p_1 + p_2(i - k) + p_3(i - k)^2$ is the background function, np is the number of peaks in the fitting interval and functions f_{ij} are given by

$$f_{ij} = p_{2+2j} \exp\left[-1/2(i - p_{3+2j})^2/w_j^2\right]$$

$$\text{for} \quad p_{3+2j} - l_j \leq i \leq p_{3+2j} + h_j$$

$$f_{ij} = p_{2+2j} \exp\left[1/2\, l_j(2i - 2p_{3+2j} + l_j)/w_j^2\right]$$

$$\text{for} \quad i < p_{3+2j} - l_j$$

$$f_{ij} = p_{2+2j} \exp\left[1/2\, h_j(2p_{3+2j} - 2i + h_j)/w_j^2\right]$$

$$\text{for} \quad i > p_{3+2j} + h_j$$

Parameters p_1, p_2 and p_3 define the continuum, p_{2+2j} and p_{3+2j} are the j-th peak height and centroid, respectively and w_j, l_j and h_j are the shape defining parameters. Fig. 76 depicts the result of fitting a multiplet in

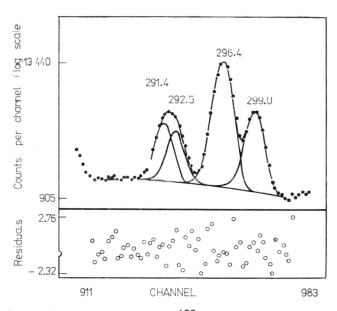

Fig. 76. Fitting of a multiplet in the ^{177m}Lu spectrum using the SAMPO program [78].

421

a ^{177m}Lu spectrum. SAMPO carries out minimization by an iteration gradient method with variable metric. Other programs, e.g. GAMANAL, employ a non-linear least squares method for iteration.

e) Determination of Gamma-Ray Energies and Intensities

The last step in reduction of spectral data is calculation of the exact energies and emission intensities from the measured peak positions and areas. The energy dependence of the pulse height (given by the peak position) is almost linear with modern spectrometers. Small non-linearity of the calibration curve can readily be fitted by a second or third order polynomial. Analytical applications of gamma-ray spectrometry usually do not require so precise determination of the energies as in nuclear spectroscopy.

On the contrary, the analyst requires adequate calibration of the detector efficiency and suitable interpretation of peak intensities. Methods for the calculation of gamma emission rates from observed peak areas usually require a separate efficiency calibration curve for every sample geometry and type encountered. This approach is often very accurate but it is not flexible with respect to changes in sample shape, strength and composition. Because of difficulties in mathematical modelling of the efficiency curve, many programs employ semi-empirical models. For example, the GAMANAL program uses only a single efficiency curve plus other variables that describe the detector, source (sample) and the sample distance from the detector.

4.4.3 INTERPRETATION OF SPECTRA AND CALCULATION OF FINAL RESULTS

4.4.3.1 Identification of Radionuclides

Programs assigning the evaluated peaks to radionuclides contain a library of principal data (half-lives, energies and intensities of gamma-radiation) for a large number of radionuclides; further they may include data on relationships between parent and daughter nuclides, etc. This library can be used both for radionuclide identification and for quantitative interpretation of peak intensities.

Assuming good energy calibration of the spectrum, the identification chiefly depends on agreement of the experimental spectral peak energy with

the energy of a gamma-line contained in the library. In the absence of interferences, this simple method is sufficient. Most programs for activation analysis and other applications, where interferences do occur, must contain additional tests based on further information (presence of other associated gamma
rays of the assumed radionuclide, a sufficiently long half-life with respect to
the sample decay time, the probability of production of the particular radionuclide by the nuclear reactions considered, etc.).

4.4.3.2 Quantitative Calculations in the Presence of Interferences in the Spectrum

In spite of high resolution of semiconductor detectors and of the use of
complex peak-finding and fitting procedures, interferences are encountered
in gamma-ray spectra (e.g. two nuclides exhibit peaks with virtually identical
energies) and programs must consider them. The problem of interferences
is often solved by constructing an interference matrix, whose elements are
described by the peaks found and by the nuclides that are assumed to be
present. The GAMANAL [82] program considers the interference matrix as
a set of linear equations,

$$Y_i = \sum_{j=1}^{j-m} A_j \cdot x_{ij} \qquad (4.43)$$

where Y_i is the intensity of the i-th peak (photon/min), x_{ij} is the branching
intensity of the j-th component in the i-th peak and A_j is the disintegration
rate for the j-th component, i.e. the number which is actually to be found.
The matrix is then solved by the least squares method; larger computers
are mostly required.

4.4.3.3 Calculation of Element Concentrations in Multielement Activation Analysis

In analytical applications of gamma-ray spectrometry, especially in instrumental neutron activation analysis (INAA) and in other types of activation analysis, too, all steps can be carried out be means of a computer
program. In addition to subroutines for data reduction, the main program
involves identification and interpretation of gamma-ray spectra and a subroutine that converts the data obtained into element concentration values

for the given sample. This subroutine differs for relative method of activation analysis (i.e. an appropriate standard is irradiated and measured for each element; sometimes standards containing several elements can be used) and for the comparator method, in which only one (or two or three) standard-comparator is used and the results for many elements are obtained on the basis of knowledge of the nuclear constants and experimentally determined measuring and irradiation parameters. Absolute activation analysis that does not require standards at all (only measured values of thermal and epithermal neutron fluxes) has recently been perfected. To achieve higher accuracy of multielement analysis by INNA, a computer oriented standardization method introducing generalized k_0 factors has been developed [85]. This method is an intermediate between the single comparator and absolute methods, at the same time eliminating some of their disadvantages.

Computer codes for activation analysis require a great deal of input data, especially nuclear (activation cross-sections, flux density of the activating particles, decay characteristics, etc.). Further, data on samples (sample numbers and masses), irradiation, decay and measuring times, and geometrical factors of the measurement must also be supplied. The programs usually contain corrections for interference contributions from other types of nuclear reactions as well. The final analytical results then contain the element concentrations, statistical counting errors and sometimes also the total error of the analytical procedure together with the detection limits for the determination of many elements by an activation analysis method used.

This chapter concentrates on discussion of the main features of computer technique in gamma-ray spectrometry with semiconductor detectors and in its applications. Programs for gamma-ray spectrometry with scintillation detectors are now used rather rarely and the main principles can be found in the work by Crouthamel [73]. Laboratories all over the world employ various kinds of program codes for gamma-ray spectrometry which differ in details, but which have many of the common features discussed in this chapter. The programs are tailored for particular experimental conditions in the individual laboratories. The comparison of programs for reduction of gamma-ray spectral data, carried out in a number of laboratories, has shown that simple algorithms used in systems with small computers often yield better results than much more complicated algorithms employed in large computers. The experimenter's experience often plays a decisive role.

424

5 Appendices

5.1 Appendix A

Block Schemes of Principal Matrix Operations

Symbols: A — matrix with dimensions $(n \times m)$
$A_{i,j}$ — matrix element
i — row index $(i = 1, 2, ..., n)$
j — column index $(j = 1, 2, ..., m)$
matrix with dimensions $(1 \times m)$ = row vector
matrix with dimensions $(n \times 1)$ = column vector
element of a column vector $= B_{i,1} = B_i$

1) Filling of matrices

a) with zeros $\qquad A = 0$

$i = 1, n$
$\quad j = 1, m$
$\qquad A_{i,j} = 0$

b) with units $\qquad A = 1$

$i = 1, n$
$\quad j = 1, m$
$\qquad A_{i,j} = 1$

c) with a unit matrix $\quad A = E$
$\quad (n = m)$

$i = 1, n$
$\quad j = 1, m$
$\qquad A_{i,j} = 0$
$\quad A_{i,i} = 1$

d) by assigning $\quad A = B$

$i = 1, n$

$j = 1, m$
$A_{i,j} = B_{i,j}$

2) Addition and subtraction
of matrices
$A = B \pm C$
$(A = A \pm C)$

$i = 1, n$

$j = 1, m$
$A_{i,j} = B_{i,j} \pm C_{i,j}$

3) Scalar multiplication
$A = (k) . C$
$(A = (k) . A)$

$i = 1, n$

$j = 1, m$
$A_{i,j} = k . C_{i,j}$

4) Vector multiplication
$A = B \times C$
$(n, m) = (n, l) \times (l, m)$

$i = 1, n$

$j = 1, m$

$A_{i,j} = 0$

$k = 1, l$
$A_{i,j} = A_{i,j} + B_{i,k} . C_{k,j}$

5) Transposition
$A = B^t$
$(n, m) (m, n)$

$i = 1, n$

$j = 1, m$
$A_{i,j} = B_{j,i}$

426

5.2 Appendix B

Block Scheme of Matrix Inversion by the Gauss Elimination Method

The given matrix, **B**, and auxiliary matrix **D** occupied by a unit matrix are rearranged together. Using the elimination method, a unit matrix is formed from matrix **B** and matrix **D**, is then inverted with respect to **B**, and finally is transferred back into matrix **B**.

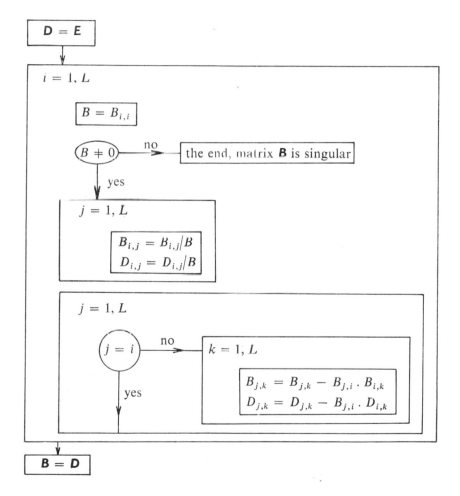

Refinement of Inverted Matrix

If inverted matrix B^{-1} is to be refined, the following method can be used. Original matrix B and inverted matrix D are retained. Their vector product would yield a unit matrix, if the inverted matrix were determined precisely. When inverted matrix D is multiplied by the difference between the unit matrix and product $B \times D$, correction Q is obtained. This procedure can be repeated until product $B \times D$ is identical with the unit matrix. Another criterion for termination of iterations can be used in practice when a unit matrix cannot be attained because of rounding-off errors; then the minimum of the sum of the squares of the elements of matrix H is sought.

The refining algorithm itself requires two more matrices (Q, H) with the same dimensions $(L \times L)$. The block scheme given below does not include testing, for simplicity. The refining procedure is repeated three times.

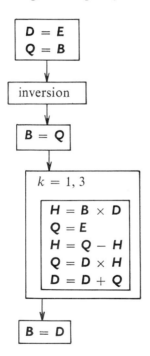

5.3 Appendix C

Block Scheme of Linear Regression — procedure a)

Matrix dimensions: $F(L, N)$, $G(N, L)$, $B(L, L)$, $A(L)$, $C(L)$, $X(N)$, $Y(N)$, $W(N)$, $V(N)$, $Z(N)$, $R(N)$, $T(1, N)$

Input: N, X_i, Y_i, W_i, L, f_j

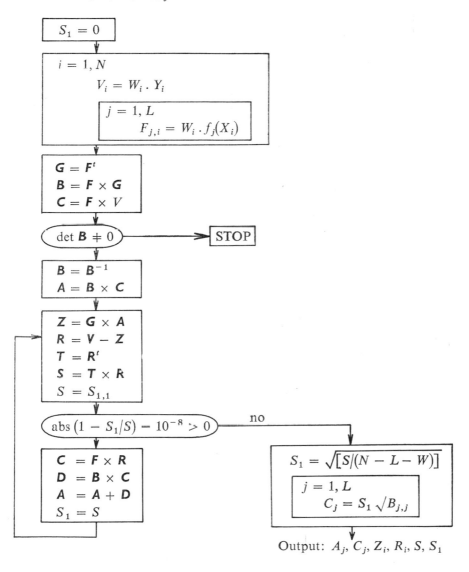

$$S_1 = 0$$

$i = 1, N$
$$V_i = W_i \cdot Y_i$$

$j = 1, L$
$$F_{j,i} = W_i \cdot f_j(X_i)$$

$$G = F^t$$
$$B = F \times G$$
$$C = F \times V$$

$\det B \neq 0$ → STOP

$$B = B^{-1}$$
$$A = B \times C$$

$$Z = G \times A$$
$$R = V - Z$$
$$T = R^t$$
$$S = T \times R$$
$$S = S_{1,1}$$

$\text{abs}(1 - S_1/S) - 10^{-8} > 0$ — no →

$$C = F \times R$$
$$D = B \times C$$
$$A = A + D$$
$$S_1 = S$$

$$S_1 = \sqrt{[S/(N - L - W)]}$$

$j = 1, L$
$$C_j = S_1 \sqrt{B_{j,j}}$$

Output: A_j, C_j, Z_i, R_i, S, S_1

Block Scheme of Linear Regression — procedure b)
Matrix dimensions: $X(N)$, $Y(N)$, $W(N)$, $Z(N)$, $R(N)$, $B(L, L)$, $D(L,L)$, $A(L)$, $C(L)$, $F(L)$, $G(1, L)$

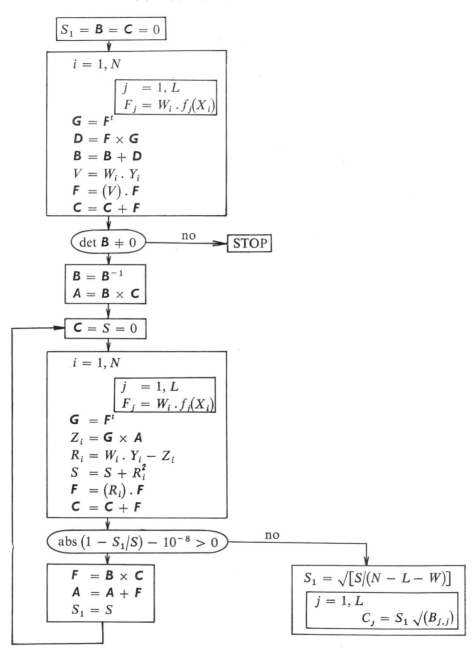

5.4 Appendix D

Approximation of the Critical Values of the Student t-Distribution and the Grubbs T-Distribution

The critical values of the Student t-distribution can be approximated by a polynomial in the form,

$$t(\alpha, v) = \sum_{i=0}^{k} a_i/v^i \qquad k = 5 \text{ or } 6$$

and those for the Grubbs distribution by the polynomial,

$$T(\alpha, n) = \sum_{i=0}^{k} b_i/(\sqrt{n})^i \qquad k = 3 \text{ or } 4$$

Using the Horner scheme, these polynomials can be transcribed in a very economic form,

$$t(\alpha, v) = (((((a_6/v + a_5)/v + a_4)/v + a_3)/v + a_2)/v + a_1)/v + a_0$$

or

$$T(\alpha, n) = (((b_4/\sqrt{(n)} + b_3)/\sqrt{(n)} + b_2)\sqrt{(n)} + b_1)/\sqrt{(n)} + b_0$$

Therefore, a single-row function can be used instead of extensive tables in programs requiring these critical values.

The precision of the approximation (to four decimal places) is sufficient for most purposes; for lower values of α, more polynomial terms must be used to attain a comparable precision.

Table of coefficients for approximation of the Student t-distribution

$\alpha =$	0.1	0.05	0.01
a_0	1.64485	1.95996	2.57583
a_1	1.52392	2.37365	4.91391
a_2	1.41574	2.78267	8.94301
a_3	1.02112	2.89696	10.70767
a_4	0.30292	0.44565	20.24612
a_5	0.40519	2.24731	-12.91220
a_6	—	—	29.18241

Table of coefficients for approximation of the Grubbs T-distribution

$\alpha =$	0.1	0.05	0.01
b_0	3.6374	3.9254	4.3773
b_1	-7.4490	-8.7140	-9.3382
b_2	11.5940	17.7670	20.6360
b_3	-9.3295	-24.1976	-38.2479
b_4	—	11.2685	26.1920

5.5 Appendix E

Approximation of the Critical Values of the Non-Central Student Distribution

The dependence of the critical values of the non-central Student distribution on parameter d, which expresses the shift of the arithmetic mean by a systematic error, was approximated by

$$t(\alpha, v, d) = t(\alpha, v, \infty) + B\Big[1 - \exp\Big(-\sum_{i=1}^{3} b_i d\Big)\Big]$$

where $t(\alpha, v, \infty)$ is the limit of $t(\alpha, v, d)$ for $d \to \infty$ and coefficient B equals the difference, $t(\alpha, v, 0) - t(\alpha, v, d)$. The $t(\alpha, v, \infty)$ values are tabulated below, together with parameters B, b_1, b_2 and b_3 for significance levels $\alpha = 0.1$, 0.05 and 0.01 and for the individual numbers of degrees of freedom, v. The approximations are precise to three decimal places.

The limits of the interval over which the integral of the Student distribution equals $(1 - \alpha)$ can be determined using the approximated values as:

left-hand limit, $-t(\alpha, v, d)$

right-hand limit, $+t(\alpha, v, d) + 2d$

Table of coefficients

$\alpha = 0.1$	$t(\alpha, \nu, \infty)$	B	b_1	b_2	b_3
1	3.0777	3.2361	0.3092	-0.0022	-0.0033
2	1.8856	1.0344	0.9894	-0.0127	-0.0276
3	1.6377	0.7156	1.4396	0.0561	-0.0661
4	1.5332	0.5986	1.7173	0.1921	-0.1147
5	1.4759	0.5392	1.8967	0.3513	-0.1632
6	1.4398	0.5034	2.0189	0.5096	-0.2072
7	1.4149	0.4797	2.1062	0.6570	-0.2449
8	1.3968	0.4627	2.1713	0.7901	-0.2769
9	1.3830	0.4501	2.2215	0.9080	0.3009
10	1.3722	0.4403	2.2615	1.0119	-0.3204
11	1.3634	0.4325	2.2941	1.1031	-0.3351
12	1.3562	0.4261	2.3214	1.1830	-0.3458
13	1.3502	0.4208	2.3444	1.2539	-0.3539
14	1.3450	0.4163	2.3643	1.3159	-0.3587
15	1.3406	0.4124	2.3818	1.3707	0.3612
16	1.3368	0.4091	2.3972	1.4190	-0.3618
17	1.3334	0.4062	2.4111	1.4618	-0.3608
18	1.3304	0.4037	2.4234	1.5006	-0.3594
19	1.3277	0.4014	2.4348	1.5337	-0.3554
20	1.3253	0.3994	2.4455	1.5620	-0.3492
21	1.3232	0.3976	2.4553	1.5878	-0.3432
22	1.3212	0.3959	2.4624	1.6194	-0.3443
23	1.3195	0.3944	2.4705	1.6412	-0.3383
24	1.3178	0.3930	2.4780	1.6612	-0.3322
25	1.3163	0.3918	2.4848	1.6800	-0.3265
30	1.3104	0.3868	2.5146	1.7451	-0.2894
35	1.3062	0.3834	2.5396	1.7745	-0.2419
40	1.3031	0.3808	2.5567	1.8042	-0.2098
45	1.3006	0.3788	2.5717	1.8199	-0.1759
50	1.2987	0.3772	2.5857	1.8221	-0.1373
60	1.2958	0.3748	2.5983	1.8674	-0.1173
80	1.2922	0.3719	2.6277	1.8553	-0.0217
100	1.2901	0.3702	2.6415	1.8661	0.0222
150	1.2872	0.3679	2.6621	1.8717	0.0899
200	1.2858	0.3667	2.6611	1.9325	0.0675
∞	1.2816	0.3633	2.7011	1.8906	0.2312

$\alpha = 0.05$	$t(\alpha, v, \infty)$	B	b_1	b_2	b_3
1	6.3138	6.3925	0.1565	−0.0001	−0.0006
2	2.9200	1.3827	0.7323	0.0052	−0.0181
3	2.3534	0.8291	1.2360	0.0390	−0.0482
4	2.1318	0.6446	1.5954	0.1264	−0.0875
5	2.0150	0.5555	1.8493	0.2493	−0.1316
6	1.9432	0.5037	2.0333	0.3872	−0.1760
7	1.8946	0.4700	2.1708	0.5269	−0.2177
8	1.8595	0.4465	2.2765	0.6614	−0.2554
9	1.8331	0.4290	2.3600	0.7873	−0.2887
10	1.8125	0.4157	2.4273	0.9033	−0.3176
11	1.7959	0.4051	2.4827	1.0095	−0.3424
12	1.7823	0.3965	2.5293	1.1055	−0.3630
13	1.7709	0.3894	2.5685	1.1944	−0.3819
14	1.7613	0.3835	2.6025	1.2736	−0.3965
15	1.7530	0.3784	2.6322	1.3455	−0.4083
16	1.7459	0.3740	2.6584	1.4105	−0.4178
17	1.7396	0.3702	2.6817	1.4696	−0.4253
18	1.7341	0.3669	2.7023	1.5244	−0.4322
19	1.7291	0.3639	2.7212	1.5727	−0.4361
20	1.7247	0.3612	2.7383	1.6167	−0.4379
21	1.7207	0.3589	2.7538	1.6576	−0.4398
22	1.7171	0.3567	2.7680	1.6954	−0.4409
23	1.7139	0.3548	2.7806	1.7324	−0.4439
24	1.7109	0.3530	2.7915	1.7696	−0.4482
25	1.7081	0.3514	2.8030	1.7976	−0.4458
30	1.6973	0.3450	2.8494	1.9114	−0.4300
35	1.6896	0.3405	2.8884	1.9672	−0.3867
40	1.6839	0.3372	2.9164	2.0170	−0.3562
45	1.6794	0.3347	2.9391	2.0523	−0.3262
50	1.6759	0.3327	2.9560	2.0879	−0.3066
60	1.6706	0.3296	2.9754	2.1741	−0.3074
80	1.6641	0.3259	3.0153	2.2005	−0.2189
100	1.6602	0.3237	3.0363	2.2326	−0.1778
150	1.6551	0.3208	3.0517	2.3447	−0.1885
200	1.6525	0.3194	3.0661	2.3671	−0.1601
∞	1.6449	0.3151	3.0891	2.5446	−0.1742

$\alpha = 0.01$	$t(\alpha, v, \infty)$	B	b_1	b_2	b_3
1	31.8205	31.8362	0.0314	0.0000	0.0000
2	6.9646	2.9603	0.3382	0.0060	−0.0035
3	4.5407	1.3002	0.7762	0.0321	−0.0215
4	3.7469	0.8571	1.1881	0.0763	−0.0487
5	3.3649	0.6672	1.5347	0.1433	−0.0813
6	3.1427	0.5648	1.8176	0.2312	−0.1178
7	2.9980	0.5015	2.0478	0.3340	−0.1564
8	2.8965	0.4589	2.2365	0.4450	−0.1956
9	2.8214	0.4284	2.3929	0.5593	−0.2342
10	2.7638	0.4055	2.5240	0.6736	−0.2713
11	2.7181	0.3877	2.6352	0.7852	−0.3066
12	2.6810	0.3735	2.7304	0.8935	−0.3400
13	2.6503	0.3620	2.8128	0.9969	−0.3712
14	2.6245	0.3523	2.8848	1.0946	−0.3994
15	2.6025	0.3442	2.9477	1.1895	−0.4280
16	2.5835	0.3373	3.0040	1.2764	−0.4518
17	2.5669	0.3313	3.0540	1.3593	−0.4746
18	2.5524	0.3261	3.0986	1.4391	−0.4971
19	2.5395	0.3215	3.1391	1.5131	−0.5171
20	2.5280	0.3174	3.1763	1.5802	−0.5330
21	2.5176	0.3137	3.2097	1.6463	−0.5499
22	2.5083	0.3104	3.2403	1.7090	−0.5651
23	2.4999	0.3075	3.2685	1.7655	−0.5786
24	2.4922	0.3048	3.2943	1.8211	−0.5920
25	2.4851	0.3023	3.3179	1.8754	−0.6072
30	2.4573	0.2927	3.4157	2.0927	−0.6506
35	2.4377	0.2861	3.4856	2.2684	−0.6890
40	2.4233	0.2812	3.5447	2.3731	−0.6788
45	2.4121	0.2775	3.5757	2.5477	−0.7626
50	2.4033	0.2745	3.6236	2.5591	−0.6898
60	2.3901	0.2702	3.6579	2.8030	−0.8166
80	2.3739	0.2648	3.7298	2.9493	−0.7890
100	2.3642	0.2617	3.7688	3.0666	−0.7922
150	2.3515	0.2575	3.7897	3.4317	−1.0310
200	2.3451	0.2555	3.8830	3.0971	−0.5543
∞	2.3263	0.2495	3.7162	4.9692	−2.4449

5.6 Appendix F

Iterative Methods for Solution of Equations

There are several methods for finding an unknown root of an equation, x, from implicit equation $g(x) = 0$.
(x_i is the i-th estimate of root x)

1) Newton method:

$$x_{i+1} = x_i - \frac{g(x_i)}{g'(x_i)}$$

2) Tchebyshev method:

$$x_{i+1} = x_i - \frac{g(x_i)}{g'(x_i)} - \frac{g''(x_i) \cdot [g(x_i)]^2}{2[g'(x_i)]^3}$$

3) Aitken method of accelerating convergence:

$$x_{3i+3} = x_{3i} - \frac{(x_{3i+1} - x_{3i})^2}{x_{3i+2} - 2x_{3i+1} + x_{3i}}$$

4) "Regula falsi" method:

$$x_0 = a - \frac{g(a)}{g(b) - g(a)} (b - a)$$

$$x_{i+1} = x_i - \frac{x_s - x_i}{g(x_s) - g(x_i)} \cdot g(x_i)$$

5.7 Appendix G

Block Scheme of the Gauss-Newton Method

Matrix dimensions: $X(N)$, $Y(N)$, $Z(N)$, $R(N)$, $B(L, L)$, $D(L, L)$, $A(L)$, $C(L)$, $F(L)$, $G(1, L)$

Input: data: N, X_i, Y_i, initial estimate: L, A_j

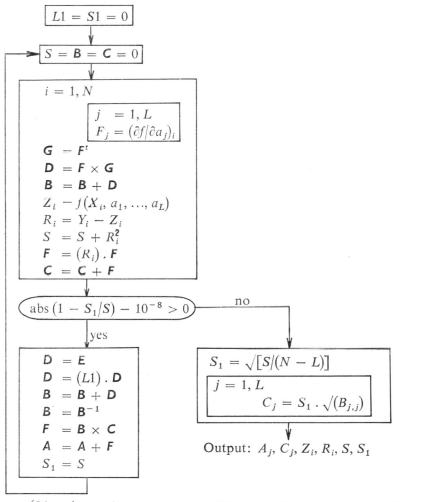

$$L1 = S1 = 0$$

$$S = B = C = 0$$

$$i = 1, N$$

$$j = 1, L$$
$$F_j = (\partial f / \partial a_j)_i$$

$$G = F^t$$
$$D = F \times G$$
$$B = B + D$$
$$Z_i = f(X_i, a_1, \ldots, a_L)$$
$$R_i = Y_i - Z_i$$
$$S = S + R_i^2$$
$$F = (R_i) \cdot F$$
$$C = C + F$$

$$\text{abs}(1 - S_1/S) - 10^{-8} > 0 \qquad \text{no}$$

yes

$$D = E$$
$$D = (L1) \cdot D$$
$$B = B + D$$
$$B = B^{-1}$$
$$F = B \times C$$
$$A = A + F$$
$$S_1 = S$$

$$S_1 = \sqrt{[S/(N - L)]}$$

$$j = 1, L$$
$$C_j = S_1 \cdot \sqrt{(B_{j,j})}$$

Output: A_j, C_j, Z_i, R_i, S, S_1

($L1 = \lambda$, see the Marquardt modification of the $G - N$ method)

REFERENCES

1. Malissa H.: *Z. Anal. Chem.* **256**, 7 (1971).
2. Eckschlager K., Štěpánek V.: Information Theory as Applied to Chemical Analysis, Wiley — Interscience, New York 1979.
3. Brillouin L.: Science and Information Theory, Academic Press, New York 1962.
4. Shannon C. E., Weaver W.: The Mathematical Theory of Information, The University of Illinois Press, Urbana 1949.
5. Cleij P., Dijkstra A.: *Z. Anal. Chem.* **298**, 97 (1979).
6. Liteanu C., Rica I.: *Anal. Chem.* **51**, 1986 (1979).
7. Danzer K., Eckschlager K.: *Talanta* **25**, 725 (1975).
8. Ziegler E.: Computer in der instrumentellen Analytik, Akademische Verlagsgesellschaft, Frankfurt/Main 1973.
9. Massart D. L., Dijkstra A., Kaufman L.: Evaluation and Optimization of the Laboratory Methods and Analytical Procedures, Elsevier, Amsterdam 1978.
10. Varmuza K.: Pattern Recognition in Chemistry, Springer Verlag, Berlin-Heidelberg-New York 1980.
11. Kowalski B. R.: *Anal. Chem.* **52**, 119 R (1980); **54**, 232 R (1982).
12. Štrouf O., Fusek J.: *Collect. Czech. Chem. Commun.* **44**, 1370 (1979).
13. Schoenfeld P. S., De Voe J. R.: *Anal. Chem.* **48**, 403 R (1976).
14. Anderson R. E.: *Chromatographia* **2**, 105 (1972).
15. Ziegler E., Henneberg D., Schomburg G.: *Anal. Chem.* **42**, 51 A (1970).
16. Fok J. S., Abrahamson E. A.: *Chromatographia* **7**, 423 (1974).
17. 3352B Laboratory Data System, Hewlett-Packard Bull. 3352B.
18. Varian Research Notes, Sept. 1972.
19. The Perkin-Elmer Gas Chromatography Data System Design and Performance, presented at the Pittsburgh Conference 1971.
20. Auswertung von Gas-Chromatographen mit Prozessrechner. Programmsystem PRAG 320, Siemens techn. paper.
21. Kelly P. C., Horlick G.: *Anal. Chem.* **45**, 518 (1973).
22. Kishimoto K., Musha S.: *J. Chromat. Sci.* **9**, 608 (1971).
23. Chesler S. N., Cram S. P.: *Anal. Chem.* **44**, 2240 (1972).
24. Sutre P., Malengé J. P.: *Chromatographia* **5**, 141 (1972).
25. Craven D. A., Everett E. S., Rubel M.: *J. Chromat. Sci.* **9**, 541 (1971).
26. Hoeschele Jr. D. F.: Analog-to-Digital and Digital-to-Analog Conversion Techniques, J. Wiley, New York 1968.
27. Savitzki A., Golay M. J. E.: *Anal. Chem.* **36**, 1627 (1964).
28. Hershey H. C., Zakin J. L., Simha R.: *IEC Fundamentals* **3**, 413 (1967).
29. Milne W. E.: Numerical Calculus, Princeton University Press, Princeton 1949.
30. Fozart A., Franses J. J., Wyatt A.: *Chromatographia* **5**, 377 (1972).
31. Charrier G., Dupuis M. C., Merlivat J. C., Pons J., Sigelle R.: *Chromatographia* **5**, 119 (1972).
32. Hegedüs L. L., Petersen E. E.: *J. Chromatog. Sci.* **9**, 551 (1971).

33. System IV a New Computing Integrator for Chromatography, Autolab Technical Bull. 105—72.
34. Keller W. D., Lusebrink T. R., Sederholm C. H.: *J. Chem. Phys.* **44**, 782 (1966).
35. Papoušek D., Plíva J.: *Collect. Czech. Chem. Commun.* **30**, 3007 (1965).
36. Pitha J., Jones R. N.: *Canadian J. Chem.* **44**, 3031 (1966).
37. Farrar T. C., Becker E. D.: Pulse and Fourier Transformation NMR, Academic Press, London 1971.
38. Computers in Chemistry and Instrumentation, Vol. 7, Infrared Correlation and Fourier Transformation Spectroscopy, M. Dekker, New York 1977.
39. Hieftje G. M., Powell L. A.: *Anal. Chim. Acta* **100**, 313 (1978).
40. König H.: Spectra Analysis Automation Review, Chimia Ger. Blendax Werke, Schneider and Co., Mainz 1977.
41. Clerc J. T.: *Chimia* **31**, 353 (1977).
42. Norman Jones R., Ed.: Computer Programs for Infrared Spectrophotometry, National Research Council of Canada, 1977.
43. Corio P. L.: Structure of High-Resolution NMR Spectra, Academic Press, New York 1966.
44. Pople J. A., Schneider W. G., Bernstein H. J.: High-Resolution Nuclear Magnetic Resonance, McGraw-Hill, New York 1959.
45. Lincoln S. F.: Kinetic Applications of NMR Spectroscopy; Progress in Reaction Kinetics, Pergamon Press, Oxford 1977.
46. Matherny M.: *Wiss. Z. Karl Marx Univ. Leipzig, Naturwiss.* **28**, 449 (1979).
47. Matherny M.: *Kém. Közl.* **48**, 363 (1977); **48**, 365 (1977).
48. Matherny M.: *Anal. Chim. Acta, Computer Techniques and Optimization,* **112**, 277 (1979).
49. Matherny M.: *Chim. Analyt.* **21**, 339 (1976); **21**, 1053 (1976).
50. Matherny M., Florián K.: Zborník ved. prác VŠT, Košice 1977, Vol. 2, 183.
51. Florián K.: *Kém. Közl.* **48**, 347 (1977).
52. Florián K., Pliešovská N.: *Chem. Zvesti* **31**, 204 (1977).
53. Plško E.: *Collect. Czech. Chem. Commun.* **30**, 1246 (1965).
54. Matherny M.: *Chem. Zvesti* **24**, 112 (1970).
55. Matherny M.: *Spectroscopy Letters* **6**, 711 (1973).
56. Florián K., Matherny M., Lavrin A.: *Chem. Zvesti* **27**, 623 (1973).
57. Matherny M.: *Z. Anal. Chem.* **271**, 101 (1974).
58. Matherny M., Ondáš J.: *Anal. Chim. Acta, Computer Techniques and Optimization* **133**, 51 (1981).
59. Kaiser H.: *Spectrochim. Acta* **3**, 159 (1947).
60. Török T., Zimmer K.: Quantitative Evaluation of Spectrograms by Means of *l*-transformation, Akadémiai Kiadó, Budapest 1972.
61. Plško E.: *Chem. Zvesti* **23**, 150 (1969).
62. Churchill J. R.: *Ind. Eng. Chem., Anal. Ed.,* **16**, 653 (1944).
63. Florián K., Matherny M.: *Spectrochim. Acta* **33 B**, 429 (1978).
64. Kaiser H.: *Spectrochim. Acta* **3**, 297 (1947).
65. Matherny M.: *Kém. Közl.* **52**, 49 (1979).

66. Prokofiev V. K.: Spektrální analysa kovů, SNTL, Praha 1954.
67. Burrington R. S., May D. C.: Handbook of Probability and Statistics with Tables, 2nd Ed., McGraw-Hill, New York 1970.
68. Hald A.: Statistical Theory with Engineering Application, J. Wiley, New York 1952.
69. Riesel G.: *Z. Angew. Geol.* **21**, 513 (1975).
70. Mai H., Stahlberg U.: *Jenaer Rundschau* **21**, 316 (1976).
71. Malínek M.: *Canad. Appl. Spectrosc.* **20**, 68 (1975).
72. Matherny M.: *Spectroscopy Letters* **5**, 227 (1972).
73. Crouthamel C. E., Adams F., Adams R.: Applied Gamma-Ray Spectrometry, Pergamon Press, Oxford 1970.
74. De Soete D., Gijbels R., Hoste J.: Neutron Activation Analysis, Wiley-Interscience, New York 1972.
75. Mariscotti M. A: *Nucl. Instr. Methods* **50**, 309 (1967).
76. Barness V.: *IEEE Trans. on Nucl. Sci.* **15**, 437 (1968).
77. Felawka L. T. et al.: AECL - 4217 1973, Atomic Energy of Canada.
78. Routti J. T., Prussin S.G.: *Nucl. Instr. Meth.* **72**, 125 (1969).
79. Phillips G., Marlow K. W.: *Nucl. Instr. Meth.* **137**, 525 (1976).
80. Black W. W.: *Nucl. Instr. Meth.* **71**, 318 (1969).
81. Op De Beeck J.: *Atomic Energy Review* **13**, 743 (1975).
82. Gunnink R., Niday J. B.: Computerized Quantitative Analysis by Gamma-Ray Spectrometry, UCRL - 51061, Vol. 1—4, Lawrence Livermore Laboratory 1971.
83. Baedecker P. A.: *Anal. Chem.* **43**, 405 (1971.
84. Hertogen J., De Dourder J., Gijbels R.: *Nucl. Instr. Meth.* **115**, 197 (1974).
85. Simonits A., Moens L., De Corte F., De Wispelaere A., Elek A., Hoste J.: *J.Radioanal. Chem.* **60**, 461 (1980).

Subject Index

Absorbance, 164

Absorption-stat, 175

Accuracy, 260

Activator, in catalytic reaction, 41, 57, 58

—, in catalytic reaction, determination of, 63, 64, 226

Alcohols, determination of, 225

Algorithm, 295

Amino acid, determination of, 225

—, oxidase, determination of, 223

Ammonia, determination of in water, 191

Analysis, activation, 423, 424

—, qualitative, 257

—, quantitative, 257

—, structural, spectroscopic, 376ff, 382—384

Biamperostat, 176

Biological clock, 43

Bipotentiometry, 169ff

Brdička reaction, 112

Bunching, 364

Calibration, in emission spectroscopy, 392ff

Catalase, determination of, 223

Catalysis, mechanism of, 36ff

Catalyst, definition of, 33

—, limit of determination of, 153, 154

—, organic, 44

Cell, electrochemical, optically transparent, 166

Cell, electrochemical, thin-layer, 166

Cellulase, determination of, 221

Chelation, 40

Choline esterase, determination of, 221

Chymotrypsine, determination of, 221

Coefficient, correlation, 322,395

Coenzyme, determination of, 63, 64

Collections of spectra, 379, 380

Complex, activated, 27

—, charge-transfer, 37

—, polydentate, 88

Complex formation, 87ff

Computation, in real time, 287

Computer, classification of, 291

—, in chemical laboratory, 284ff

—, off-line, 286ff

—, on-line, 287ff

Conductometry, in kinetic analyses, 171

Confidence interval, 271, 319, 329, 337

Constant, Michaelis, 55ff, 66, 70

—, rate, 26,81

Convertor, analog-to-digital, 361ff, 386, 398, 412

—, voltage-to-frequency, 362

Convolution, 417

Copper, determination of in serum, 210

Correlation, 321, 322

Coulometry, in kinetic analyses, 171

Current, catalytic, 110ff

—, —, of hydrogen, 111ff

—, kinetic, 107ff

Curve, density, characteristic, 388

—, fitting, 308, 338ff, 365

441